T0350774

STOCHASTIC STRUCTURAL DYNAMICS

STOCHASTIC STRUCTURAL DYNAMICS
APPLICATION OF FINITE ELEMENT METHODS

Cho W. S. To

Professor of Mechanical Engineering
University of Nebraska-Lincoln, USA

WILEY

This edition first published 2014
© 2014 John Wiley & Sons, Ltd

Registered office
John Wiley & Sons Ltd, The Atrium, Southern Gate, Chichester, West Sussex, PO19 8SQ, United Kingdom

For details of our global editorial offices, for customer services and for information about how to apply for permission to reuse the copyright material in this book please see our website at www.wiley.com.

Library of Congress Cataloging-in-Publication Data

To, Cho W. S.
    Stochastic structural dynamics : application of finite element methods / Cho W. S. To. – First edition.
        pages cm
    Includes bibliographical references and index.
    ISBN 978-1-118-34235-0 (hardback)
    1. Structural dynamics–Engineering.   2. Finite element method.   3. Stochastic analysis.   I. Title.
    TA654.T59 2014
    624.1'70151922–dc23
                                                                                    2013023517

A catalogue record for this book is available from the British Library.

ISBN 9781118342350

Set in 10/12pt Times by Aptara Inc., New Delhi, India

Printed and bound in Singapore by Markono Print Media Pte Ltd

1   2014

Lidong Leighton and Lizhen Jane

Contents

Preface

Stochastic structural dynamics is concerned with the studies of dynamics of structures and structural systems that are subjected to complex excitations treated as random processes. In engineering practice, many structures and structural systems cannot be dealt with analytically and therefore the versatile numerical analysis techniques, the finite element methods (FEM) are employed.

The parallel developments of the FEM in the 1950's and the engineering applications of stochastic processes in the 1940's provided a combined numerical analysis tool for the studies of dynamics of structures and structural systems under random loadings. In the open literature, there are books on statistical dynamics of structures and books on structural dynamics with chapter(s) dealing with random response analysis. However, a systematic treatment of stochastic structural dynamics applying the FEM seems to be lacking. The present book is believed to be the first relatively in-depth and systematic treatment on the subject. It is aimed at advanced and specialist level. It is suitable for classes taken by master's degree level post-graduate students and specialists.

The present book has seven chapters and ten appendices. Chapter 1 introduces the displacement-based FEM, element equations of motion for temporally and spatially stochastic systems, hybrid stress-based element equations of motion, incremental variational principle and mixed formulation-based nonlinear element matrices, constitutive relations and updating of configurations and stresses.

Chapter 2 is concerned with the spectral analysis and response statistics of linear structural systems. It includes evolutionary spectral analysis, evolutionary spectra of engineering structures, modal analysis and time-dependent response statistics, and response statistics of engineering structures.

Direct integration methods for linear structural systems are presented in Chapter 3. The stochastic central difference method with time co-ordinate transformation and its application, extended stochastic central difference method for narrow-band excitations, stochastic Newmark family of algorithms, and their applications to plate structures are presented in this chapter.

Modal analysis and response statistics of quasi-linear structural systems are covered in Chapter 4. Modal analysis of temporally stochastic quasi-linear systems and the bi-modal approach are included. Response analysis of plate structures by the Melosh-Zienkiewicz-Cheung bending plate element, and the high precision triangular plate element are presented.

Chapter 5 is concerned with the application of the direct integration methods for response statistics of quasi-linear structural systems. Recursive covariance matrices of displacements of cantilever pipes containing turbulent fluids and subjected to modulated white noise as well as narrow-band random excitations are derived in this chapter.

Direct integration methods for temporally stochastic nonlinear structural systems subjected to stationary and nonstationary random excitations are presented in Chapter 6. A brief introduction to the statistical linearization techniques is included. Symplectic members of the deterministic and stochastic versions of the Newmark family of algorithms are identified. The stochastic central difference method with time co-ordinate transformation and adaptive time schemes are introduced and applied to the computation of large responses of plate and shell structures.

Chapter 7 is concerned with the presentation of the direct integration methods for temporally and spatially stochastic nonlinear structural systems. The stochastic FEM or probabilistic FEM is introduced. The stochastic central difference method for temporally and spatially stochastic structural systems subjected to stationary and nonstationary random excitations are developed. Application of the method to spatially homogeneous and non-homogeneous shell structures is made.

Finally, a word about symbols is in order. Mathematically, random variables and random processes are different. But without ambiguity the same symbols for random variables and processes are applied in the present book, unless it is stated otherwise.

Acknowledgments

Thanks are due to the author's several former graduate students, Gregory Zidong Chen, Sherwin Xingling Dai, Derick Hung, Meilan Liu, Irewole Raphael Orisamolu, and Bin Wang who provided various drawings in this book.

The author would like to express his sincere thanks to Paul Petralia, Senior Editor and his project team members, Tom Carter, Sandra Grayson, Anna Smart, and Liz Wingett.

Finally, the author would also like to thank Elsevier Science for permission to reproduce the following figures. Figures 1.1 and 1.2 are from To, C. W. S. (1979): Higher order tapered beam finite elements for vibration analysis, *Journal of Sound and Vibration*, **63(1)**, 33-50. Figures 1.3 and 1.4 are from To, C. W. S. and Liu, M. L. (1994): Hybrid strain based three-node flat triangular shell elements, *Finite Elements in Analysis and Design*, **17**, 169-203. Figures 2.1 through 2.3 are from To, C. W. S. (1982): Nonstationary random responses of a multi-degree-of-freedom system by the theory of evolutionary spectra, *Journal of Sound and Vibration*, **83(2)**, 273-291. Figures 2.12, 2.13, and 2B.1 are from To, C. W. S. (1984): Time-dependent variance and covariance of responses of structures to non-stationary random excitations, *Journal of Sound and Vibration*, **93(1)**, 135-156. Figures 2.14 through 2.19 are from To, C. W. S. and Wang, B. (1993): Time-dependent response statistics of axisymmetrical shell structures, *Journal of Sound and Vibration*, **164(3)**, 554-564. Figures 2.20 through 2.26 are from To, C. W. S. and Wang, B. (1996): Nonstationary random response of laminated composite structures by a hybrid strain-based laminated flat triangular shell finite element, *Finite Elements in Analysis and Design*, **23**, 23-35. Figures 3.2 through 3.10 are from To, C. W. S. and Liu, M. L. (1994): Random responses of discretized beams and plates by the stochastic central difference method with time co-ordinate transformation, *Computers and Structures*, **53(3)**, 727-738. Figures 3.11 through 3.15, and Figures 5.9 through 5.14 are from Chen, Z. and

To, C. W. S. (2005): Responses of discretized systems under narrow band nonstationary random excitations, *Journal of Sound and Vibration*, **287**, 433-458. Figures 4.1 through 4.7 are from To, C. W. S. and Orisamolu, I. R. (1987): Response of discretized plates to transversal and in-plane non-stationary random excitations, *Journal of Sound and Vibration*, **114(3)**, 481-494. Figures 6.1 through 6.7, and 6.10 through 6.13 are from To, C. W. S. and Liu, M. L. (2000): Large nonstationary random responses of shell structures with geometrical and material nonlinearities, *Finite Elements in Analysis and Design*, **35**, 59-77.

1

Introduction

The parallel developments of the finite element methods (FEM) in the 1950's [1, 2] and the engineering applications of the stochastic processes in the 1940's [3, 4] provided a combined numerical analysis tool for the studies of dynamics of structures and structural systems under random loadings. There are books on statistical dynamics of structures [5, 6] and books on structural dynamics with chapter(s) dealing with random response analysis [7, 8]. In addition, there are various monographs and lecture notes on the subject. However, a systematic treatment of the stochastic structural dynamics applying the FEM seems to be lacking. The present book is believed to be the first relatively in-depth and systematic treatment of the subject that applies the FEM to the field of stochastic structural dynamics.

Before the introduction to the concept and theory of stochastic quantities and their applications with the FEM in subsequent chapters, the two FEM employed in the investigations presented in the present book are outlined in this chapter. Specifically, Section 1.1 is concerned with the derivation of the temporally stochastic element equation of motion applying the displacement formulation. The consistent element stiffness and mass matrices of two beam elements, each having two nodes are derived. One beam element is uniform and the other is tapered. The corresponding temporally and spatially stochastic element equation of motion is derived in Section 1.2. The element equations of motion based on the mixed formulation are introduced in Section 1.3. Consistent element matrices for a beam of uniform cross-sectional area are obtained. This beam element has two nodes, each of which has two degrees-of-freedom (dof). This beam element is applied to show that stiffness matrices derived from the displacement and mixed formulations are identical. The incremental variational principle and element matrices based on the mixed formulation for nonlinear

Stochastic Structural Dynamics: Application of Finite Element Methods, First Edition. C.W.S. To.
© 2014 John Wiley & Sons, Ltd. Published 2014 by John Wiley & Sons, Ltd.

structures are presented in Section 1.4. Section 1.5 deals with constitutive relations and updating of configurations and stresses. Closing remarks for this chapter are provided in Section 1.6.

1.1 Displacement Formulation-Based Finite Element Method

Without loss of generality and as an illustration, the displacement formulation based element equations of motion for temporally stochastic linear systems are presented in this section. These equations are similar in form to those under deterministic excitations. It is included in Sub-section 1.1.1 while application of the technique for the derivation of element matrices of a two-node beam element of uniform cross-section is given in Sub-section 1.1.2. The tapered beam element is presented in Sub-section 1.1.3.

1.1.1   Derivation of element equations of motion

The Rayleigh-Ritz (RR) method approximates the displacement by a linear set of admissible functions that satisfy the geometric boundary conditions and are p times differentiable over the domain, where p is the number of boundary conditions that the displacement must satisfy at every point of the boundary of the domain. The admissible functions required by the RR method are constructed employing the finite element displacement method with the following steps:

(a) idealization of the structure by choosing a set of imaginary reference or node points such that on joining these node points by means of imaginary lines a series of finite elements is formed;

(b) assigning a given number of dof, such as displacement, slope, curvature, and so on, to every node point; and

(c) constructing a set of functions such that every one corresponds to a unit value of one dof, with the others being set to zero.

Having constructed the admissible functions, the element matrices are then determined. For simplicity, the damping matrix of the element will be disregarded. Thus, in the following the definition of consistent element mass and stiffness matrices in terms of deformation patterns usually referred to as shape functions is given.

Assuming the displacement $u(x,t)$ or simply u at the point x (for example, in the three-dimensional case it represents the local co-ordinates r, s and t at the point) within the e'th element is expressed in matrix form as

$$u(x,t) = N(x)q(t) , \tag{1.1}$$

where $N(x)$ or simply N is a matrix of element shape functions, and $q(t)$ or q a matrix of nodal dof with reference to the local axes, also known as the vector of nodal displacements or generalized displacements.

The matrix of strain components ε thus takes the form

$$\varepsilon = Bq , \qquad (1.2)$$

where B is a differential of the shape function matrix N.

The matrix of stress components σ is given by

$$\sigma = D\varepsilon , \qquad (1.3)$$

where D is the elastic matrix.

Substituting Eq. (1.2) into (1.3) gives

$$\sigma = DBq . \qquad (1.4)$$

In order to derive the element equations of motion for a conservative system, the Hamilton's principle can be applied

$$L = T - (U + W) , \qquad (1.5)$$

where T and $(U + W)$ are the kinetic and potential energies, respectively.

It may be appropriate to note that for a non-conservative system or system with non-holonomic boundary conditions, the modified Hamilton's principle [9] or the virtual power principle [10, 11] may be applied. Non-holonomic systems are those with constraint equations containing velocities which cannot be integrated into relations in co-ordinates or displacements only. An example of a non-holonomic system is the bicycle moving down an inclined plane in which enforcing no slipping at the contact point gives rise to non-holonomic constraint equations. Another example is a disk rolling on a horizontal plane. In this case enforcing no slipping at the contact point also give rise to non-holonomic constraint equations.

The kinetic energy density of the element is defined as

$$dT = \tfrac{1}{2} \rho \, \dot{u}^T \dot{u} \, dV \qquad (1.6)$$

where ρ is the density of the material, dV is the incremental volume, and the over-dot denotes the differentiation with respect to time t.

By making use of Eq. (1.6), the kinetic energy of the element becomes

$$T = \frac{1}{2} \iiint_V \rho \, \dot{u}^T \dot{u} \, dV . \qquad (1.7)$$

The strain energy density for a linear elastic body is defined as

$$dU = \frac{1}{2}\varepsilon^T \sigma \, dV = \frac{1}{2}\varepsilon^T D\varepsilon \, dV \,. \tag{1.8}$$

The potential energy for a linearly elastic body can be expressed as the sum of internal work, the strain energy due to internal stress, and work done by the body forces and surface tractions. Thus,

$$U + W = \iiint_V dU - \iiint_V u^T \overline{Q} \, dV - \iint_S u^T \overline{Y} \, dS, \tag{1.9}$$

where S now is the surface of the body on which surface tractions $\overline{Y}$ are prescribed. The last two integrals on the right-hand side (rhs) of Eq. (1.9) represent the work done by the external random forces, the body forces $\overline{Q}$ and surface tractions $\overline{Y}$. In the last equation the over-bar of a letter designates the quantity is specified.

Applying Eq. (1.8), the total potential of the element from Eq. (1.9) becomes

$$U + W = \frac{1}{2}\iiint_V \varepsilon^T D\varepsilon \, dV \quad - \iiint_V u^T \overline{Q} \, dV$$
$$- \iint_S u^T \overline{Y} \, dS \,. \tag{1.10}$$

Substituting Eqs. (1.7) and (1.10) into (1.5), the functional of a linearly elastic element,

$$L = \frac{1}{2}\iiint_V \left(\rho \, \dot{u}^T \dot{u} - \varepsilon^T D\varepsilon \, + 2u^T \overline{Q}\right) dV$$
$$+ \iint_S u^T \overline{Y} \, dS \,. \tag{1.11}$$

On substituting Eqs. (1.1) through (1.3) into the last equation and using the matrix relation $(XY)^T = Y^T X^T$, the Lagrangian becomes

$$L = \frac{1}{2}\iiint_V \left(\rho \dot{q}^T N^T N\dot{q} - q^T B^T DB\,q \right.$$
$$\left. + 2q^T N^T \overline{Q}\right) dV + \iint_S q^T N^T \overline{Y} \, dS \,. \tag{1.12}$$

Applying Hamilton's principle, it leads to

$$
\int_{t_1}^{t_2} \left(\delta\dot{q}^T \iiint_V \rho N^T N dV \dot{q} - \delta q^T \iiint_V B^T D B dV q \right.
$$

$$
\left. + \delta q^T \iiint_V N^T \overline{Q} dV + \delta q^T \iint_S N^T \overline{Y} dS \right) dt = 0 . \tag{1.13}
$$

Integrating the first term inside the brackets on the left-hand side (lhs) of Eq. (1.13) by parts with respect to time t results

$$
\int_{t_1}^{t_2} \delta\dot{q}^T \iiint_V \rho N^T N dV \dot{q} \, dt
$$

$$
= \left[\delta q^T \iiint_V \rho N^T N dV \dot{q} \right]_{t_1}^{t_2} \tag{1.14}
$$

$$
- \int_{t_1}^{t_2} \delta q^T \iiint_V \rho N^T N dV \ddot{q} \, dt .
$$

According to Hamilton's principle, the tentative displacement configuration must satisfy given conditions at times t_1 and t_2, that is,

$$
\delta q(t_1) = 0 , \qquad \delta q(t_2) = 0 .
$$

Hence, the first term on the rhs of Eq. (1.14) vanishes.

Substituting Eq. (1.14) into (1.13) and rearranging, it becomes

$$
\int_{t_1}^{t_2} \delta q^T \left(\iiint_V \rho N^T N dV \ddot{q} + \iiint_V B^T D B dV q \right.
$$

$$
\left. - \iiint_V N^T \overline{Q} dV - \iint_S N^T \overline{Y} dS \right) dt = 0 . \tag{1.15}
$$

As the variations of the nodal displacements δq are arbitrary, the expressions inside the parentheses must be equal to zero in order that Eq. (1.15) is satisfied. Therefore, the equation of motion for the e'th element in matrix form is

$$m\ddot{q} + kq = f, \tag{1.16}$$

where the element mass and stiffness matrices are defined, respectively as

$$m = \iiint_V \rho N^T N \, dV, \quad k = \iiint_V B^T D B \, dV,$$

and the element random load matrix

$$f = \iiint_V N^T \overline{Q} \, dV + \iint_S N^T \overline{Y} \, dS.$$

Applying the generalized co-ordinate form of displacement model the displacement can be expressed as

$$u = \Phi \zeta, \tag{1.17}$$

where Φ is a matrix of function of variables x and ζ is the vector of generalized co-ordinates, also known as generalized displacement amplitudes. The coefficient matrix may be determined by introducing the nodal co-ordinates successively into Eq. (1.17) such that the vector u and matrix Φ become the nodal displacement vector q and coefficient matrix C, respectively. That is,

$$q = C\zeta. \tag{1.18}$$

Hence, the generalized displacement amplitude vector

$$\zeta = C^{-1} q, \tag{1.19}$$

where C^{-1} is the inverse of the coefficient matrix also known as the transformation matrix and is independent of the variables x.

Substituting Eq. (1.19) into (1.17) one has

$$u = \Phi C^{-1} q. \tag{1.20}$$

Comparing Eqs. (1.1) and (1.20), one has the shape function matrix

$$N = \Phi C^{-1}. \tag{1.21}$$

On application of Eqs. (1.16) and (1.21), the element mass, stiffness and load matrices can be evaluated.

To provide a more concrete illustration of the shape function matrix and a better understanding of the steps in the derivation of element mass and stiffness matrices, a uniform beam element is considered in the next sub-section.

1.1.2   Mass and stiffness matrices of uniform beam element

The uniform beam element considered in this sub-section has two nodes, each of which has two dof. The latter include nodal transverse displacement, and rotation or angular displacement about an axis perpendicular to the plane containing the beam and the transverse displacement. For simplicity, the theory of the Euler beam is assumed. The cross-sectional area A and second moment of area I are constant. Let ρ and E be the density and modulus of elasticity of the beam. The bending beam element is shown in Figure 1.1 where the edge displacements and angular displacements are included. The convention adopted in the figure is sagging being positive.

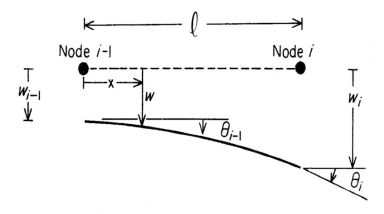

Figure 1.1   Uniform beam element with edge displacements.

Applying Eq. (1.17) so that the transverse displacement at a point inside the beam element can be written as

$$u = w = \Phi\,\zeta, \qquad \Phi = [\,1 \quad x \quad x^2 \quad x^3\,] \qquad\qquad (1.22\text{a, b})$$

$$\zeta^T = \begin{bmatrix} \zeta_1 & \zeta_2 & \zeta_3 & \zeta_4 \end{bmatrix}. \qquad\qquad (1.22\text{c})$$

Consider the nodal values. At $x = 0$, $w = w_{i-1}$ and $\theta = \partial w/\partial x = \theta_{i-1}$ so that upon application of Eq. (1.22a) one has

$$w_{i-1} = [1 \quad 0 \quad 0 \quad 0]\zeta, \qquad \theta_{i-1} = [0 \quad 1 \quad 0 \quad 0]\zeta. \qquad (1.23a, b)$$

Similarly, at $x = \ell$, $w = w_i$ and $\theta = \theta_i$ so that upon application of Eq. (1.22a) it leads to

$$w_i = \begin{bmatrix} 1 & \ell & \ell^2 & \ell^3 \end{bmatrix}\zeta, \qquad \theta_i = \begin{bmatrix} 0 & 1 & 2\ell & 3\ell^2 \end{bmatrix}\zeta. \qquad (1.23c, d)$$

Re-writing Eq. (1.23) in matrix form as in Eq. (1.18), one has

$$q = \begin{Bmatrix} w_{i-1} \\ \theta_{i-1} \\ w_i \\ \theta_i \end{Bmatrix} = \begin{bmatrix} 1 & 0 & 0 & 0 \\ 0 & 1 & 0 & 0 \\ 1 & \ell & \ell^2 & \ell^3 \\ 0 & 1 & 2\ell & 3\ell^2 \end{bmatrix} \begin{Bmatrix} \zeta_1 \\ \zeta_2 \\ \zeta_3 \\ \zeta_4 \end{Bmatrix}. \qquad (1.24)$$

Thus, the inverse of matrix C becomes

$$C^{-1} = \frac{1}{\ell^3} \begin{bmatrix} \ell^3 & 0 & 0 & 0 \\ 0 & \ell^3 & 0 & 0 \\ -3\ell & -2\ell^2 & 3\ell & -\ell^2 \\ 2 & \ell & -2 & \ell \end{bmatrix}. \qquad (1.25)$$

Making use of Eqs. (1.22b) and (1.25), the shape function matrix by Eq. (1.21) is obtained as

$$N = \begin{bmatrix} N_{11} & N_{12} & N_{13} & N_{14} \end{bmatrix}, \qquad (1.26)$$

in which

$$N_{11} = 1 - 3\xi^2 + 2\xi^3, \qquad N_{12} = x\left(1 - 2\xi + \xi^2\right), \qquad \xi = \frac{x}{\ell},$$

$$N_{13} = 3\xi^2 - 2\xi^3, \qquad N_{14} = -x\left(\xi - \xi^2\right).$$

Substituting Eq. (1.26) into the equation for element mass matrix defined in Eq. (1.16), one can show that

$$m = \rho A \int_0^\ell N^T N dx = \frac{\rho A \ell}{420} \begin{bmatrix} 156 & 22\ell & 54 & -13\ell \\ \cdot & 4\ell^2 & 13\ell & -3\ell^2 \\ symmetric & \cdot & 156 & -22\ell \\ \cdot & \cdot & \cdot & 4\ell^2 \end{bmatrix}. \quad (1.27)$$

Similarly, the element stiffness matrix is obtained as

$$k = \frac{EI}{\ell^6} \int_0^\ell B^T B dx = \frac{2EI}{\ell^3} \begin{bmatrix} 6 & 3\ell & -6 & 3\ell \\ \cdot & 2\ell^2 & -3\ell & \ell^2 \\ symmetric & \cdot & 6 & -3\ell \\ \cdot & \cdot & \cdot & 2\ell^2 \end{bmatrix}, \quad (1.28)$$

in which

$$B = \frac{\partial^2 N}{\partial x^2} = \begin{bmatrix} B_{11} & B_{12} & B_{13} & B_{14} \end{bmatrix}, \qquad B_{11} = 12x - 6\ell,$$

$$B_{12} = 6x\ell - 4\ell^2, \qquad B_{13} = -12x + 6\ell, \qquad B_{14} = 6x\ell - 2\ell^2.$$

1.1.3 Mass and stiffness matrices of higher order taper beam element

The tapered beam element considered in this sub-section has two nodes, each of which has four dof. The latter include nodal displacement, rotation or angular displacement, curvature, and shear dof. This is the higher order tapered beam element first developed and presented by the author [12].

The tapered beam element of length ℓ, shown in Figure 1.2, is assumed to be of homogeneous and isotropic material. Its cross-sectional area and second moment of area are, respectively given by

$$A(x) = c_1 b(s)d(x), \qquad I(x) = c_2 b(s)d^3(x), \quad (1.29)$$

where c_1 and c_2 depend on the shape of the beam cross-section. For an elliptic-type closed curve cross-section, they are given by [13]

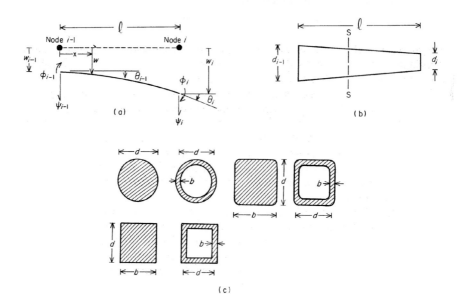

Figure 1.2   Linearly tapered beam element: (a) beam element with edge forces; (b) tapered beam element; (c) cross-section at section S-S in (b).

$$c_1 = \frac{\Gamma\left(\dfrac{1}{\mu_1} + 1\right)\Gamma\left(\dfrac{1}{\mu_2} + 1\right)}{\Gamma\left(\dfrac{1}{\mu_1} + \dfrac{1}{\mu_2} + 1\right)}, \quad c_2 = \frac{\Gamma\left(\dfrac{1}{\mu_1} + 1\right)\Gamma\left(\dfrac{3}{\mu_2} + 1\right)}{12\Gamma\left(\dfrac{1}{\mu_1} + \dfrac{3}{\mu_2} + 1\right)}, \quad (1.30a, b)$$

in which $\Gamma(.)$ is the gamma function, and μ_1 and μ_2 are real positive numbers which need not be integers. When $\mu_1 = \mu_2 = 1$, the cross-section is a triangle and in this case the factor $1/12$ in c_2 should be replaced by $1/9$. When $\mu_1 = \mu_2 = 2$, the cross-section is an ellipse. As μ_1 and μ_2 each approaches infinity, it is a rectangle.

The cross-sectional dimensions, $b(x)$ and $d(x)$, vary linearly along the length of the element so that

$$b(x) = b_{i-1}\left[1 + (\alpha - 1)\frac{x}{\ell}\right], \quad d(x) = d_{i-1}\left[1 + (\beta - 1)\frac{x}{\ell}\right], \quad (1.31a, b)$$

where $\alpha = b_i / b_{i-1}$ and $\beta = d_i / d_{i-1}$ are the taper ratios for the beam element.

Substituting Eq. (1.31) into (1.29) leads to

$$A(x) = A_{i-1}\left(1 + \gamma_1 \xi + \gamma_2 \xi^2\right), \tag{1.32a}$$

$$I(x) = I_{i-1}\left(1 + \delta_1 \xi + \delta_2 \xi^2 + \delta_3 \xi^3 + \delta_4 \xi^4\right), \tag{1.32b}$$

$$\xi = \frac{x}{\ell}, \quad \gamma_1 = (\alpha - 1) + (\beta - 1), \quad \gamma_2 = (\alpha - 1)(\beta - 1),$$

$$\delta_1 = (\alpha - 1) + 3(\beta - 1), \quad \delta_2 = 3(\alpha - 1)(\beta - 1) + 3(\beta - 1)^2,$$

$$\delta_3 = 3(\alpha - 1)(\beta - 1)^2 + (\beta - 1)^3, \quad \delta_4 = (\alpha - 1)(\beta - 1)^3,$$

A_{i-1} and I_{i-1} are respectively the cross-sectional area and second moment of area associated with Node i - 1.

It should be noted that in applying Eq. (1.32) to hollow beams, of square or circular cross-section, for instance, either the ratio b/d must be small or the ratio b/d must be constant because in Eq. (1.29) for a square hollow cross-section c_1 = 4 and $c_2 = (2/3)[1 + (b/d)^2]$, and for a circular hollow cross-section $c_1 = \pi$ and $c_2 = (\pi/8)[1 + (b/d)^2]$.

With the cross-sectional area and second moment of area defined, the element mass and stiffness matrices can be derived accordingly. To this end let the transverse displacement of the beam element be

$$w = \sum_{j=1}^{8} \zeta_j x^{j-1}, \quad \text{or} \quad w = \Phi \zeta, \tag{1.33}$$

where the row and column vectors are respectively

$$\Phi = \begin{bmatrix} 1 & x & x^2 & x^3 & x^4 & x^5 & x^6 & x^7 \end{bmatrix}, \quad \zeta = \begin{bmatrix} \zeta_1 & \zeta_2 & \zeta_3 & \zeta_4 & \zeta_5 & \zeta_6 & \zeta_7 & \zeta_8 \end{bmatrix}^T.$$

Equation (1.33) can be identified as Eq. (1.17) in which the displacement function u is replaced by w. Thus, the nodal displacement vector in Eq. (1.18) for the present tapered beam element becomes

$$q = \begin{bmatrix} w_{i-1} & \theta_{i-1} & \phi_{i-1} & \psi_{i-1} & w_i & \theta_i & \phi_i & \psi_i \end{bmatrix}^T, \quad \phi_i = \frac{\partial^2 w_i}{\partial x^2}, \quad \psi_i = \frac{\partial^3 w_i}{\partial x^3}.$$

The corresponding coefficient matrix in Eq. (1.18) is obtained as [12]

$$
C = \begin{bmatrix}
1 & 0 & 0 & 0 & 0 & 0 & 0 & 0 \\
0 & 1 & 0 & 0 & 0 & 0 & 0 & 0 \\
0 & 0 & 2 & 0 & 0 & 0 & 0 & 0 \\
0 & 0 & 0 & 6 & 0 & 0 & 0 & 0 \\
1 & \ell & \ell^2 & \ell^3 & \ell^4 & \ell^5 & \ell^6 & \ell^7 \\
0 & 1 & 2\ell & 3\ell^2 & 4\ell^3 & 5\ell^4 & 6\ell^5 & 7\ell^6 \\
0 & 0 & 2 & 6\ell & 12\ell^2 & 20\ell^3 & 30\ell^4 & 42\ell^5 \\
0 & 0 & 0 & 6 & 24\ell & 60\ell^2 & 120\ell^3 & 210\ell^4
\end{bmatrix} .
\qquad (1.34)
$$

The inverse of matrix C can be found to be [12]

$$
C^{-1} = \begin{bmatrix}
1 & 0 & 0 & 0 & 0 & 0 & 0 & 0 \\
0 & 1 & 0 & 0 & 0 & 0 & 0 & 0 \\
0 & 0 & \dfrac{1}{2} & 0 & 0 & 0 & 0 & 0 \\
0 & 0 & 0 & \dfrac{1}{6} & 0 & 0 & 0 & 0 \\
-35/\ell^4 & -20/\ell^3 & -5/\ell^2 & -2/3\ell & 35/\ell^4 & -15/\ell^3 & 5/2\ell^2 & -1/6\ell \\
84/\ell^5 & 45/\ell^4 & 10/\ell^3 & 1/\ell^2 & -84/\ell^5 & 39/\ell^4 & -7/\ell^3 & 1/2\ell^2 \\
-70/\ell^6 & -36/\ell^5 & -15/2\ell^4 & -2/3\ell^3 & 70/\ell^6 & -34/\ell^5 & 13/2\ell^4 & -1/2\ell^3 \\
20/\ell^7 & 10/\ell^6 & 2/\ell^5 & 1/6\ell^4 & -20/\ell^7 & 10/\ell^6 & -2/\ell^5 & 1/6\ell^4
\end{bmatrix} .
$$

With this inverse matrix and operating on Eq. (1.21) one can obtain the shape function matrix for the present higher order tapered beam element as

$$
N = \begin{bmatrix} N_{11} & N_{12} & N_{13} & N_{14} & N_{15} & N_{16} & N_{17} & N_{18} \end{bmatrix} ,
\qquad (1.35)
$$

where the shape functions are defined by

$$
N_{11} = 1 - 35\xi^4 + 84\xi^5 - 70\xi^6 + 20\xi^7 ,
$$

$$N_{12} = \xi\ell\left(1 - 20\xi^3 + 45\xi^4 - 36\xi^5 + 10\xi^6\right),$$

$$N_{13} = \frac{1}{2}(\xi\ell)^2\left(1 - 10\xi^2 + 20\xi^3 - 15\xi^4 + 4\xi^5\right),$$

$$N_{14} = \frac{1}{6}(\xi\ell)^3\left(1 - 4\xi + 6\xi^2 - 40\xi^3 + \xi^4\right),$$

$$N_{15} = 35\xi^4 - 84\xi^5 + 70\xi^6 - 20\xi^7,$$

$$N_{16} = \xi\ell\left(39\xi^4 - 15\xi^3 - 34\xi^5 + 10\xi^6\right),$$

$$N_{17} = \frac{1}{2}(\xi\ell)^2\left(5\xi^2 - 14\xi^3 + 13\xi^4 - 4\xi^5\right),$$

$$N_{18} = \frac{1}{6}(\xi\ell)^3\left(3\xi^2 - \xi - 3\xi^3 + \xi^4\right).$$

By making use of Eqs. (1.28) and (1.16), one can find the mass and stiffness matrices of the tapered beam element. These element matrices are given in Appendix 1A.

1.2 Element Equations of Motion for Temporally and Spatially Stochastic Systems

The displacement based FEM presented in Section 1.1 can straightforwardly be extended to temporally and spatially stochastic systems. Without loss of generality and for easy understanding in the following presentation the notation applied in the last section is adopted in this section.

Consider now the elastic matrix in Eq. (1.2) is replaced by the following spatially stochastic elastic matrix,

$$D_h = D + \tilde{D}, \tag{1.36}$$

in which D is the deterministic elastic matrix while the second term on the rhs is the spatially stochastic component of the elastic matrix whose ensemble average is zero such that the element stiffness matrix with spatially stochastic elastic component becomes

$$k_h = k + \tilde{k}, \tag{1.37}$$

where the element stiffness matrices associated with the deterministic and spatially stochastic components are, respectively

$$k = \iiint_V B^T D B \, dV , \qquad \tilde{k} = \iiint_V B^T \tilde{D} B \, dV . \qquad (1.38a, b)$$

To provide a simple example, suppose the modulus of elasticity of the material is spatially stochastic such that it can be written as

$$E_h = E + \tilde{E} , \qquad (1.39)$$

where E is the deterministic component of the modulus of elasticity whereas the second term on the rhs of Eq. (1.39) is the spatially stochastic component of the modulus of elasticity with zero ensemble in the spatial domain.

With reference to Eq. (1.38b), the spatially stochastic component of the stiffness matrix can be written as

$$k_h = k + \tilde{k} = k + r\varphi , \qquad (1.40)$$

where the spatially stochastic component of the stiffness matrix is $r\varphi$.

Substituting Eq. (1.40) into (1.16), the element equations of motion for the temporally and spatially stochastic system becomes

$$m\ddot{q} + (k + r\varphi)q = f . \qquad (1.41)$$

For systems with other spatially stochastic material properties, similar element equations of motion can be obtained accordingly. Note that the spatially stochastic matrix r can have large stochastic variation. This is different from that applying the SFEM or PFEM in which the spatially stochastic variation is limited to a small quantity.

Applying Eq. (1.41) for the entire system, the assembled equation of motion can be constructed in the usual manner.

1.3 Hybrid Stress-Based Element Equations of Motion

The main objective of this section is to provide the element equations of motion by applying the hybrid stress FEM pioneered by Pian [14]. In addition, it is shown by way of derivation of the element mass and stiffness matrices that the hybrid stress-based FEM can give results identical to those obtained by the displacement formulation-based FEM. The hybrid stress-based formulation is presented in Sub-section 1.3.1 whereas the derivation of the element matrices is included in Sub-section 1.3.2.

1.3.1   Derivation of element equations of motion

The Hellinger-Reissner's variational principle is adopted in this sub-section

$$
\pi_{HR} = \iiint_V \left(\sigma^T \mathcal{L} u - \frac{1}{2} \sigma^T C \sigma \right) dV - \iiint_V b^T u \, dV
$$
$$
- \iint_{S_u} (\Gamma \sigma)^T (u - \bar{u}) dS - \iint_{S_t} \tau^T u \, dS ,
$$

(1.42)

where σ is the stress vector, u is the displacement vector, C is the compliance matrix, b is the body force vector, τ is the prescribed traction vector on boundary S_t, $\bar{u}$ is the prescribed displacement on boundary S_u, $\mathcal{L}$ is the linear differential operator to derive strain from displacement, and Γ is the linear differential operator to evaluate surface traction from stress.

In dynamic problems, one can introduce the kinetic energy term to Eq. (1.42) such that a new functional is formed

$$
T - \pi_{HR} = \frac{1}{2} \iiint_V \rho \dot{u}^T \dot{u} \, dV - \iiint_V \left(\sigma^T \mathcal{L} u - \frac{1}{2} \sigma^T C \sigma \right) dV
$$
$$
+ \iiint_V b^T u \, dV + \iint_{S_u} (\Gamma \sigma)^T (u - \bar{u}) dS + \iint_{S_t} \tau^T u \, dS .
$$

(1.43)

It is observed that the lhs of Eq. (1.43) can be identified as the Lagrangian in Eq. (1.11). Therefore, Hamilton's principle can be applied. A formal presentation is included in Section 1.4 for nonlinear dynamic problems.

Returning to the linear element equation of motion, the assumed displacement field and assumed stress field are, respectively, given by

$$
u = Nq , \qquad \sigma = P\beta , \qquad \text{(1.44a, b)}
$$

where β, different from that in Eq. (1.31b), is the vector of stress parameters, q and N are defined in Eq. (1.1), and P is the stress shape function matrix.

Substituting Eq. (1.44) into (1.43) and some manipulation, one has

$$
T - \pi_{HR} = \pi_{DHR} = \frac{1}{2} \dot{q}^T m \dot{q} - \frac{1}{2} \beta^T H \beta - \beta^T G q - f^T q , \quad \text{(1.45)}
$$

where π_{DHR} is similar to L in Eq. (1.11), f the nodal force vector, and

$$H = \int\int\int_V P^T C P \, dV, \quad G = \int\int\int_V P^T \mathcal{L} N \, dV, \quad m = \int\int\int_V \rho \, N^T N \, dV,$$

in which H and G are known as, respectively, the generalized stiffness matrix and leverage matrix.

For π_{DHR} in Eq. (1.45) to have an extremum value, one has

$$\delta\pi_{DHR} = \frac{\partial\pi_{DHR}}{\partial t}\delta t + \frac{\partial\pi_{DHR}}{\partial\beta}\delta\beta + \frac{\partial\pi_{DHR}}{\partial q}\delta q = 0, \qquad (1.46)$$

in which it is understood that the division of a quantity by a vector is not permitted in matrix operations. However, it is used as an abbreviation for the partial differentiation of all the elements or entries in the vector concerned. Since δt, $\delta\beta$, and δq are all arbitrary, Eq. (1.46) holds only if the following equations are satisfied

$$\frac{\partial\pi_{DHR}}{\partial t} = 0, \quad \frac{\partial\pi_{DHR}}{\partial\beta} = 0, \quad \frac{\partial\pi_{DHR}}{\partial q} = \frac{\partial\pi_{DHR}}{\partial t}\left(\frac{\partial t}{\partial q}\right) = 0. \quad (1.47\text{a, b, c})$$

When Eq. (1.47a) is satisfied, it simultaneously satisfies Eq. (1.47c). Thus, Eqs. (1.47b) and (1.47a) give, respectively

$$-H\beta - Gq = 0, \qquad \ddot{q}^T m - \beta^T G - f^T = 0. \qquad (1.48\text{a, b})$$

From Eq. (1.48a), one has

$$\beta = -H^{-1}Gq. \qquad (1.49)$$

Substituting Eq. (1.49) into the transpose of Eq. (1.48b) yields

$$m\ddot{q} + G^T H^{-1} G q - f = 0, \qquad m^T = m.$$

This equation becomes Eq. (1.16) with the definition,

$$k = G^T H^{-1} G. \qquad (1.50)$$

Once the nodal displacement vector is determined, it is substituted into Eq. (1.49) which is substituted, in turn, into Eq. (1.44b) to recover the stress vector.

1.3.2   Mass and stiffness matrices of uniform beam element

To provide an understanding of the steps involved in the derivation of element mass and stiffness matrices by applying the hybrid stress or mixed formulation, and for illustration as well as for simplicity, the beam element of uniform cross-sectional area A and length ℓ as shown in Figure 1.1 is considered in this sub-section. Its material is assumed to be isotropic and homogeneous. It has two nodes, each of which has two dof as in Sub-section 1.1.2. Thus, the shape function matrix in Eq. (1.44a) is identical to that in Eq. (1.26). The assumed stress shape functions and stress parameters are related to the stress by the following equation

$$\sigma = P\beta = [\,1 \quad x \quad x^2\,]\beta\,, \qquad \beta = \begin{bmatrix} \beta_1 & \beta_2 & \beta_3 \end{bmatrix}^T\,.$$

Applying the definitions in Eq. (1.45), the element mass matrix is identical to that given by Eq. (1.27) since it is the same beam element with identical shape functions. Similarly, the generalized stiffness matrix becomes

$$H = \frac{1}{EI}\int_0^\ell \begin{pmatrix} 1 \\ x \\ x^2 \end{pmatrix}\begin{bmatrix} 1 & x & x^2 \end{bmatrix} dx = \frac{\ell}{60EI}\begin{bmatrix} 60 & 30\ell & 20\ell^2 \\ & 20\ell^2 & 15\ell^3 \\ symmetric & & 12\ell^4 \end{bmatrix}. \quad (1.51)$$

The inverse of the generalized stiffness matrix is found as

$$H^{-1} = \frac{3EI}{\ell}\begin{bmatrix} 3 & -12/\ell & 10/\ell^2 \\ & 64/\ell^2 & -60/\ell^3 \\ symmetric & & 60/\ell^4 \end{bmatrix}. \quad (1.52)$$

The leverage matrix is

$$G = \int_0^\ell \begin{pmatrix} 1 \\ x \\ x^2 \end{pmatrix}\frac{\partial^2 N}{\partial x^2} dx = \begin{bmatrix} 0 & -1 & 0 & 1 \\ 1 & 0 & -1 & \ell \\ \ell & \ell^2/6 & -\ell & 5\ell^2/6 \end{bmatrix}. \quad (1.53)$$

By making use of Eqs. (1.52), (1.53), (1.50), and after some manipulation, the element stiffness matrix is found to be identical to that given by Eq. (1.28).

Now, consider a lower order stress shape function matrix so that

$$\sigma = P\beta = [\,1 \quad x\,]\beta\,, \qquad \beta = \begin{bmatrix} \beta_1 & \beta_2 \end{bmatrix}^T\,.$$

In this case, the generalized stiffness matrix becomes

$$H = \frac{1}{EI} \int_0^\ell \begin{pmatrix} 1 \\ x \end{pmatrix} \begin{bmatrix} 1 & x \end{bmatrix} dx = \frac{\ell}{6EI} \begin{bmatrix} 6 & 3\ell \\ 3\ell & 2\ell^2 \end{bmatrix}. \tag{1.54}$$

The inverse of this matrix is

$$H^{-1} = \frac{2EI}{\ell^3} \begin{bmatrix} 2\ell^2 & -3\ell \\ -3\ell & 6 \end{bmatrix}. \tag{1.55}$$

The corresponding leverage matrix is

$$G = \int_0^\ell \begin{pmatrix} 1 \\ x \end{pmatrix} \frac{\partial^2 N}{\partial x^2} dx = \begin{bmatrix} 0 & -1 & 0 & 1 \\ 1 & 0 & -1 & \ell \end{bmatrix}. \tag{1.56}$$

By making use of Eqs. (1.55), (1.56), and (1.50), one arrives at the identical element stiffness matrix defined by Eq. (1.28) which agrees with that presented in [15, 16]. This confirms the fact that the displacement and hybrid stress formulations are equivalent [17].

   The two and three stress parameters of the assumed stress fields give an identical element stiffness matrix because they satisfy the Tong, Pian and Chen condition [18, 19],

$$n_s \geq n_d - n_r, \tag{1.57}$$

where n_s is the number of assumed stress modes, n_d is the number of generalized displacements, and n_r is the number of zero-eigenvalues or rigid-body modes. In the three stress parameter case, the number of assumed stress modes $n_s = 3$, the number of generalized displacements $n_d = 4$, and the number of rigid-body modes $n_r = 2$. Therefore, Eq. (1.57) is satisfied. Similarly, for the two stress parameter case, $n_s = 2$, $n_d = 4$, and $n_r = 2$. Thus, Eq. (1.57) is also satisfied and therefore identical element stiffness matrix is obtained.

1.4 Incremental Variational Principle and Mixed Formulation-Based Nonlinear Element Matrices

   In this section the incremental mixed formulation for nonlinear element equations of motion and derivation of the nonlinear mass and stiffness matrices for lower order flat triangular shell elements are presented. The formulation and derivation of element matrices closely follow those presented in [20, 21]. As the detailed presentation has been provided in the latter reference, the following is an outline of [21] with some changes of symbols in the present section.

In addition to other advantages over the displacement formulation, the two main ones are: (a) the elements derived are free of the locking phenomenon, and (b) the strains and stresses are continuous through the application of Eq. (1.72).

In the following, Sub-section 1.4.1 deals with the formulation and linearization of the nonlinear incremental variational principle while Sub-section 1.4.2 includes an outline of the derivation of shell element matrices.

1.4.1   Incremental variational principle and linearization

Consider the dynamic counterpart of the Hellinger-Reissner variational principle for every element of the nonlinear structural systems [22] where the kinetic energy of the system is defined by

$$\delta \int_{t_1}^{t_2} \left(T - \pi_{HR} \right) d\tau = 0 , \qquad (1.58)$$

π_{HR} is the Hellinger-Reissner's functional, and the remaining symbols have

$$T = \frac{1}{2} \int\int\int_V \rho \, \dot{u}^T \dot{u} \, dV , \qquad (1.59)$$

their usual meaning. Note that the integrand in Eq. (1.58) is π_{DHR} in Eq. (1.45).

For nonlinear dynamic systems the incremental approach is adopted in the response analysis. Therefore, in the following the incremental variational principle and linearization are first presented.

As pointed out in [20, 21], the fundamental difficulty in any nonlinear analysis is the unknown configuration of a body at time $(t+\Delta t)$. In obtaining an approximate solution, the static and kinematic variables in the current configuration C^t of the incremental formulation are assumed to be known. Their values in an unknown neighbouring configuration $C^{t+\Delta t}$ at a later time $(t+\Delta t)$ are determined from the known solutions. In the present study the starting point of such an incremental analysis is the incremental or modified Hellinger-Reissner variational principle. It has two independently assumed fields, the incremental generalized displacement and the incremental strain fields. The incremental Hellinger-Reissner variational principle can be written as [21]

$$\Delta\pi_{HR}(\Delta u, \Delta e) = \frac{1}{2} \int_{V^t} \left[-(\Delta e)^T D(\Delta e) + 2\sigma^T(\Delta e^u) \right.$$

$$\left. + 2(\Delta e)^T D(\Delta e^u) \right] dV - W^{t+\Delta t} ,$$

(1.60)

where the integration is performed over the reference volume V^t,

  Δu is the vector of assumed incremental displacement,

  Δe is the vector of assumed incremental Green strain,

  Δe^u is the vector of incremental Washizu strain calculated from the vector Δu, see Eq. (1.61) in the following,

  D is the time-dependent material elastic matrix so that $\Delta S = D\Delta e$ with ΔS being the incremental second Piola-Kirchhoff (PK2) stress vector, see later in the following,

  σ is the Cauchy or true stress vector at time t, and

  $W^{t+\Delta t}$ is the work-equivalent term corresponding to prescribed body-force and surface traction in configuration $C^{t+\Delta t}$.

In Eq. (1.60), the component form of the incremental Washizu strain vector Δe^u is given by

$$\Delta e^u_{ij} = \Delta\varepsilon_{ij} + \Delta\eta_{ij} ,$$

$$\Delta\varepsilon_{ij} = \frac{1}{2}\left(\Delta u_{i,j} + \Delta u_{j,i}\right), \quad \Delta\eta_{ij} = \frac{1}{2}\Delta u_{k,i}\Delta u_{k,j} ,$$

(1.61)

where the Einstein summation convention for indices has been adopted for the integer k and the differentiation is with respect to the reference co-ordinates, x_i^t with $i = 1, 2, 3$, at time t.

In the hybrid strain or mixed formulation FEM it is assumed that within every element one can write

$$\Delta u = N\Delta q , \quad \Delta e = P\Delta\beta ,$$

(1.62a, b)

where Δq is the vector of incremental generalized nodal displacements and $\Delta\beta$ the vector of incremental strain parameters, while N and P are the matrices of displacement interpolation functions and strain interpolation functions. Substituting Eq. (1.62) into Eq. (1.60) gives

$$\Delta\pi_{HR}(\Delta q, \Delta\beta) = U - W^{t+\Delta t}(\Delta q) ,$$

(1.63)

where

$$U = \frac{1}{2} \int\limits_{V_e} \Big[-(\Delta\beta)^T P^T D P (\Delta\beta) + 2\sigma^T B_L \Delta q$$

$$+ (\Delta q)^T (B_{NL})^T \sigma_1{}^T B_{NL} (\Delta q) + 2(\Delta\beta)^T P^T D B_L \Delta q$$

$$+ 2(\Delta e)^T D(\Delta\eta) \Big] dV ,$$

V_e is the volume of an element at the reference configuration,
σ_1 is the matrix that contains the Cauchy stress components at t,
B_L is the linear strain-displacement matrix, and
B_{NL} is the nonlinear strain-displacement matrix.

The last term within the square brackets in the last equation contains the third order product of Δq and $\Delta\beta$ and therefore will be disregarded when linearizing Eq. (1.60) or Eq. (1.63).

Note that in arriving at Eq. (1.63) the following two relations have been applied

$$\Delta\varepsilon = (B_L)(\Delta q), \quad \sigma^T \Delta\eta = \frac{1}{2}(\Delta q)^T (B_{NL})^T (\sigma_1)^T (B_{NL})(\Delta q). \quad (1.64a, b)$$

To proceed further one can define

$$H = \int\limits_{V_e} P^T D P \, dV , \qquad G = \int\limits_{V_e} P^T D B_L \, dV , \qquad (1.65a, b)$$

$$k_{NL} = \int\limits_{V_e} (B_{NL})^T \sigma_1 B_{NL} \, dV , \qquad F_1 = \int\limits_{V_e} (B_L)^T \sigma \, dV , \qquad (1.65c, d)$$

and substitutes Eq. (1.65) into Eq. (1.63) to give

$$\Delta\pi_{HR}(\Delta q, \Delta\beta) = \sum_{i=1}^{5} \pi_e^{(i)} , \qquad (1.66)$$

in which

$$\pi_e^{(1)} = -\frac{1}{2}(\Delta\beta)^T H(\Delta\beta) , \quad \pi_e^{(2)} = F_1(\Delta q) , \quad \pi_e^{(3)} = \frac{1}{2}(\Delta q)^T k_{NL}(\Delta q) ,$$

$$\pi_e^{(4)} = (\Delta\beta)^T G(\Delta q) , \qquad \pi_e^{(5)} = -[F(t+\Delta t)]^T (\Delta q) ,$$

with $F(t+\Delta t)$, the externally generalized nodal force vector in $C^{t+\Delta t}$ associated with the $W^{t+\Delta t}(\Delta q)$ term on the rhs of Eq. (1.63).

Applying stationarity to Eq. (1.66) with respect to $\Delta\beta$ yields

$$\Delta\beta = H^{-1}G(\Delta q) . \tag{1.67}$$

Substituting Eq. (1.67) into Eq. (1.66) and applying stationarity with respect to Δq, it results in the following equilibrium equation

$$\left(k_L + k_{NL}\right)(\Delta q) = F(t+\Delta t) - F_1 , \tag{1.68}$$

where the linear or small displacement element stiffness matrix

$$k_L = G^T H^{-1}G , \tag{1.69}$$

and the matrix k_{NL} on the lhs of Eq. (1.68) and defined by Eq. (1.65c) is the element initial stress stiffness matrix.

On the rhs of Eq. (1.68), F_1 is the pseudo-force vector. One may rewrite $F(t+\Delta t) - F_1$ as $\Delta F + F(t) - F_1$. Thus, $F(t+\Delta t) - F_1$ consists of ΔF, the incremental external force from the current time t to the next time step $(t+\Delta t)$, and $F(t) - F_1$, the equilibrium imbalance at time t. If the equilibrium at time t is satisfied in an average sense, $F(t) - F_1$ vanishes and $F(t+\Delta t) - F_1$ reduces to the increment of external force ΔF from t to $(t+\Delta t)$. Note that Eq. (1.60) differs from those presented by Boland and Pian [23], and Saleeb $et\ al.$ [24]. The variational principle applied in Refs. [23, 24] was based on the hybrid stress formulation and it contained one additional term, which is, in the present notation,

$$\int_{V^t} \left[(\Delta e)^T D\left(e - e^u\right) \right] dV \tag{1.70}$$

where Δe and D are the same quantities in Eq. (1.60), but e is the vector of Almansi strain accumulated from Δe. The vector e^u is the Almansi strain vector calculated from the total displacements u. In other words,

$$e^u = \tfrac{1}{2}\left(u_{i,j} + u_{j,i} - u_{k,i}u_{k,j}\right) .$$

The aforementioned additional term, Eq. (1.70), accounts for the compatibility mismatch due to inaccurate total displacements and strains. Boland and Pian [23] argued that the term should not be expected to vanish. On the other hand, the numerical experiments of Saleeb $et\ al.$ [24] showed that, though totally discarding the term resulted in convergence difficulties, including the term in only the first iteration of every load step yielded essentially the same results as those having the term under all circumstances. In the present formulation, the compatibility mismatch term vanishes as a consequence of the presently employed hybrid strain formulation. Specifically, $(\Delta e)^T D(e \text{-} e^u)$ results in third and fourth order products of strain terms after applying the linearization to the incremental variational principle. Thus, these third and fourth product terms are eliminated after linearization.

The above element level formulas are applied to every element. Once all the element matrices have been determined, they are assembled to form the global equilibrium equation. This equation is then solved for the displacement increments.

Finally, at the element level the incremental Washizu strain can be obtained by using Eqs. (1.62b) and (1.67) as

$$\Delta e = P(\Delta\beta) = PH^{-1}G(\Delta q) ,\qquad(1.71)$$

whereas the incremental PK2 stress can be found by the following equation

$$\Delta S = D\Delta e .\qquad(1.72)$$

1.4.2   Linear and nonlinear element stiffness matrices

This section is concerned with the presentation of an outline of the derivation of element stiffness matrices. The element geometry and co-ordinate systems are introduced first. The assumed incremental displacement field and the assumed incremental strain field within an element are considered subsequently. These are followed by the derivation of the linear and initial stress stiffness matrices. A simplified version of the stiffness matrices from those derived with the nonlinear formulation is included.

1.4.2.1 Element geometry and co-ordinate systems

The finite elements under consideration are the three-node flat triangular shell finite elements. An example is shown in Figure 1.3. The three nodes are located at the three corners of the mid-surface of the element. A local rectangular co-ordinate system is attached to Node 1, with its r-axis coinciding with the side 1-2, its t-axis being parallel to the normal of the element and its s-axis perpendicular to the r-t plane. With such a co-ordinate system, the r and s co-ordinates of Nodes 1, 2, and 3 are: $(0,0)$, $(r_2 ,0)$ and (r_3 ,s_3), respectively. A director orthogonal frame with components d_r, d_s and d at any points on the mid-surface is also considered. In the undeformed configuration, the director d

coincides with the normal to the mid-surface of the shell element. However, as the shell deforms, the director is, in general, not normal to the mid-surface. Thus, the director orthogonal frame differs from point to point and from the rectangular co-ordinate system. The director orthogonal frame serves as a basis of measuring and recording the change of orientation of points situated on the mid-surface. As indicated in Figure 1.3 the nodal dof are:

u being the displacement in the r-direction and should not be confused with the assumed displacement vector defined in Eq. (1.62a),

v being the displacement in the s-direction,

w being the displacement in the t-direction,

θ_r being the rotation component about the r-axis,

θ_s being the rotation component about the s-axis, and

θ_t being the rotation about the t-axis, or the drilling dof (ddof).

The right-handed screw rule is adopted in the present formulation.

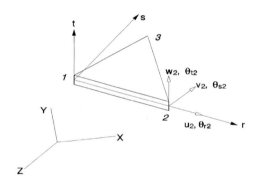

Figure 1.3   Shell element in local and global co-ordinate systems.

1.4.2.2 Assumed incremental displacement field within an element

The isoparametric approach is employed. The local co-ordinates (r,s,t) of an arbitrary point within the element can be written as

$$\begin{pmatrix} r^t \\ s^t \\ t^t \end{pmatrix} = \sum_{i=1}^{3} \xi_i \begin{pmatrix} \bar{r}_i^t \\ \bar{s}_i^t \\ 0 \end{pmatrix} + \eta^t \sum_{i=1}^{3} \xi_i d_i^t , \qquad (1.73)$$

where the superscript t denotes time t and the over bars indicate variables that are defined on the mid-surface. Recall that rotational or angular displacements as well as directors and their increments are defined on the mid-surface. Therefore, over-bars will not be placed on them for conciseness. The symbol d_i^t, with $i = 1, 2, 3$, denotes the director of Node i at time t, while ξ_i are the natural or area co-ordinates of the triangle satisfying

$$0 \le \xi_i \le 1 , \quad \sum_{i=1}^{3} \xi_i = 1 , \qquad (1.74a,b)$$

and η^t is the co-ordinate along the director direction satisfying

$$-\frac{h^t}{2} \le \eta^t \le \frac{h^t}{2} \qquad (1.74c)$$

with h^t being the thickness of the shell at time t. Unless stated otherwise, h^t is considered constant over the entire element. The first summation of Eq. (1.73) represents the position of the mid-surface while the second summation indicates that the director orthogonal frame is interpolated in exactly the same way as the mid-surface r and s co-ordinates. This scheme of interpolation is commonly known as the continuum consistent interpolation [25].

The incremental displacements of any point within the element from time t to $t+\Delta t$ can be expressed as

$$\begin{pmatrix} \Delta u^t \\ \Delta v^t \\ \Delta w^t \end{pmatrix} = \sum_{i=1}^{3} \xi_i \begin{pmatrix} \Delta \bar{u}_i^t \\ \Delta \bar{v}_i^t \\ \Delta \bar{w}_i^t \end{pmatrix} + \eta^t \sum_{i=1}^{3} \xi_i \left(\Delta d_i^t \right) . \qquad (1.75)$$

In this equation the first summation term contains incremental displacements of any point located on the mid-surface. The second summation term includes the change in orientation of director of any point on the mid-surface which is interpolated from Δd_i^t, the increment of director of Node i from time t to $t+\Delta t$. By following the steps in Refs. [20, 21], one can obtain the shape function matrix through the following equation

$$\begin{bmatrix} \Delta u^t & \Delta v^t & \Delta w^t \end{bmatrix}^T =$$

$$N \begin{bmatrix} \Delta \bar{u}_1^{(0)} & \Delta \theta_1^{(0)} \dots \Delta \bar{u}_3^{(0)} & \Delta \theta_3^{(0)} \end{bmatrix}^T , \qquad (1.76)$$

where the displacement shape function matrix for the shell element is

$$N = \begin{bmatrix} N_{11} & N_{12} & N_{13} \end{bmatrix}_{3\times18} , \tag{1.77}$$

in which the incremental displacement row vectors

$$\Delta\bar{u}_i^{(0)} = \begin{bmatrix} \Delta\bar{u}_i^t & \Delta\bar{v}_i^t & \Delta\bar{w}_i^t \end{bmatrix}, \quad \Delta\theta_i^{(0)} = \begin{bmatrix} \Delta\theta_{ri}^t & \Delta\theta_{si}^t & \Delta\theta_{ti}^t \end{bmatrix}, \quad i = 1,2,3,$$

and the 3×6 sub-matrix N_{1i} is defined by

$$N_{1i} = \begin{bmatrix} \xi_i & 0 & 0 & \eta^t \xi_i \Lambda_{i(11)}^t & \eta^t \xi_i \Lambda_{i(12)}^t & \bar{p}_i \\ 0 & \xi_i & 0 & \eta^t \xi_i \Lambda_{i(21)}^t & \eta^t \xi_i \Lambda_{i(22)}^t & \bar{q}_i \\ 0 & 0 & \xi_i & \eta^t \xi_i \Lambda_{i(31)}^t - \bar{p}_i & \eta^t \xi_i \Lambda_{i(32)}^t - \bar{q}_i & 0 \end{bmatrix}, \tag{1.78}$$

where the 3×2 matrix Λ_i^t is

$$\Lambda_i^t = \begin{bmatrix} \Lambda_{i(11)}^t & \Lambda_{i(12)}^t \\ \Lambda_{i(21)}^t & \Lambda_{i(22)}^t \\ \Lambda_{i(31)}^t & \Lambda_{i(32)}^t \end{bmatrix} = -\Omega_i^t \left(\Gamma_i^s \right)^t , \tag{1.79}$$

where the skew-symmetric matrix

$$\Omega_i^t = \begin{bmatrix} 0 & -d_{ti}^t & d_{si}^t \\ d_{ti}^t & 0 & -d_{ri}^t \\ -d_{si}^t & d_{ri}^t & 0 \end{bmatrix} , \tag{1.80}$$

and $(\Gamma_i^s)^t$ consists of the first two columns of the exponential mapping Γ_i^t which is an orthogonal matrix associated with Node i at time t. The exponential mapping satisfies the following relation

$$d_i^t = \Gamma_i^t e_3 , \tag{1.81}$$

with e_3 being the unit vector along the t-axis. That is, $e_3 = [0,0,1]^T$. The lhs of Eq. (1.81) is the current position of the director at Node i and is known from configuration updating [20, 21].

That is, the known director is given by

$$d_i^t = R(\Delta\theta) d_i^{t-\Delta t} ,$$ (1.82)

where the transformation matrix

$$R(\Delta\theta) = \cos(|\Delta\theta|) I_3 + \frac{\sin(|\Delta\theta|)}{|\Delta\theta|}(\Delta\theta^s) ,$$ (1.83)

in which $\Delta\theta = (\Delta\theta_i)^{t-\Delta t}$, and $|\cdot|$ denotes the magnitude of the enclosed vector, I_3 is the unity matrix of order 3 while $\Delta\theta^s$ is a skew-symmetric matrix constructed from $\Delta\theta$,

$$\Delta\theta^s = \begin{bmatrix} 0 & -\Delta\theta_t & \Delta\theta_s \\ \Delta\theta_t & 0 & -\Delta\theta_r \\ -\Delta\theta_s & \Delta\theta_r & 0 \end{bmatrix} .$$ (1.84)

Applying Eq. (1.81) the matrix Γ_i^t can be determined and therefore Λ_i^t in Eq. (1.79) can be found.

Substituting Eq. (1.79) into (1.78) the displacement shape functions can be obtained. Note that in Eq. (1.78)

$$\bar{p}_1 = (a_{31}\xi_3 - a_{12}\xi_2)\xi_1 , \qquad \bar{p}_2 = (a_{12}\xi_1 - a_{23}\xi_3)\xi_2 ,$$

$$\bar{p}_3 = (a_{23}\xi_2 - a_{31}\xi_1)\xi_3 , \qquad \bar{q}_1 = (b_{31}\xi_3 - b_{12}\xi_2)\xi_1 ,$$

$$\bar{q}_2 = (b_{12}\xi_1 - b_{23}\xi_3)\xi_2 , \qquad \bar{q}_3 = (b_{23}\xi_2 - b_{31}\xi_1)\xi_3 ,$$

with the coefficients on the rhs being defined as

$$2a_{12} = \ell_{12}\cos\gamma_{12} , \qquad 2b_{12} = \ell_{12}\sin\gamma_{12} ,$$

$$2a_{23} = \ell_{23}\cos\gamma_{23} , \qquad 2b_{23} = \ell_{23}\sin\gamma_{23} ,$$

$$2a_{31} = \ell_{31}\cos\gamma_{31} , \qquad 2b_{31} = \ell_{31}\sin\gamma_{31} ,$$

in which, as shown in Figure 1.4, ℓ_{ij} is the length of the side of the triangular element joining Nodes i and j.

In closing, it should be noted that because of the incorporation of the ddof in the foregoing formulation, the displacement shape function matrix defined by Eq. (1.78) makes the present finite element sub-parametric since different schemes are employed for the geometry and displacements.

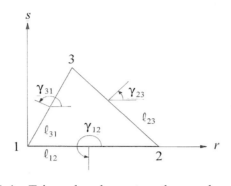

Figure 1.4   Triangular element on the *r-s* plane of the
local co-ordinate system.

1.4.2.3 Assumed incremental strain field within an element
The assumed strain field for any point within the element is defined as

$$\left[\Delta\varepsilon_r \ \ \Delta\varepsilon_s \ \ \Delta\varepsilon_{rs} \ \ \Delta\varepsilon_{st} \ \ \Delta\varepsilon_{tr} \ \ \Delta\varepsilon^s_{rs}\right]^T =$$

$$P\left[\Delta\beta_1 \ \ \Delta\beta_2 \ \ \cdots \ \ \Delta\beta_9 \ \ \Delta\beta_{10}\right]^T, \tag{1.85}$$

where the strain shape function matrix is defined by

$$P = \begin{bmatrix} \left[P_5^{(1)} \quad P_5^{(2)}\right]_{5\times9} & (0)_{5\times1} \\ [0]_{1\times9} & 1 \end{bmatrix}_{6\times10}, \tag{1.86a}$$

in which the sub-matrix $[P_5^{(1)} \ \ P_5^{(2)}]_{5\times9}$ is the strain shape function matrix
without the ddof. The two component sub-matrices are

$$P_5^{(1)} = \begin{bmatrix} 1 & 0 & 0 & \eta' & 0 & 0 \\ 0 & 1 & 0 & 0 & \eta' & 0 \\ 0 & 0 & 1 & 0 & 0 & \eta' \\ 0 & 0 & 0 & 0 & 0 & 0 \\ 0 & 0 & 0 & 0 & 0 & 0 \end{bmatrix}_{5\times6}, \tag{1.86b}$$

$$P_5^{(2)} = \begin{bmatrix} 0 & 0 & 0 \\ 0 & 0 & 0 \\ 0 & 0 & 0 \\ -s_3(1-2\xi_2) & s_3(2\xi_2+2\xi_3-1) & 0 \\ -r_3(1-2\xi_2) & r_{32}(2\xi_2+2\xi_3-1) & r_2(1-2\xi_3) \end{bmatrix}_{5\times3} , \quad (1.86c)$$

with $r_{32} = r_3 - r_2$, and the superscript s on the lhs of Eq. (1.85) denoting the skew-symmetric part of the strain tensor. Note that the incremental assumed strain components $\Delta\varepsilon_{rs}$, $\Delta\varepsilon_{st}$, and $\Delta\varepsilon_{tr}$ employ the engineering definition. Of the strain parameters, $\Delta\beta_1$ through $\Delta\beta_3$ are associated with membrane strains, $\Delta\beta_4$ through $\Delta\beta_6$ bending strains, $\Delta\beta_7$ through $\Delta\beta_9$ transversal strains, and $\Delta\beta_{10}$ the skew-symmetric strain component.

1.4.2.4 *Linear element stiffness matrix k_L*

With reference to Eq. (1.68), the element stiffness matrix is the sum of linear and initial stress stiffness matrices, k_L and k_{NL}. The outline of the derivation of the linear stiffness matrix k_L is presented in the following.

Recall that the element linear stiffness matrix is defined by Eq. (1.69) in which the inverse of the generalized stiffness matrix H and the leverage matrix G are required. These, in turn, require the strain shape function matrix P, and the strain and displacement matrix B_L for a given material elastic matrix D. Since P is obtained by using Eq. (1.86a) and therefore B_L is required.

Symbolically, one can write

$$B_L = \begin{bmatrix} B_{L_1} & B_{L_2} & B_{L_3} \end{bmatrix}, \quad (1.87)$$

where the sub-matrices are defined by

$$B_{L_i} = B_{L_{mi}} + B_{L_{bi}} + B_{L_{si}} + B_{L_{ti}}, \quad i = 1, 2, 3, \quad (1.88a)$$

in which the first, second, third, and fourth terms on the rhs are respectively associated with the membrane, bending, shear, and torsional components of the element linear stiffness matrix. The membrane and bending parts are

$$B_{L_{mi}} = \begin{bmatrix} B_{Li}^m \\ [0]_{3\times6} \end{bmatrix}, \quad B_{L_{bi}} = \begin{bmatrix} B_{Li}^b \\ [0]_{3\times6} \end{bmatrix}, \quad (1.88b, c)$$

in which $[0]_{3 \times 6}$ is a 3×6 null matrix and

$$
B_{Li}^m = \begin{bmatrix}
\xi_{i,r} & 0 & 0 & 0 & 0 & \bar{p}_{i,r} \\
0 & \xi_{i,s} & 0 & 0 & 0 & \bar{q}_{i,s} \\
\xi_{i,s} & \xi_{i,r} & 0 & 0 & 0 & \bar{p}_{i,s}+\bar{q}_{i,r}
\end{bmatrix},
$$

$$
B_{Li}^b = \begin{bmatrix}
0 & 0 & 0 & \eta^t \xi_{i,r} \Lambda_{i(11)}^t & \eta^t \xi_{i,r} \Lambda_{i(12)}^t & 0 \\
0 & 0 & 0 & \eta^t \xi_{i,s} \Lambda_{i(21)}^t & \eta^t \xi_{i,s} \Lambda_{i(22)}^t & 0 \\
0 & 0 & 0 & b_{34}^b & b_{35}^b & 0
\end{bmatrix},
$$

where the elements or entries have been defined in Eq. (1.78) except

$$
b_{34}^b = \eta^t \left(\xi_{i,r} \Lambda_{i(21)}^t + \xi_{i,s} \Lambda_{i(11)}^t \right), \qquad b_{35}^b = \eta^t \left(\xi_{i,r} \Lambda_{i(22)}^t + \xi_{i,s} \Lambda_{i(12)}^t \right),
$$

$$
\xi_{1,r} = -1/r_2^t, \quad \xi_{1,s} = \left(r_3^t - r_2^t \right) \left(r_2^t s_3^t \right)^{-1}, \quad \xi_{2,r} = 1/r_2^t,
$$

$$
\xi_{2,s} = -r_3^t \left(r_2^t s_3^t \right)^{-1}, \quad \xi_{3,r} = 0, \quad \xi_{3,s} = 1/s_3^t,
$$

$$
\bar{p}_{i,s} = \frac{\partial \bar{p}_i}{\partial s}, \qquad \bar{p}_{i,r} = \frac{\partial \bar{p}_i}{\partial r}, \qquad \bar{q}_{i,s} = \frac{\partial \bar{q}_i}{\partial s}, \qquad \bar{q}_{i,r} = \frac{\partial \bar{q}_i}{\partial r}.
$$

The shear component on the rhs of Eq. (1.88a) is defined as

$$
B_{L_{si}} = \begin{bmatrix} [0]_{3 \times 6} \\ B_{Li}^s \end{bmatrix}, \qquad B_{Li}^s = \begin{bmatrix}
0 & 0 & \xi_{i,s} & b_{44}^s & b_{45}^s & 0 \\
0 & 0 & \xi_{i,r} & b_{54}^s & b_{55}^s & 0 \\
0 & 0 & 0 & 0 & 0 & 0
\end{bmatrix}, \qquad (1.88\text{d, e})
$$

where

$$
b_{44}^s = \eta^t \xi_{i,s} \Lambda_{i(31)}^t - \bar{p}_{i,s} + \xi_i \Lambda_{i(21)}^t,
$$

$$b_{45}^s = \eta' \xi_{i,s} \Lambda_{i(32)}^t - \bar{q}_{i,s} + \xi_i \Lambda_{i(22)}^t , \qquad b_{54}^s = \eta' \xi_{i,r} \Lambda_{i(31)}^t - \bar{p}_{i,r} + \xi_i \Lambda_{i(11)}^t ,$$

$$b_{55}^s = \eta' \xi_{i,r} \Lambda_{i(32)}^t - \bar{q}_{i,r} + \xi_i \Lambda_{i(12)}^t .$$

The torsional component by the displacement formulation is defined by

$$
B_{L_t} = \frac{1}{2} \left[\frac{r_3 - r_2}{r_2 s_3}, \frac{1}{r_2}, 0, 0, 0, b_6^d, -\frac{r_3}{r_2 s_3}, -\frac{1}{r_2}, \right.
$$

$$
\left. 0, 0, 0, b_{12}^d, \frac{1}{s_3}, 0, 0, 0, 0, b_{18}^d \right]_{1 \times 18} , \tag{1.88f}
$$

in which

$$b_6^d = 2\xi_1 + \left(\bar{p}_{1,s} - \bar{q}_{1,r} \right) , \qquad b_{12}^d = 2\xi_2 + \left(\bar{p}_{2,s} - \bar{q}_{2,r} \right) ,$$

$$b_{18}^d = 2\xi_3 + \left(\bar{p}_{3,s} - \bar{q}_{3,r} \right) .$$

With Eqs. (1.86a) and (1.87) defined, the element generalized stiffness matrix and the leverage matrix can be evaluated. Thus, the linear element stiffness matrix k_L can be determined. It is convenient to express it as

$$k_L = k_{L_m} + k_{L_b} + k_{L_s} + k_t , \tag{1.89}$$

where the component matrices are understood to be of the same order and in proper locations inside the matrices. They are defined by

$$k_{L_m} = \left(G_{L_m} \right)^T H^{-1} G_{L_m} , \qquad k_{L_b} = \left(G_{L_b} \right)^T H^{-1} G_{L_b} , \tag{1.90a, b}$$

$$k_{L_s} = \left(G_{L_s} \right)^T H^{-1} G_{L_s} , \qquad k_t = \tfrac{1}{2} G_e h^t \int_A \left(B_{L_t} \right)^T B_{L_t} \, dA , \tag{1.90c, d}$$

in which G_e is the shear modulus of elasticity of the material, H is evaluated by Eq. (1.65a), and G by Eq. (1.65b) such that

$$G_{L_m} = \int_V P^T D B_{L_m} \, dV, \qquad G_{L_b} = \int_V P^T D B_{L_b} \, dV, \qquad (1.91\text{a, b})$$

$$G_{L_s} = \int_V P^T D B_{L_s} \, dV, \qquad B_{L_m} = \begin{bmatrix} B_{L_{m1}} & B_{L_{m2}} & B_{L_{m3}} \end{bmatrix}, \qquad (1.91\text{c, d})$$

$$B_{L_b} = \begin{bmatrix} B_{L_{b1}} & B_{L_{b2}} & B_{L_{b3}} \end{bmatrix}, \; B_{L_s} = \begin{bmatrix} B_{L_{s1}} & B_{L_{s2}} & B_{L_{s3}} \end{bmatrix}, \; B_{L_t} = \begin{bmatrix} B_{L_{t1}} & B_{L_{t2}} & B_{L_{t3}} \end{bmatrix}, \quad (1.91\text{e, f, g})$$

where the sub-matrices have been defined in Eqs. (1.88a) through (1.88d). The order of every matrix in Eqs. (1.91c) through (1.91g) is appropriately adjusted to 6 × 18. For completeness, the explicit expressions for the element stiffness matrix defined by Eq. (1.89) are presented in Appendix 1B.

1.4.2.5 Element initial stress stiffness matrix k_{NL}

To obtain the element initial stress stiffness matrix k_{NL}, one first has the nonlinear strain-displacement matrix B_{NL} as

$$B_{NL} = \begin{bmatrix} \dfrac{\partial N_{1i}}{\partial r} & \dfrac{\partial N_{1i}}{\partial s} & \dfrac{\partial N_{1i}}{\partial \eta} \end{bmatrix}_{9 \times 18}, \qquad (1.92)$$

where N_{1i} with $i = 1, 2, 3$, are defined in Eq. (1.78). For instance,

$$\frac{\partial N_{1i}}{\partial \eta} = \frac{\partial N_{1i}}{\partial \eta'} = \begin{bmatrix} 0 & 0 & 0 & \xi_i \, \Lambda^t_{i(11)} & \xi_i \, \Lambda^t_{i(12)} & 0 \\ 0 & 0 & 0 & \xi_i \, \Lambda^t_{i(21)} & \xi_i \, \Lambda^t_{i(22)} & 0 \\ 0 & 0 & 0 & \xi_i \, \Lambda^t_{i(31)} & \xi_i \, \Lambda^t_{i(32)} & 0 \end{bmatrix}.$$

It is noted that the partial differentiation in Eq. (1.92) is with respect to the local co-ordinates at time t or in configuration C^t. According to Eq. (1.65c), the evaluation of k_{NL} requires the Cauchy stress matrix σ_1 in addition to the nonlinear strain-displacement matrix B_{NL}. Hence, the Cauchy stress matrix

$$\sigma_1 = \begin{bmatrix} \sigma_{11} I_3 & \left(\sigma_{12} + \overset{s}{\sigma}_{12} \right) I_3 & \sigma_{31} I_3 \\ \left(\sigma_{12} - \overset{s}{\sigma}_{12} \right) I_3 & \sigma_{22} I_3 & \sigma_{23} I_3 \\ \sigma_{31} I_3 & \sigma_{23} I_3 & O_3 \end{bmatrix} \qquad (1.93)$$

with O_3 being a 3×3 null matrix. Note that because of σ_{12}^s, matrix σ_1 is no longer symmetric. Consequently, k_{NL} is non-symmetric. However, in the present study σ_{12}^s is disregarded since its contribution has been assumed to be small. Further, in [26] Bufler has shown that for conservative systems tangent stiffness matrices are symmetric. Thus, Eq. (1.93) can be reduced to

$$
\sigma_1 = \begin{bmatrix} \sigma_{11} I_3 & \sigma_{12} I_3 & \sigma_{31} I_3 \\ \sigma_{12} I_3 & \sigma_{22} I_3 & \sigma_{23} I_3 \\ \sigma_{31} I_3 & \sigma_{23} I_3 & O_3 \end{bmatrix}. \tag{1.94}
$$

Equations (1.71), (1.72) and (1.86a) indicate that, the membrane components of the incremental PK2 stress, ΔS are constant within an element. The bending components vary in the thickness direction only. However, the transverse components vary linearly within the element. Consequently, the PK2 and Cauchy stresses at time t, S and σ are functions of mid-surface position. Their updating requires storage of stresses at the three nodes.

Returning to the integration of Eq. (1.65), if one considers σ as mid-surface position dependent, it would produce very tedious expressions. Therefore, in the present study the transverse stress components of σ are considered constant in the element. That is, all the stress components of σ are computed and updated only at the centroid of every element. The justification for such an operation is that the membrane and bending stresses in shell structures are usually dominant. Such an approach reduces substantially the computation efforts involved and requires much less computer storage space. Note that, in general, every σ_{ij} of Eqs. (1.93) and (1.94) is a combination of membrane, bending and transverse shear components. At any point, including the centroid of the element, membrane and transverse shear components are obtained by setting η^t to zero for the purpose of evaluating the nonlinear stiffness matrix k_{NL}. The stress at the top surface of the shell element can be obtained by substituting $\eta^t = h^t/2$. The difference between the stresses at the top and the mid-surface of the element, after being divided by $h^t/2$, is the slope of bending stress component. Then, σ_{ij} is re-written as

$$
\sigma_{ij} = \sigma_{ij(0)} + \eta^t \, d\sigma_{ij}, \qquad d\sigma_{ij} = \frac{2}{h^t}\left(\sigma_{ij(+)} - \sigma_{ij(0)}\right), \tag{1.95a, b}
$$

where $\sigma_{ij(0)}$ and $\sigma_{ij(+)}$ denote stresses at the middle and top surfaces of the shell element, respectively while $d\sigma_{ij}$ is the slope of the bending stress. Equation (1.95) is also valid for plastic deformation. In this case, $d\sigma_{ij}$ becomes zero because the non-layered approach has been employed in the present study. Of course, for the layered approach, $d\sigma_{ij}$ is not zero in general.

With B_{NL} and σ_1 now defined in Eqs. (1.92) and (1.94), they are substituted into Eq. (1.65c) so that k_{NL} can be obtained after evaluating the required integration. By applying the algebraic manipulation package MAPLE, explicit expressions for k_{NL} have been determined [20]. For brevity, these explicit expressions are not presented in the present book.

In closing, it should be mentioned that recovery of the linear element stiffness matrices from the nonlinear formulation developed in the foregoing has been made and presented [20, 21]. These recovered linear element stiffness matrices do not contain the directors and they have their usefulness to be employed for cases where the directors are not uniquely defined or are difficult to determine, typically at joints and discontinuities of shell structures. For reference the element stiffness matrices derived in the foregoing are called the director version of element stiffness matrices or simply the director version.

1.4.2.6 Consistent element mass matrices

In Refs. [20, 21] it was shown that the derivation of the consistent element mass matrix consists of three parts. The first part is concerned with the derivation of the translational component of the consistent element mass matrix m_u, the second part the rotational component without the ddof component of the consistent mass matrix m_r, and the third part the ddof component m_t. Symbolically, the consistent element mass matrix can be written as

$$m = m_u + m_r + m_t , \qquad (1.96)$$

where the first two terms on the rhs are defined by

$$m_u = \int_{A^t} \rho^t \, h^t \, N_u^T \, N_u \, dA , \qquad m_r = \int_{A^t} J^t \, N_r^T \, N_r \, dA . \qquad (1.97a, b)$$

This particular element is identical in both the director version and the non-director or simplified version provided in [27]. Note that the displacement shape function and angular displacement shape function matrices, and J^t in Eq. (1.97) are defined, respectively, by

$$
N_u = \begin{bmatrix} N_{u1} & N_{u2} & N_{u3} \end{bmatrix}, \qquad
N_{ui} =
\begin{bmatrix}
\xi_i & 0 & 0 & 0 & 0 & \bar{p}_i \\
0 & \xi_i & 0 & 0 & 0 & \bar{q}_i \\
0 & 0 & \xi_i & -\bar{p}_i & -\bar{q}_i & 0 \\
0 & 0 & 0 & 0 & 0 & 0 \\
0 & 0 & 0 & 0 & 0 & 0 \\
0 & 0 & 0 & 0 & 0 & 0
\end{bmatrix},
\qquad (1.98\text{a, b})
$$

$$
N_r = \begin{bmatrix} N_{r1} & N_{r2} & N_{r3} \end{bmatrix}, \qquad
N_{ri} =
\begin{bmatrix}
0 & 0 & 0 & 0 & 0 & 0 \\
0 & 0 & 0 & 0 & 0 & 0 \\
0 & 0 & 0 & 0 & 0 & 0 \\
0 & 0 & 0 & -\xi_i \Lambda^t_{i(21)} & -\xi_i \Lambda^t_{i(22)} & 0 \\
0 & 0 & 0 & \xi_i \Lambda^t_{i(11)} & \xi_i \Lambda^t_{i(12)} & 0 \\
0 & 0 & 0 & 0 & 0 & 0
\end{bmatrix},
\qquad (1.98\text{c, d})
$$

and $J^t = \rho^t (h^t)^3 / (12)$.

It is not difficult to recognize that the first three rows of Eq. (1.98b) are in fact constructed by applying Eq. (1.77) with the shell thickness co-ordinate η^t being set to zero because the displacement field is defined in terms of the mid-surface displacements.

It is observed that in the foregoing the derivation of consistent element mass matrices is with respect to the reference configuration C^t. Therefore, the mass matrices have to be updated at every time step since the density, thickness of the shell and element geometry change from one time step to another for large deformation problems. This approach is different from those in [28, 29], for example. In the latter two references the mass matrix was defined with respect to the undeformed configuration and was kept constant throughout the entire analysis. As pointed out in [20], for the updated Lagrangian formulation a consistent element mass matrix that was defined with respect to the reference state was obtained in [30]. However, numerical results were not included in the latter reference. In the present study, the option of updating the mass matrix has been kept as in Refs. [20, 21]. Comparisons of numerical results using constant or updated mass matrices have been investigated and reported in [31].

For completeness, explicit expressions for the consistent mass matrix of the triangular shell element are included in Appendix 1C.

1.5 Constitutive Relations and Updating of Configurations and Stresses

The constitutive relations are first introduced in Sub-sections 1.5.1 and 1.5.2. These relations are for linear, elastic and isotropic materials with small or finite strains, and elasto-plastic materials with isotropic strain hardening. The latter case also includes deformations of small and finite strains. The updating of configurations and stresses are dealt with in Sub-section 1.5.3.

1.5.1   Elastic materials

The constitutive fourth order tensor for homogeneous, isotropic and linearly elastic materials undergoing deformation of small strain is

$$D_{ijkl} = \frac{E}{1+v}\left[\frac{v}{1-2v}\delta_{ij}\delta_{kl} + \frac{1}{2}\left(\delta_{ik}\delta_{jl} + \delta_{il}\delta_{jk}\right)\right] \tag{1.99}$$

where E is the Young's modulus of elasticity, v Poisson's ratio and δ_{ij} the Kronecker delta. Equation (1.99) can be cast in matrix form as

$$D^c = \begin{bmatrix} c_{11} & c_{22} & c_{22} & 0 & 0 & 0 \\ c_{22} & c_{11} & c_{22} & 0 & 0 & 0 \\ c_{22} & c_{22} & c_{11} & 0 & 0 & 0 \\ 0 & 0 & 0 & c_{33} & 0 & 0 \\ 0 & 0 & 0 & 0 & c_{33} & 0 \\ 0 & 0 & 0 & 0 & 0 & c_{33} \end{bmatrix} \tag{1.100}$$

with the elements of this matrix being defined by

$$c_{11} = \frac{E}{1+v}\left(\frac{1-v}{1-2v}\right), \quad c_{22} = \frac{E}{1+v}\left(\frac{v}{1-2v}\right), \quad c_{33} = \frac{1}{2}\left(\frac{E}{1+v}\right).$$

The stress and strain vectors accompanying Eq. (1.100) are

$$\sigma = \left[\sigma_{11}, \sigma_{22}, \sigma_{33}, \sigma_{12}, \sigma_{23}, \sigma_{31}\right]^T ,$$

$$\varepsilon = \left[\varepsilon_{11}, \varepsilon_{22}, \varepsilon_{33}, \varepsilon_{12}, \varepsilon_{23}, \varepsilon_{31}\right]^T .$$

(1.101a, b)

To include the skew-symmetric component of stress and strain tensors, D^c in Eq. (1.00) needs to be expanded to the following 7×7 matrix

$$D^s = \begin{bmatrix} (D^c)_{6 \times 6} & (0)_{6 \times 1} \\ [0]_{1 \times 6} & c_{33} \end{bmatrix} ,$$

(1.102)

where $(0)_{6 \times 1}$ and $[0]_{1 \times 6}$ are null matrices of orders 6×1 and 1×6, respectively. The accompanying stress and strain vectors are

$$\sigma = \left[\sigma_{11}, \sigma_{22}, \sigma_{33}, \sigma_{12}, \sigma_{23}, \sigma_{31}, \sigma_{12}^s\right]^T ,$$

$$\varepsilon = \left[\varepsilon_{11}, \varepsilon_{22}, \varepsilon_{33}, \varepsilon_{12}, \varepsilon_{23}, \varepsilon_{31}, \varepsilon_{12}^s\right]^T ,$$

(1.103)

and the superscript s denotes the skew-symmetric components.

Finally, Eq. (1.102) should be reduced from general 3D applications to plate or shell analyses by imposing zero normal stress condition $\sigma_{33} = 0$. Hughes and Liu [32] proposed performing a transformation defined as

$$D = (T_c)^T D^s (T_c) ,$$

(1.104)

where the transformation matrix

$$T_c = \begin{bmatrix} 1 & 0 & 0 & 0 & 0 & 0 \\ 0 & 1 & 0 & 0 & 0 & 0 \\ t_1 & t_2 & t_4 & t_5 & t_6 & t_7 \\ 0 & 0 & 1 & 0 & 0 & 0 \\ 0 & 0 & 0 & 1 & 0 & 0 \\ 0 & 0 & 0 & 0 & 1 & 0 \\ 0 & 0 & 0 & 0 & 0 & 1 \end{bmatrix} , \quad t_k = -\frac{D_{3k}^s}{D_{33}^s} ,$$

(1.105a, b)

in which D^{s}_{3k} is the element located in the third row and k'th column of D^{s} of Eq. (1.102).

For finite strain deformations, Refs. [22, 33, 34] suggested adding to Eq. (1.99) the following term,

$$D^{h}_{ijkl} = -\frac{1}{2}\left(\sigma_{ik}\delta_{jl} + \sigma_{jk}\delta_{il} + \sigma_{il}\delta_{jk} + \sigma_{jl}\delta_{ik}\right). \qquad (1.106)$$

Note that this term is a consequence of transforming the Jaumann stress rate to the incremental PK2 stress. If cast into matrix form with Eq. (1.103) as the accompanying stress and strain vectors, Eq. (1.106) becomes

$$D^{r} = \begin{bmatrix} D^{\beta}_{11} & D^{\beta}_{12} \\ D^{\beta}_{21} & D^{\beta}_{22} \end{bmatrix}, \quad D^{\beta}_{11} = -2\begin{bmatrix} \sigma_{11} & 0 & 0 \\ 0 & \sigma_{22} & 0 \\ 0 & 0 & \sigma_{33} \end{bmatrix}, \qquad (1.107)$$

where

$$D^{\beta}_{12} = -\begin{bmatrix} \sigma_{12} & 0 & \sigma_{13} & 0 \\ \sigma_{12} & \sigma_{23} & 0 & 0 \\ 0 & \sigma_{23} & \sigma_{13} & 0 \end{bmatrix}, \quad D^{\beta}_{21} = \left(D^{\beta}_{12}\right)^{T},$$

$$D^{\beta}_{22} = -\frac{1}{2}\begin{bmatrix} \sigma_{11}+\sigma_{22} & \sigma_{13} & \sigma_{23} & 0 \\ \sigma_{13} & \sigma_{22}+\sigma_{33} & \sigma_{12} & 0 \\ \sigma_{23} & \sigma_{12} & \sigma_{11}+\sigma_{33} & 0 \\ 0 & 0 & 0 & 0 \end{bmatrix}.$$

The transformation rule of Eq. (1.104) applies to $(D^{s} + D^{r})$, except that σ_{33} in Eq. (1.107) is zero.

Other approaches in dealing with finite strain problems include performing numerical integration on rate constitutive equations [32, 35], or on the Jaumann rate [28]. The approach in the present study is, however, simpler and more direct, compared with those in Refs. [28, 32, 35].

1.5.2   Elasto-plastic materials with isotropic strain hardening

Notable large strain elasto-plastic deformation theories include those by, for examples, Green and Naghdi [36], Nemat-Nasser [37] and Lee [38]. It seems that most of the controversies arise in cases where both the elastic and plastic parts of strain are large. However, if confined to applications of small elastic, but large plastic strain (thus, large total strain), it can be shown that all the different theories reduce to the so-called J_2 flow theory of plasticity [28, 22, 39, 33-35]. As the material of interest in the present study is metal with small elastic, but large plastic strain, in what follows, the J_2 flow theory is considered.

The small strain formulation of J_2 flow theory involves the stress deviator σ^D and the J_2 invariant of stresses. In Cartesian co-ordinates, they are defined as in [33],

$$\sigma_{ij}^D = \sigma_{ij} - \frac{1}{3}\sigma_{kk}\delta_{ij} , \quad J_2 = \frac{1}{2}\sigma_{ij}^D\sigma_{ij}^D . \tag{1.108}$$

Consequently, the constitutive fourth order tensor is now given by

$$D_{ijkl}^{ep} = D_{ijkl} - \left(\frac{E}{1+\nu}\right)\left(\frac{\alpha}{E^P}\right)\sigma_{ij}^D\sigma_{kl}^D , \tag{1.109}$$

where

$$E^P = \frac{2}{3}(\sigma^e)^2 \left(\frac{E - \dfrac{1-2\nu}{3}E_T}{E - E_T}\right) , \tag{1.110}$$

in which E_T is the tangent modulus, ν Poisson's ratio, and $(\sigma^e)^2 = 3J_2$ the square of the effective stress while α, not to be confused with that in Eq. (1.31a), is a parameter having the value of either zero or unity. When $\alpha = 0$, it is associated with elastic loading or any unloading, and when $\alpha = 1$, it is associated with plastic loading. Whether the material is undergoing plastic loading or not, it can be determined through the following conditions

$$\alpha = \begin{cases} 1 & \dot{J}_2 = \sigma_{ij}^D\dot{\sigma}_{ij} \geq 0 \quad \wedge \quad J_2 = (J_2)_{max} \\ 0 & \dot{J}_2 < 0 \quad \vee \quad J_2 < (J_2)_{max} \end{cases} \tag{1.111}$$

with $\wedge$ denoting the logical "and" and $\vee$ the logical "or".

The above small strain formulation can be extended to finite strain cases in a similar approach to that of Sub-section 1.5.1. This is achieved by adding Eq. (1.106) to (1.109). Then, the matrix form of the constitutive tensor becomes

$$D^{ep} = D^s + D^r + D^\alpha ,\tag{1.112}$$

where D^s and D^r are defined in Eqs. (1.102) and (1.107), respectively.

The term D^α denotes the component matrix associated with α in Eq. (1.109). It is defined by

$$D^\alpha = -\lambda \begin{vmatrix} D_{11}^\alpha & D_{12}^\alpha \\ D_{21}^\alpha & D_{22}^\alpha \end{vmatrix},\tag{1.113}$$

where

$$D_{11}^\alpha = \begin{bmatrix} \sigma_{11}^D\sigma_{11}^D & \sigma_{11}^D\sigma_{22}^D & \sigma_{11}^D\sigma_{33}^D \\ \sigma_{22}^D\sigma_{11}^D & \sigma_{22}^D\sigma_{22}^D & \sigma_{22}^D\sigma_{33}^D \\ \sigma_{33}^D\sigma_{11}^D & \sigma_{33}^D\sigma_{22}^D & \sigma_{33}^D\sigma_{33}^D \end{bmatrix}, \quad D_{21}^\alpha = \begin{bmatrix} \sigma_{12}\sigma_{11}^D & \sigma_{12}\sigma_{22}^D & \sigma_{12}\sigma_{33}^D \\ \sigma_{23}\sigma_{11}^D & \sigma_{23}\sigma_{22}^D & \sigma_{23}\sigma_{33}^D \\ \sigma_{31}\sigma_{11}^D & \sigma_{31}\sigma_{22}^D & \sigma_{31}\sigma_{33}^D \\ \sigma_{12}^s\sigma_{11}^D & \sigma_{12}^s\sigma_{22}^D & \sigma_{12}^s\sigma_{33}^D \end{bmatrix},$$

$$D_{12}^\alpha = \left(D_{21}^\alpha\right)^T, \quad D_{22}^\alpha = \begin{bmatrix} \sigma_{12}\sigma_{12} & \sigma_{12}\sigma_{23} & \sigma_{12}\sigma_{31} & \sigma_{12}\sigma_{12}^s \\ \sigma_{23}\sigma_{12} & \sigma_{23}\sigma_{23} & \sigma_{23}\sigma_{31} & \sigma_{23}\sigma_{12}^s \\ \sigma_{31}\sigma_{12} & \sigma_{31}\sigma_{23} & \sigma_{31}\sigma_{31} & \sigma_{31}\sigma_{12}^s \\ \sigma_{12}^s\sigma_{12} & \sigma_{12}^s\sigma_{23} & \sigma_{12}^s\sigma_{31} & \sigma_{12}^s\sigma_{12}^s \end{bmatrix},$$

$$\lambda = \left(\frac{E}{1+v}\right)\frac{\alpha}{E^p} .$$

Equation (1.112) is the elasto-plastic material matrix for general 3D problems. When reducing the general 3D theory to thin or moderately thick plates or shells two aspects are to be noted. First, it is usually assumed that the effects of transverse shear stresses on plastic behaviors can be disregarded [39]. Such an assumption leads to a simplified matrix D^α in which three of the four sub-matrices are

$$
D_{21}^\alpha = \begin{bmatrix} \sigma_{12}\sigma_{11}^D & \sigma_{12}\sigma_{22}^D & \sigma_{12}\sigma_{33}^D \\ 0 & 0 & 0 \\ 0 & 0 & 0 \\ \overset{s}{\sigma}_{12}\sigma_{11}^D & \overset{s}{\sigma}_{12}\sigma_{22}^D & \overset{s}{\sigma}_{12}\sigma_{33}^D \end{bmatrix}, \tag{1.114a}
$$

$$
D_{12}^\alpha = \left(D_{21}^\alpha\right)^T, \quad D_{22}^\alpha = \begin{bmatrix} \sigma_{12}\sigma_{12} & 0 & 0 & \sigma_{12}\overset{s}{\sigma}_{12} \\ 0 & 0 & 0 & 0 \\ 0 & 0 & 0 & 0 \\ \overset{s}{\sigma}_{12}\sigma_{12} & 0 & 0 & \overset{s}{\sigma}_{12}\overset{s}{\sigma}_{12} \end{bmatrix}, \tag{1.114b, c}
$$

and the sub-matrix $D^\alpha{}_{11}$ remains unchanged. Second, the transformation rule of Eq. (1.104) is applied to D^{ep}. In doing so, one can only set σ_{33} to zero while $\sigma^D{}_{33}$ is not zero.

It may be appropriate to point out that in arriving at Eqs. (1.112) and (1.114), the von Mises yield criterion has been employed. The von Mises criterion is considered presently because the materials of interest in the present investigation are metals. It can be expressed in terms of stresses or stress resultants. The first approach allows for the spread of plasticity over the thickness of plates and shells, and is termed the *layered approach*. This approach has been adopted in [39]. The second approach, the *non-layered approach*, on the other hand, employs yield functions that are in terms of stress resultants. This latter approach assumes that at a point the entire cross-section becomes plastic simultaneously. Therefore, compared with the non-layered approach, the layered approach seems to be more realistic but requires a larger number of algebraic manipulations in forming the element stiffness matrix. However, Robinson [40] showed that the discrepancy between the two approaches was insignificant, while Ref. [25] seemed to suggest the application of a large number of elements with the non-layered approach.

In the present investigation, the non-layered approach is employed for two main considerations. First, it requires less computation to evaluate element stiffness matrices. Second, the D^{ep} matrix can be written in a simple and concise way, which enables one to obtain explicit expressions for the stiffness matrix that are of a manageable size. The second main consideration is satisfied because, in the non-layered approach, stress distribution along the thickness direction is simple to formulate. Prior to the occurrence of plasticity at a point, the membrane and bending stresses are, respectively, uniformly and linearly distributed along the thickness. If the yield criterion is satisfied, the membrane stress remains uniform over the thickness. But bending stresses become symmetric about the mid-surface. That is, the bending stresses takes equal magnitudes but opposite signs below and above the mid-surface. Consequently, the yield criterion should be expressed in terms of stress resultants. Such attempts include those proposed in Refs. [41, 40, 42, 43].

In 1948, Ilyushin applied the von Mises yield criterion to thin shells [42]. The idea was further developed by Shapiro [43]. Simo and Kennedy [41] then extended the Ilyushin-Shapiro two-surface yield condition to nonlinear shell analysis. The yield conditions of Simo and Kennedy [41] were written in terms of membrane forces and bending moments, and are capable of reflecting the coupling effect of membrane forces and bending moments on plastic behaviours. Because the yield conditions included two additional parameters that were deformation-path dependent, Simo and Kennedy [41] have constructed a complex return mapping algorithm, to bring stress points outside of the yield surface onto the surface. This technique, however, requires a very significant amount of computational effort. On the other hand, Robinson [40] had shown that the Ilyushin-Shapiro yield condition reduced to a non-parametric form without loss of accuracy and generality. Consequently, the corresponding return mapping algorithm is simpler to construct and is less expensive computationally. In what follows, the yield condition of Robinson [40] and its corresponding return mapping are introduced. The yield condition in Ref. [40] shall henceforth be called Ilyushin's or the Ilyushin-Robinson yield condition.

If σ_y is the yield stress of the material in simple tension and h the thickness of the shell at time t with the superscript t being disregarded in all the quantities considered here for conciseness, one defines the dimensionless membrane forces n_{11}^{P}, n_{22}^{P} and n_{12}^{P}, and the dimensionless bending moments m_{11}^{P}, m_{22}^{P} and m_{12}^{P} as

$$n_{ij}^P = \frac{N_{ij}^P}{\sigma_y h}, \qquad m_{ij}^P = \frac{4M_{ij}^P}{\sigma_y h^2}, \qquad (1.115\text{a, b})$$

with the membrane forces N_{ij}^P and bending moments M_{ij}^P being related to stress components across the cross-section as

$$N_{ij}^P = \int_{-h/2}^{h/2} \sigma_{ij}\, d\eta, \quad M_{ij}^P = \int_{-h/2}^{h/2} \sigma_{ij}\eta\, d\eta. \qquad (1.116\text{a, b})$$

Therefore, the membrane forces and bending moments are defined over a cross-section with thickness h and unity width. The Ilyushin yield condition proposed by Robinson [40] is stated as

$$Q_t^P + Q_m^P + \frac{\left| Q_{tm}^P \right|}{\sqrt{3}} \le 1, \qquad (1.117)$$

where

$$Q_t^P = \left(n_{11}^P\right)^2 + \left(n_{22}^P\right)^2 - n_{11}^P n_{22}^P + 3\left(n_{12}^P\right)^2,$$

$$Q_m^P = \left(m_{11}^P\right)^2 + \left(m_{22}^P\right)^2 - m_{11}^P m_{22}^P + 3\left(m_{12}^P\right)^2,$$

$$Q_{tm}^P = n_{11}^P m_{11}^P + n_{22}^P m_{22}^P - \frac{1}{2}n_{11}^P m_{22}^P - \frac{1}{2}n_{22}^P m_{11}^P + 3 n_{12}^P m_{12}^P.$$

It should be noted that in the above yield condition transverse shear stresses have been disregarded. Robinson has shown that Eq. (1.117) is a very good approximation to the exact criterion and superior to the other linear approximations.

The return mapping follows that described in [28] for general 3D problems. Noting that in the present study membrane and bending stresses are uncoupled, it is proposed to split the return mapping into two portions, one for the membrane stress and the other for bending stress. The portion for membrane stress is, in fact, a return mapping for plane stress problems. The portion for bending stress is identical to a return mapping for plate bending problems.

Assuming that $\Delta\varepsilon^{ep}$ is the elasto-plastic strain increment which is divided into sub-increments $\Delta(\Delta\varepsilon^{ep})$ according to [29]

$$\Delta(\Delta\varepsilon^{ep}) = \gamma / \lambda_{ep} \qquad (1.118)$$

with the elasto-plastic parameters

$$\gamma = \frac{\Delta\varepsilon^{ep}}{\Delta\varepsilon} \ , \quad \lambda_{ep} = 1 + \frac{\gamma}{30} \ . \qquad (1.119a, b)$$

Therefore, the parameter γ indicates the elasto-plastic portion of total strain increment $\Delta\varepsilon$. In Eqs. (1.118) and (1.119a, b), in fact, $\Delta\varepsilon$ is divided into 30 equal sub-increments, λ_{ep} of which correspond to $\Delta\varepsilon^{ep}$, the elasto-plastic portion of the strain increment. For each of the λ_{ep} sub-increments of the elasto-plastic strain, $\Delta(\Delta\varepsilon^{ep})$, the updating formula is

$$\sigma_m \ \leftarrow \ \sigma_m \ + \ D_3^{ep} \, \Delta(\Delta\varepsilon^{ep}) \ , \qquad (1.120)$$

where $\leftarrow$ denotes "assign to". For the membrane stress portion, σ_m and $\Delta(\Delta\varepsilon^{ep})$ are understood to be the membrane stresses and membrane strain sub-increment, respectively. For the bending moment portion, σ_m and $\Delta(\Delta\varepsilon^{ep})$ are understood to be the bending moments and bending curvature sub-increment, respectively. The 3×3 matrix D_3^{ep} is obtained through the following relation

$$D_3^{ep} = \left(T_{c4}\right)^T D_4^{ep} \left(T_{c4}\right) , \qquad (1.121)$$

where the 4×4 matrix D_4^{ep} is defined as

$$D_4^{ep} = D_4^c + D_4^r + D_4^\alpha , \qquad (1.122)$$

in which square matrices D_4^r and D_4^α are of order 4 and can, in fact, be formed by two steps. First, they are constructed by selecting the first 4 rows and columns of D^r in Eq. (1.107), and D^α in Eq. (1.113), respectively. Second, the stresses σ_{ij}, and so on in Eqs. (1.107) and (1.113) are replaced with appropriate σ_m. The matrix D_4^c, is defined as

$$D_4^c = \lambda_\ell \begin{bmatrix} 1-v & v & v & 0 \\ v & 1-v & v & 0 \\ v & v & 1-v & 0 \\ 0 & 0 & 0 & \dfrac{1-2v}{2} \end{bmatrix} \qquad (1.123)$$

with the subscript $\ell = 1, 2$ such that λ_1 is associated with membrane stresses and λ_2 bending moments. These parameters are defined as

$$\lambda_1 = \frac{E}{(1+v)(1-2v)} \ , \qquad \lambda_2 = \frac{Eh^3}{12(1+v)(1-2v)} \ .$$

It is observed that with $\ell = 1$, Eq. (1.123) is just the matrix formed by selecting the first four rows and first four columns of D^c in Eq. (1.100).

In closing this sub-section, it should be mentioned that the transformation matrix T_{c4} in Eq. (1.121) is given by

$$T_{c4} = \begin{bmatrix} 1 & 0 & 0 \\ 0 & 1 & 0 \\ t_1 & t_2 & t_4 \\ 0 & 0 & 1 \end{bmatrix} \ , \qquad t_k = \frac{\left(D_4^{ep}\right)_{3k}}{\left(D_4^{ep}\right)_{33}} \ , \qquad (1.124a, b)$$

where $(D_4^{ep})_{3k}$ is the element in the third row and k'th column of D_4^{ep}.

1.5.3   Configuration and stress updatings

The updating of configuration and stresses at every time step is required in problems with large deformations. Therefore, in this sub-section steps of updating of configuration and stresses are provided.

1.5.3.1 Updating of configuration

This consists of the updating of mid-surface co-ordinates and directors. Mid-surface co-ordinates are updated by adding mid-surface displacements to mid-surface co-ordinates of the reference configuration. To this end, one makes use of Eqs. (1.73) and (1.75) with $\eta^t = 0$ such that

$$\begin{pmatrix} \bar{r}^{t+\Delta t} \\ \bar{s}^{t+\Delta t} \\ \bar{t}^{t+\Delta t} \end{pmatrix} = \begin{pmatrix} \bar{r}^t \\ \bar{s}^t \\ \bar{t}^t \end{pmatrix} + \begin{pmatrix} \Delta\bar{u}^t \\ \Delta\bar{v}^t \\ \Delta\bar{w}^t \end{pmatrix} \ . \qquad (1.125)$$

Similar updating for the global co-ordinates can be made. But for brevity, those in terms of global co-ordinates are not presented here.

In the present study, updating procedures for directors follow those proposed in Refs. [24, 44], for example. Assuming that incremental rotations $(\Delta\theta_i)^t$, $i = 1,2,3$, have been determined at the current time step t, the directors at the next time step or new position are obtained by making use of Eq. (1.82),

$$d_i^{t+\Delta t} = R(\Delta\theta)\, d_i^t \,, \tag{1.126}$$

where the transformation matrix $R(\Delta\theta)$ or simply written as R is defined by Eq. (1.83) except that now it is in configuration C^t so that $\Delta\theta = (\Delta\theta_i)^t$.

Having updated the directors in configuration $C^{t+\Delta t}$ and since the director angular velocity as well as the director angular acceleration in C^t are known, the director angular velocity and director angular acceleration fields are updated applying the following scheme [44]

$$\omega^{t+\Delta t} = \frac{\gamma\,\Delta\theta}{\beta\,\Delta t} - R\left[\omega^t + \left(\frac{\gamma}{\beta} - 2\right)\left(\omega^t + \frac{\Delta t\,\dot\omega^t}{2}\right)\right],$$

$$\dot\omega^{t+\Delta t} = \frac{\omega^{t+\Delta t}}{\gamma\,\Delta t} - R\left[\frac{\omega^t}{\gamma\,\Delta t} + \left(\frac{1}{\gamma} - 1\right)\dot\omega^t\right], \tag{1.127a, b}$$

in which $\omega^t = d^t \times d(d^t)/dt$, β and γ are the parameters of the Newmark family of algorithms, not to be confused with that in Eqs. (1.31b) and (1.44b), and (1.119b). It should be noted that the numerical integration scheme in Eq. (1.127) is applied to the updating of the translational velocity and acceleration,

$$\dot u^{t+\Delta t} = \frac{\gamma\,\Delta u}{\beta\Delta t} - \left[\dot u^t + \left(\frac{\gamma}{\beta} - 2\right)\left(\dot u^t + \frac{\Delta t\,\ddot u^t}{2}\right)\right],$$

$$\ddot u^{t+\Delta t} = \frac{\dot u^{t+\Delta t}}{\gamma\Delta t} - \left[\frac{\dot u^t}{\gamma\Delta t} + \left(\frac{1}{\gamma} - 1\right)\ddot u^t\right], \tag{1.128a, b}$$

where Δu is understood to be the incremental displacement vector at time t.

For the trapezoidal rule in which $\beta = 1/4$ and $\gamma = \frac{1}{2}$, Eq. (1.127) reduces to the following

$$\omega^{t+\Delta t} = \frac{2\,\Delta\theta}{\Delta t} - R\,\omega^t \,,$$

$$\dot\omega^{t+\Delta t} = \frac{2\,\omega^{t+\Delta t}}{\Delta t} - R\left(\frac{2\,\omega^t}{\Delta t} + \dot\omega^t\right), \tag{1.129a, b}$$

Other quantities such as the translational velocity and acceleration can be similarly obtained but are not included presently for brevity.

It is noted that in the limiting case of small rotations, $\Delta\theta \to 0$. Consequently, $\cos|\Delta\theta| \to 1$, $\sin(|\Delta\theta|)/|\Delta\theta| \to 1$ and $R \to I_3$. Naturally, in this case, the directors needs not be updated and the updating scheme only has to use Eq. (1.128) for the translational velocity and acceleration vectors.

It should also be noted that the foregoing updating scheme, Eqs. (1.126) through (1.129) and so on, has embodied one implicit condition [24, 44]. That is, the incremental rotation vector $\Delta\theta$ is perpendicular to the director's reference position d_i^t. The physical interpretation is that the incremental rotational component along the director does not have any effects on the re-orientation of the director. Only those incremental rotational components lying on the plane perpendicular to the director can have an effect on bringing the director to a new position. Fox and Simo [45] proposed replacing the exponential mapping Γ_i^t of Eq. (1.81) with another mapping that in fact contained the product of two exponential mappings. One of these mappings was constructed from the ddof and the director. However, in [45] it was shown that such a new mapping was not uniquely defined. To preserve uniqueness *drill rotation constraint* has to be applied. This no doubt complicates the computation. It may be appropriate to note that no numerical results were included in [45]. As Ref. [46], the precursor of [45], pointed out, the aim was to identify the independent rotation field, which is the ddof in the present study, with the rotation of the continuum. This aim is realized in the foregoing formulation and derivation.

The present study adopts Eqs. (1.126) and (1.127) as the director updating scheme for two main considerations. First, it has a good physical basis. Second, it involves relatively less computational efforts.

Finally, as a part of configuration updating, the updating of density and thickness are included in the present investigation. Such an updating requires the calculation of the *relative* deformation gradient [22] which is defined as

$$
\mathscr{F}_t^{t+\Delta t} =
\begin{bmatrix}
\dfrac{\partial(\Delta u^t)}{\partial r^t} & \dfrac{\partial(\Delta u^t)}{\partial s^t} & \dfrac{\partial(\Delta u^t)}{\partial t^t} \\[3mm]
\dfrac{\partial(\Delta v^t)}{\partial r^t} & \dfrac{\partial(\Delta v^t)}{\partial s^t} & \dfrac{\partial(\Delta v^t)}{\partial t^t} \\[3mm]
\dfrac{\partial(\Delta w^t)}{\partial r^t} & \dfrac{\partial(\Delta w^t)}{\partial s^t} & \dfrac{\partial(\Delta w^t)}{\partial t^t}
\end{bmatrix},
\tag{1.130}
$$

where the displacement increments Δu^t, Δv^t, and Δw^t are defined by Eq. (1.75). Equation (1.130) expresses the deformations of the body in $C^{t+\Delta t}$ with respect to the reference configuration C^t. Thus, the density and thickness relations of the shell element become, respectively,

$$\rho^{t+\Delta t} = \frac{\rho^t}{\det\!\left(\mathscr{F}_t^{t+\Delta t}\right)}, \qquad h^{t+\Delta t} = \frac{h^t \det\!\left(\mathscr{F}_t^{t+\Delta t}\right) A^t}{A^{t+\Delta t}}, \qquad (1.131\text{a, b})$$

where $\det(\cdot)$ denotes the *determinant of*, and the area of the shell element at t, A^t is given by

$$A^t = \frac{r_2^t\, s_3^t}{2}. \qquad (1.131\text{c})$$

1.5.3.2 Updating of Stresses

After solving the nodal displacement increments, the incremental strain and stress increments can be recovered by applying Eqs. (1.71) and (1.72). These stress increments are defined with respect to C^t, and therefore they can be added to the Cauchy stress σ^t since they are referred to the same reference state. The sum of σ^t and ΔS is equal to $S^{t+\Delta t}$. That is, $S^{t+\Delta t} = \sigma^t + \Delta S$, which is the PK2 stress vector in deformation state $C^{t+\Delta t}$ measured with respect to C^t. The transformation of $S^{t+\Delta t}$ to the Cauchy stress $\sigma^{t+\Delta t}$ is [22, 28]

$$\sigma^{t+\Delta t} = \frac{1}{\det\!\left(\mathscr{F}_t^{t+\Delta t}\right)} \mathscr{F}_t^{t+\Delta t} S^{t+\Delta t} \left(\mathscr{F}_t^{t+\Delta t}\right)^T, \qquad (1.132)$$

where the deformation gradient $\mathscr{F}_t^{t+\Delta t}$ is defined by Eq. (1.130).

1.6 Concluding Remarks

While it is assumed that the readers have at least a minimum knowledge of a first course in the FEM, the basic theories and steps in the displacement formulation and the hybrid stress formulation have been included in this chapter.

To illustrate the steps in a more concrete way, the derivation of element mass and stiffness matrices for the uniform and tapered beam elements have been presented. An outline of the incremental hybrid strain or mixed formulation has also been included. Explicit expressions for the element mass and stiffness matrices have been derived and presented as appendices to this chapter.

Throughout this chapter the element equations of motion are emphasized.

No attempt has been made to include the assembled equations of motion for the structural systems. This is because assembled equations of motion are concerned with computer programming and it is a standard feature in any FEM package.

Finally, it is observed that the literature on FEM is vast and the topics presented in this chapter serve as a preparation for a better understanding of the concepts and detailed steps in subsequent chapters. However, Sections 1.4 and 1.5 contain more advanced topics and therefore they may be disregarded by the average or general readers.

References

[1] Turner, M.J., Clough, R.W., Martin, H.C. and Topp, L.J. (1956). Stiffness and deflection analysis of complex structures, *Journal of the Aeronautical Sciences*, **23**, 805-823.

[2] Zienkiewicz, O.C. and Taylor, R.L. (1988). *The Finite Element Method*, **1**, McGraw-Hill, New York.

[3] Uhlenbeck, G.E. and Ornstein, L.S. (1930). On the theory of the Brownian motion, *Physical Review*, **36**, 832-841.

[4] Rice, S.O. (1944). Mathematical analysis of random noise, *Bell System Technical Journal*, **23**, 282-332.

[5] Lin, Y.K. (1967). *Probabilistic Theory of Structural Dynamics*, McGraw-Hill, New York.

[6] Bolotin, V.V. (1984). *Random Vibration of Elastic Systems*, Martinus and Nijhoff Publishers, The Hague, The Netherlands.

[7] Clough, R.W. and Penzien, J. (1975). *Dynamics of Structures*, McGraw-Hall, New York.

[8] Paz, M. (1985). *Structural Dynamics: Theory and Computation*, Van Nostrand Reinhold, New York.

[9] Meirovitch, L. (1997). *Principles and Techniques of Vibrations*, Prentice-Hall, New Jersey.

[10] Jourdain, P.F.B. (1909). Note on an analogue of Gauss principle of least constraint, *Quarterly Journal of Pure and Applied Mathematics*, **4L**.

[11] Kane, T.R. (1961). Dynamics of nonholonomic systems, *Trans. A.S.M.E. Journal of Applied Mechanics*, **28**, 574-578.

[12] To, C.W. S. (1979). Higher order tapered beam finite elements for vibration analysis, *Journal of Sound and Vibration*, **63(1)**, 33-50.

[13] Worley, W. J. and Breuer, F. D. (1957). Areas, centroids and inertias for a family of elliptic-type closed curves, *Product Engineering*, **28**, 141-144.

[14] Pian, T.H.H. (1964). Derivation of element stiffness matrices by assumed stress distributions, *A.I.A.A. Journal*, **2**, 1333-1336.

[15] Archer, J.S. (1963). Consistent mass matrix for distributed mass sytems, *Proc. A.S.C.E. Journal of Structural Division*, **89(ST4)**, 161-178.

[16] Weaver, W. Jr. and Johnston, P. R. (1984). *Finite Elements for Structural analysis*, Prentice-Hall, New Jersey.

[17] Kang, D. S. S. (1991). C^0 continuity elements by hybrid stress method, NASA CR 189040.

[18] Tong, P. and Pian, T. H. H. (1969). A variational principle and the convergence of a finite element method based on assumed stress distribution, *Int. J. Num. Meth. Eng.*, **5**, 463-472.

[19] Pian, T. H. H. and Chen, D. P. (1983). On the suppression of zero energy deformation modes, *Int. J. Num. Meth. Eng.*, **19**, 1741-1752.

[20] Liu, M. L. and To, C. W. S. (1995). Hybrid strain based three-node flat triangular shell elements, Part I: Nonlinear theory and incremental formulation, *Computers and Structures*, **54(6)**, 1031-1056.

[21] To, C. W. S. (2010). *Nonlinear Random Vibration: Computational Methods*, Zip Publishing, Columbus, Ohio.

[22] Kleiber, M. (1989). *Incremental Finite Element Modelling in Nonlinear Solid Mechanics*, Ellis Horwood, Chichester.

[23] Boland, P. L. and Pian, T. H. H. (1977). Large deflection analysis of thin elastic structures by the assumed stress hybrid finite element method, *Computers and Structures*, **7**, 1-12.

[24] Saleeb, A. F., Chang, T. Y., Graf, W. and Yingyeunyong, S. (1990). A hybrid/mixed model for non-linear shell analysis and its applications to large rotation problems, *Int. J. Num. Meth. Eng.*, **29**, 407-446.

[25] Simo, J. C. and Fox, D. D. (1989). On a stress resultant geometrically exact shell model, Part I: Formulation and optimal parametrization, *International Journal for Numerical Methods in Engineering*, **72**, 267-304.

[26] Bufler, H. (1993). Nonlinear conservative systems and tangent operators, *Proc. 2nd Int. Conf. on Nonl. Mech.*, pp. 110-113, Aug. 23-26, Beijing.

[27] Liu, M. L., and To, C. W. S. (1995). Vibration analysis of structures by hybrid strain based three-node flat triangular shell elements, *Journal of Sound and Vibration*, **184(5)**, 801-821.

[28] Bathe, K. J. (1982). *Finite Element Procedures in Engineering Analysis*, Prentice-Hall, New York.

[29] Bathe, K. J., Wilson, E. L., and Iding, R.H. (1974). NONSAP, A structural analysis program for static and dynamic response of nonlinear systems, Report No. SESM 74-3, Struct. Eng. Lab., U. of California, Berkeley.

[30] Gadala, M.S., Dokainish, M.A. and Oravas, G.A.E. (1984). Formulation methods of geometric and material nonlinearity problems, *International Journal for Numerical Methods in Engineering*, **20**, 887-914.

[31] To, C. W. S., and Liu, M. L. (1995). Hybrid strain based three-node flat triangular shell elements, Part II: Numerical investigation of nonlinear problems, *Computers and Structures*, **54(6)**, 1057-1076.

[32] Hughes, T. J. R., and Liu, W. K. (1981). Nonlinear finite element analysis of shells, Part I: Three dimensional shells, *Computer Methods in Applied Mechanics and Engineering*, **26**, 331-362.

[33] Hutchinson, J. W. (1975). Finite strain analysis of elasto-plastic solids and structures, in *Numerical Solution of Nonlinear Structural Problems*, pp. 17-29, R.F. Hartung (edr.), A.S.M.E., Detroit.

[34] Nagtegaal, J.C., and de Jong, J.E. (1981). Some computational aspects of elastic-plastic large strain analysis. *International Journal for Numerical Methods in Engineering*, **17**,15-41.

[35] Hughes, T.J.R., and Winget, J.(1980). Finite rotation effects in numerical integration of rate type constitutive equations arising in large deformation analysis, *Int. J. for Numerical Methods in Engineering*, **15**, 1862-1867.

[36] Green, A.E., and Naghi, P.M. (1971). Some remarks on elastic-plastic deformation at finite strain, *Int. J. of Engineering Science*, **9**, 1219-1229.

[37] Nemat-Nasser, S. (1979). Decomposition of strain measures and their rates in finite deformation elastoplasticity, *International Journal of Solids and Structures*, **15**, 155-166.

[38] Lee, E. H. (1969). Elastic plastic deformation at finite strains, *Journal of Applied Mechanics*, **36**, 1-6.

[39] Owen, D. R. J. and Hinton, E. (1980). *Finite Elements in Plasticity: Theory and Practice*, Pineridge Press., Swansea.

[40] Robinson, M. (1971). A comparison of yield surfaces for thin shells, *International Journal of Mechanical Sciences,* **13**, 345-354.

[41] Simo, J. C. and Kennedy, J. G. (1992). On a stress resultant geometrically exact shell model, Part V: Nonlinear plasticity, formulation and integration algorithms, *Comput. Meth.Appl. Mech. and Eng.,* **96**, 133-171.

[42] Ilyushin, A. A. (1948). *Plasticity* (in Russian), Gostekhizdat, Moscow.

[43] Shapiro, G. S. (1961). On yield surfaces for ideally plastic shells, in *Problems of Continuum Mechanics*, pp. 414-418, S.I.A.M., Philadelphia, Pa.

[44] Simo, J. C., Rifai, M. S. and Fox, D. D.(1992). On a stress resultant geometrically exact shell model, Part VI: Conserving algorithms for nonlinear dynamics, *Int. J. Num. Meth. Eng.,* **34**, 117-164.

[45] Fox, D. D., and Simo, J. C. (1992). A drill rotation formulation for geometrically exact shells, *Computer Methods in Applied Mechanics. and Engineering.,* **98** , 329-343.

[46] Simo, J. C., Fox, D. D., and Hughes, T. J. R. (1992). Formulations of finite elasticity with independent rotations, *Computer Methods in Applied Mechanics and Engineering,* **95**, 277-288.

2
Spectral Analysis and Response Statistics of Linear Structural Systems

This chapter contains a review of the spectral analysis of stationary random responses, the evolutionary spectral analysis of nonstationary random responses, and the modal analysis for time-dependent response statistics of linear structures. Closed form solutions of evolutionary spectra and response statistics of displacements, displacements and velocities, and velocities are presented. Evolutionary spectra of engineering structures such as the mast antenna, cantilever beam, and plates idealized by the finite element method (FEM) are considered. Response statistics of mast antenna structures, truncated conical shell structures, and laminated composite plate and shell structures by the FEM are included.

2.1 Spectral Analysis

Lindberg and Olson [1] appeared to be the first to have used the FEM for the random response analysis of a multi-bay panel system. Later in 1972, Olson [2] presented a consistent FEM for the random response analysis. The method was illustrated with the computation of random response of a five-bay continuous beam. Power spectral density plots were presented. The same method was also applied to the analysis of a complex stiffened panel [3]. At about the same time, another work of applying the FEM in random vibration was presented by Newsom, Fuller and Sherrer [4]. An *ad hoc* approximation of representing the continuous field by a number of distinct concentrated forces was introduced.

In 1968 Jacobs and Lagerquist [5] applied the FEM to the sonic fatigue problem. Applying the FEM, the responses of tall structures to atmospheric turbulence which was considered as a stationary random process were investigated by Handa [6]. Random response analysis of cantilevered plate was carried out by Jones and Beadle [7]. The cantilevered plate was modeled by the displacement-based rectangular elements.

Stochastic Structural Dynamics: Application of Finite Element Methods, First Edition. C.W.S. To.
© 2014 John Wiley & Sons, Ltd. Published 2014 by John Wiley & Sons, Ltd.

Orris and Petyt [8] presented a FEM for the response analysis of periodic structures subjected to convected random pressure fields. It was shown that the problem reduced to that of finding the response of a single periodic section to a harmonic pressure wave. The inertia, stiffness and damping matrices were expressed as functions of the phase difference between the pressures at corresponding points in adjacent sections. The method was applied to a skin-rib type structure. The dimensionless power spectral density of curvature of upper skin center computed by the method was compared with the exact solution.

Application of the FEM was made by Dey [9] to the analysis of the response of discretized structures under stationary random loading. The complex matrix inversion method and the normal mode method were employed in the response computation of beam, square plate, five-bay continuous beam on equidistant simple supports, and two-dimensional framework.

Weeks and Cost [10] investigated the reliability of a composite structure that was excited by random loading. The composite structure was represented by the FE. Comparison of the computed FE results to the closed form solution was made.

Pfaffinger [11] reported a modification of the commercial FE code ADINA for analysis of discretized structures under multiple support random excitations which were assumed to be stationary Gaussian.

The stationary random response of cooling towers represented by shell FE was studied by Yang and Kapania [12].

The stationary random response analysis by applying the FEM was repeated by Elishakoff and Zhu [13]. It may be appropriate to point out that the stationary random response analysis is just a special case of the nonstationary random response analysis presented by the author [14, 15].

2.1.1   Theory of spectral analysis

The basic result underlying the spectral analysis of stationary processes is the fact that each realization of a random process can be represented as the Fourier-Stieltjes transform of the form [16, 17]

$$x(t_1) = \int_{-\infty}^{\infty} e^{i\omega t_1}\, dZ(\omega) , \qquad (2.1)$$

where $Z(\omega)$ is an orthogonal random process and $Z(\omega)$ is not differentiable, but has the property

$$dZ(\omega) = O(\sqrt{d\omega}) . \qquad (2.2)$$

In order to obtain a power spectrum for a stationary process, the average of the power at every frequency over all realizations is required. To this end, the conjugate of Eq. (2.1) has to be applied. It is

$$x(t_2) = \int_{-\infty}^{\infty} e^{-i\omega t_2} \, dZ^*(\omega) , \tag{2.3}$$

where the asterisk denotes the complex conjugate.

Taking the ensemble average of the product of Eqs. (2.1) and (2.3) results

$$\langle x(t_1)x(t_2) \rangle = \int_{-\infty}^{\infty} \int_{-\infty}^{\infty} e^{i\omega_1 t_1 - i\omega_2 t_2} \langle dZ(\omega_1)dZ^*(\omega_2) \rangle . \tag{2.4}$$

Applying the orthogonal property,

$$\langle dZ(\omega_1)dZ(\omega_2) \rangle = \delta(\omega_1 - \omega_2)\langle |dZ(\omega)|^2 \rangle \tag{2.5}$$

and writing $t = t_1 = t_2$, Eq. (2.4) becomes

$$\langle x^2(t) \rangle = \int_{-\infty}^{\infty} \langle |dZ(\omega)|^2 \rangle . \tag{2.6}$$

This enables one to define the incremental power spectrum, $dP(\omega)$, as an ensemble average of the power at every frequency. Thus, one can write

$$dP(\omega) = \langle |dZ(\omega)|^2 \rangle$$

so that

$$\langle x^2(t) \rangle = \int_{-\infty}^{\infty} dP(\omega) . \tag{2.7}$$

In this way it is useful to conceive of a power spectral density function $S(\omega)$ such that it can be related to the incremental power spectrum by the relation,

$$S(\omega)d\omega = dP(\omega) .$$

Equation (2.7) can then be written as

$$\langle x^2(t) \rangle = \int_{-\infty}^{\infty} S(\omega)d\omega . \tag{2.8}$$

2.1.2   Remarks

The above spectral description is very useful in engineering application, since it has the direct physical interpretation of being the energy density distribution over frequency. This is a special case of nonstationary random processes to be considered in the following section.

2.2 Evolutionary Spectral Analysis

The theory of evolutionary spectral density of Priestley [16, 17] and the FEM were employed by the author [14] for the nonstationary random response analysis of mast antenna structures. The nonstationary random excitations considered were modeled as products of amplitude modulating functions and zero mean Gaussian white noise processes. For stationary random response analysis the amplitude modulating functions are of unity. The digital computer program written in Fortran was general. Normal mode method was employed. Closed form transient receptance, evolutionary spectral and cross-spectral densities were presented. The explicit expressions derived are unique in that the coupling effect between various modes has been included.

2.2.1   Theory of evolutionary spectra

Priestley [16, 17] had defined a nonstationary spectral function so as to preserve a physically meaningful interpretation. It is called the evolutionary spectral density function. The important main feature of such a definition is that both stationary and nonstationary random processes can be represented by the integral equation as

$$x\left(t_1\right) = \int_{-\infty}^{\infty} \phi\left(t_1,\omega\right)dZ(\omega) \tag{2.9}$$

where $Z(\omega)$ is an orthogonal process and $\phi(t_1,\omega)$ is some suitable function. For a stationary response $x(t_1)$ one possible form of $\phi(t_1,\omega)$ is $e^{i\omega t}$. This is Eq. (2.1). For nonstationary processes the choice $\phi(t_1,\omega) = e^{i\omega t}$ is inadmissible. Another form of $\phi(t_1,\omega)$ is desired in order to preserve the notion of frequency in the resulting spectral density. Priestley introduced an *oscillatory* function $\phi(t_1,\omega)$ $= A(t_1,\omega)e^{i\omega t}$ which is a function of amplitude modulated sine and cosine waves of frequency ω. This function preserves the idea of frequency as long as the modulating function varies slowly with time. Consequently, a nonstationary random process $x(t_1)$ can be represented as

$$x(t_1) = \int_{-\infty}^{\infty} A(t_1,\omega)e^{i\omega t_1}\, dZ(\omega) \ . \tag{2.10}$$

Equation (2.10) gives an expression for $x(t_1)$ as the limit of a *sum* of sine and cosine waves with different frequencies and time-varying random amplitudes $A(t_1,\omega)dZ(\omega)$.

The conjugate to Eq. (2.10) is

$$x(t_2) = \int_{-\infty}^{\infty} A^*(t_2,\omega)e^{-i\omega t_2}\, dZ^*(\omega) \ . \tag{2.11}$$

Taking the ensemble average of the product of Eqs. (2.10) and (2.11) gives

$$\langle x(t_1)x(t_2) \rangle = \int_{-\infty}^{\infty}\int_{-\infty}^{\infty} A(t_1,\omega_1)A^*(t_2,\omega_2)e^{i\omega_1 t_1 - i\omega_2 t_2} \langle dZ(\omega_1)dZ(\omega_2) \rangle \ .$$

Applying the orthogonality property of $Z(\omega)$,

$$\langle dZ(\omega_1)dZ^*(\omega_2) \rangle = \delta(\omega_1 - \omega_2)\langle\, |\, dZ(\omega)|^2 \rangle \ ,$$

and setting $t = t_1 = t_2$, the above equation becomes

$$\langle x^2(t) \rangle = \int_{-\infty}^{\infty} |A(t,\omega)|^2\, dP(\omega) \ , \tag{2.12}$$

where the incremental power spectrum,

$$dP(\omega) = \langle\, |\, dZ(\omega)|^2 \rangle = S(\omega)d\omega \ .$$

Thus, Eq. (2.12) can be written as

$$\langle x^2(t) \rangle = \int_{-\infty}^{\infty} S(t,\omega)d\omega \ , \tag{2.13}$$

in which

$$S(t,\omega) = |A(t,\omega)|^2 S(\omega) \ .$$

This expression is called the evolutionary spectral density function.

2.2.2 Modal analysis and evolutionary spectra

Consider the governing matrix equation of motion for a discretized linear structural system under external random excitations

$$M\ddot{x} + C\dot{x} + Kx = F \tag{2.14}$$

where M, C, and K are the linear assembled mass, damping, and stiffness matrices, respectively, while F is the nonstationary random excitation vector.

Assuming that the random excitations are all nonstationary, and they are modeled as products of uniformly modulated functions and zero mean Gaussian white noise processes so that the elements of F can be written as

$$F_i = e_i w_i, \quad i = 1, 2, \cdots, n \tag{2.15}$$

where e_i is the envelope modulating function for the i'th excitation with w_i being the corresponding zero mean Gaussian white noise process.

By following the usual procedure of modal analysis as in Appendix 2A and normalizing the mode shapes, a complete set of n normalized mode shapes Ψ_m is obtained. The i'th column of the normal mode matrix Ψ_m is defined by

$$\Psi_m^{(i)} = \frac{1}{\sqrt{m_{ii}}} \Psi^{(i)}, \quad i = 1, 2, \ldots, n \tag{2.16}$$

where $\Psi^{(i)}$ is the i'th column of the modal matrix, Ψ whereas m_{ii} is the diagonal element of the matrix $\Psi^T M \Psi$ associated with mode shape $\Psi^{(i)}$. That is, $(\Psi^{(i)})^T M \Psi^{(i)} = m_{ii}$. Similarly, $(\Psi^{(i)})^T K \Psi^{(i)} = k_{ii}$. Note that because of the above definition $\Psi_m^T M \Psi_m$ is an identity matrix. Symbolically, $\Psi_m^T M \Psi_m = I$.

Introducing the transformation

$$\{x\}_{n \times 1} = \left[\Psi_m\right]_{n \times L} \{q\}_{L \times 1} \quad \text{or simply as} \quad x = \Psi_m q. \tag{2.17}$$

Premultiplying Eq. (2.14) throughout by Ψ_m^T, it leads to the uncoupled equation

$$\ddot{q}_r + 2\zeta_r \omega_r \dot{q}_r + \omega_r^2 q_r = f_r(t), \quad r = 1, 2, \cdots, L \tag{2.18}$$

in which

$$2\zeta_r \omega_r = \left(\Psi_m^{(r)}\right)^T C \Psi_m^{(r)}, \quad \omega_r^2 = \left(\Psi_m^{(r)}\right)^T K \Psi_m^{(r)}, \quad f_r(t) = \left(\Psi_m^{(r)}\right)^T F$$

and here the integer L is the total number of modes considered in the analysis. In general, L is less than n. This has to do with the fact that in structural dynamic analysis of structures discretized by the FEM the response is usually reasonably accurately represented by the first few modes even though the discretized structure may have a large number of dof. The assembled damping matrix C is assumed to be of proportional or Rayleigh type. That is, for example,

$$C = \lambda_m M + \lambda_k K , \tag{2.19}$$

where λ_m and λ_k are called the Rayleigh damping coefficients. Equation (2.19) implies that only the first two modal damping ratios can be chosen independently for the system. Note that other forms of damping equation can be constructed.

The modal response is given by

$$q_r = \int_0^t h_r(t - \tau) f_r(\tau) \, d\tau$$

where $h_r(t)$ is the impulse response function corresponding to the r'th mode. This impulse response function is defined as

$$h_r(t) = \left(\frac{1}{\Omega_r}\right) e^{-\zeta_r \omega_r t} \sin(\Omega_r t) , \qquad \Omega_r = \omega_r \sqrt{1 - \zeta_r^2} . \tag{2.20a, b}$$

Substituting Eq. (2.20a) into the modal response gives

$$q_r = \left(\frac{1}{\Omega_r}\right) \int_0^t e^{-\zeta_r \omega_r (t - \tau)} \sin\Omega_r(t - \tau) f_r(\tau) d\tau . \tag{2.21}$$

Applying the modal excitation,

$$f_r(t) = a_r(t) \eta_r(t) , \tag{2.22}$$

where $a_r(t)$ is the deterministic modal time modulating function, and $\eta_r(t)$ is the modal zero mean Gaussian white noise excitation, Eq. (2.21) can be written as

$$q_r = \left(\frac{1}{\Omega_r}\right) \int_0^t \left[e^{-\zeta_r \omega_r (t - \tau)} \sin\Omega_r(t - \tau) \right] a_r(\tau) \eta_r(\tau) d\tau . \tag{2.23}$$

The *j*'th element of the original response vector becomes

$$
x_j(t_1) = \sum_{r=1}^{L} \left(\frac{\Psi_{jr}}{\Omega_r} \right) \int_0^{t_1} e^{-\zeta_r \omega_r (t_1 - \tau_1)} \sin \Omega_r (t_1 - \tau_1) a_r(\tau_1) \eta_r(\tau_1) d\tau_1 \qquad (2.24)
$$

where Ψ_{jr} is the *j*'th element of the *r*'th mode of the discretized structure. Similarly, the *k*'th element of the original response vector becomes

$$
x_k(t_2) = \sum_{s=1}^{L} \left(\frac{\Psi_{ks}}{\Omega_s} \right) \int_0^{t_2} e^{-\zeta_s \omega_s (t_2 - \tau_2)} \sin \Omega_s (t_2 - \tau_2) a_s(\tau_2) \eta_s(\tau_2) d\tau_2 . \qquad (2.25)
$$

Taking the ensemble average of the product of $x_j(t_1)$ and $x_k(t_2)$, setting $t = t_1 = t_2$, and letting

$$
a_r(t) = E_r u_r(t) \left(e^{-\alpha_{r1} t} - e^{-\alpha_{r2} t} \right) , \qquad (2.26)
$$

where E_r, α_{r1}, and α_{r2} with $\alpha_{r1} < \alpha_{r2}$, are positive constants and $u_r(t)$ is the unit step function, one can show that

$$
\langle x_j(t) x_k(t) \rangle = \int_{-\infty}^{\infty} S_{jk}(t,\omega) d\omega , \qquad (2.27)
$$

in which the evolutionary cross-spectral density function is given as

$$
S_{jk}(t,\omega) = \sum_{i=1}^{5} S_{jk}^{(i)}(t,\omega) , \qquad (2.28)
$$

where the terms on the rhs of Eq. (2.28) are defined in Appendix 2B.

2.3 Evolutionary Spectra of Engineering Structures

In this section three cases are included to illustrate the use of the explicit evolutionary cross-spectral density function obtained. The first case is concerned with the physical model of a mast antenna structure of which the nonstationary random excitation is applied at the base [14]. The second case deals with application to a cantilever beam structure in which the nonstationary random excitation is applied at the free-end. The third case has to do with the application to a plate structure and simply-supported at all sides. The nonstationary random load is applied at the center of the plate. The results for the second and third cases are selected from Ref. [18].

2.3.1   Evolutionary spectra of mast antenna structure

The mast antenna structure is assumed to be rigidly clamped at the base. The discretized model is shown in Figure 2.1. The antenna was approximated by two discrete masses including rotary inertias attached at Node numbers 14 and 16. The mast is represented by beam finite elements, every one of which has two nodes and each node has three dof. The latter are the transversal displacement v_i, rotation θ_i about an axis perpendicular to the plane of the figure, and curvature $\partial^2 v_i / \partial x^2$, i being the nodal number . The beam and discrete mass finite elements used are identified as TB5 and DM3 in Ref. [19]. This is a special case of the tapered beam element considered in Chapter 1. Explicit expressions for element mass and stiffness matrices have been presented in Appendix 1A. The pertinent input data for the structure are included in Table 2.1. Note that the values of m_{ii} and k_{ij} in Table 2.1(b) are large because of their definitions as given below Eq. (2.16). The uniformly modulated random excitation is applied at the base of the structure so that the governing matrix equation of motion can be written as

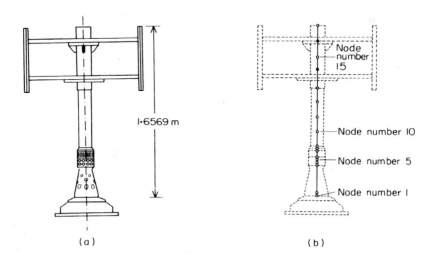

(a)   (b)

Figure 2.1   Antenna structure: (a) physical model; (b) discretized model.

Table 2.1   Input data for physical model of mast antenna structure.

(a) Nodal data and material properties of FE model

Node number	Co-ordinate (m) from base	Properties and Lumped masses
1	0.0000	Material properties:
2	0.0183	$E = 207$ GPa
3	0.2826	$\rho = 7860$ kg/m^3
4	0.3126	
5	0.3426	Lumped masses:
6	0.3726	at Node 14, $m_d = 72.2$ kg
7	0.4366	$J_d = 23.3$ kg.m^2
8	0.4620	
9	0.4874	at Node 16, $m_d = 67.8$ kg
10	0.6274	$J_d = 23.3$ kg.m^2
11	0.7674	
12	0.9074	
13	1.0474	
14	1.2299	
15	1.3474	
16	1.4649	
17	1.6569	

(b) Parameters for evolutionary spectral density plots

Values associated with first mode	Values associated with second mode	Material and structural properties
$\psi_{1,1} = 1.00$	$\psi_{1,2} = 1.00$	$m_{11} = 52122.40$
$\psi_{2,1} = 0.000346$	$\psi_{2,2} = 0.000339$	$m_{22} = 1723.58$
$\psi_{3,1} = 0.042707$	$\psi_{3,2} = 0.041578$	$\omega_1 = 12.21$ Hz
$\psi_{4,1} = 6.357140$	$\psi_{4,2} = 6.155460$	$\omega_2 = 59.43$ Hz
$\psi_{29,1} = 4.863720$	$\psi_{29,2} = 2.502030$	$k_{31} = 0.271475 \times 10^9$
$\psi_{41,1} = 14.90960$	$\psi_{41,2} = 1.87188$	$k_{41} = -0.324256 \times 10^{14}$
		$k_{51} = 0.296694 \times 10^{12}$
		$k_{61} = -0.271475 \times 10^9$

Table 2.1   Input data for physical model of mast antenna structure (continued).

(c) Element geometrical properties

Element number i	A_{i-1} of cross-section $(m^2 \times 10^{-2})$	I_{i-1} of cross-section $(m^4 \times 10^{-4})$	Taper ratio α	Taper ratio β
1	0.9330	0.5600	1.0000	1.0000
2	0.2286	0.1505	1.0000	0.5751
3	0.1734	0.0413	1.0000	1.0000
4	0.1734	0.0413	1.0000	1.0000
5	0.1734	0.0413	1.0000	1.0000
6	0.2630	0.0508	1.0000	1.0000
7	0.3770	0.0905	1.0000	1.0000
8	0.3770	0.0905	0.3685	1.0000
9	0.1431	0.0365	1.0000	1.0000
10	0.1431	0.0365	1.0000	1.0000
11	0.1431	0.0365	1.0000	1.0000
12	0.1431	0.0365	1.0000	1.0000
13	0.3770	0.0905	1.0000	1.0000
14	0.3770	0.0905	1.0000	1.0000
15	0.3770	0.0905	1.0000	1.0000
16	0.3770	0.0905	1.0000	1.0000
17	0.3770	0.0905	1.0000	1.0000

$$\begin{bmatrix} M_{yy} & M_{yx} \\ M_{xy} & M_{xx} \end{bmatrix} \begin{Bmatrix} \ddot{y} \\ \ddot{x} \end{Bmatrix} + \begin{bmatrix} C_{yy} & C_{yx} \\ C_{xy} & C_{xx} \end{bmatrix} \begin{Bmatrix} \dot{y} \\ \dot{x} \end{Bmatrix} + \begin{bmatrix} K_{yy} & K_{yx} \\ K_{xy} & K_{xx} \end{bmatrix} \begin{Bmatrix} y \\ x \end{Bmatrix} = \begin{Bmatrix} F_y \\ 0 \end{Bmatrix} \qquad (2.29)$$

where M_{xx}, C_{xx}, and K_{xx} are respectively the mass, damping, and stiffness matrices of the constrained structure, whereas x and y are the random displacement response and prescribed random displacement column vectors, respectively. The elements of the column vector F_y are the forces that are caused by the displacement vector, y.

Operating on Eq. (2.29), the second equation becomes

$$M_{xx}\ddot{x} + C_{xx}\dot{x} + K_{xx}x = F_x = -M_{xy}\ddot{y} - C_{xy}\dot{y} - K_{xy}y . \qquad (2.30)$$

With reference to Eq. (2.18) and assuming the excitation is a uniformly modulated random white noise process, one has

$$f_r(t) = \frac{1}{m_{rr}} \sum_{j=1}^{n} \psi_{jr} \left(F_{x,j} \right), \tag{2.31}$$

where m_{rr} and ψ_{jr} are, respectively, the diagonal element of the diagonalized mass matrix and element of the eigenvector associated with the r'th mode of the constrained structure in Eq. (2.30), while $F_{x,j}$ is the j'th element of force vector F_x. Note that the terms associated with the applied random velocity and acceleration are small compared with that associated with the prescribed displacement because the elements of K_{xy} are usually several orders of magnitude higher than those of M_{xy} and C_{xy}. The excitation vector, therefore reduces to

$$F_x = - K_{xy} y , \tag{2.32}$$

where in the present problem $y = [\ \hat{y}(t)\ \ 0\ ]^T$. Operating on Eq. (2.32) leads to

$$\underset{n \times 1}{F_x} = - \underset{1 \times n}{\left[k_{31}\ \ k_{41}\ \ k_{51}\ \ k_{61}\ \ 0\ 0\ \cdots\ 0 \right]^T} \hat{y}(t) , \tag{2.33}$$

in which k_{31} , ... , are the elements of the assembled stiffness matrix. In this problem the number of dof is $n = 48$ since there are 17 nodes in Figure 2.1(b) so that $n = 3 \times 17 - 3$.

Applying Eq. (2.31) gives

$$f_r(t) = - \frac{1}{m_{rr}} \left[\psi_{1r} k_{31} + \psi_{2r} k_{41} + \psi_{3r} k_{51} + \psi_{4r} k_{61} \right] \hat{y}(t) = a_r(t) \eta_r(t) ,$$

where $a_r(t)$ is defined in Eq. (2.26) while

$$\eta_r(t) = - \frac{1}{m_{rr}} \left[\psi_{1r} k_{31} + \psi_{2r} k_{41} + \psi_{3r} k_{51} + \psi_{4r} k_{61} \right] w(t) .$$

That is, $\hat{y}(t) = a_r(t) w(t)$, in which $w(t)$ is the random displacement applied at the base of the structure and approximated as a zero mean Gaussian white noise process whose spectral density is denoted by S or S_0. In other words,

$$\langle \eta_r(\tau_1)\eta_s(\tau_2) \rangle = \int_{-\infty}^{\infty} \frac{1}{m_{rr}} \left[\psi_{1r} k_{31} + \psi_{2r} k_{41} + \psi_{3r} k_{51} + \psi_{4r} k_{61} \right]$$

(2.34)

$$\times \frac{1}{m_{ss}} \left[\psi_{1s} k_{31} + \psi_{2s} k_{41} + \psi_{3s} k_{51} + \psi_{4s} k_{61} \right] e^{i\omega(\tau_1 - \tau_2)} S \, d\omega .$$

For illustration and brevity, only the evolutionary spectral density of displacement at Node 15, and cross-spectral density of the displacement responses at Nodes 15 and 11 by Eq. (2.28) for $\zeta_r = \zeta = 0.0, 0.005, 0.05,$ and 0.10 with $E_r = 9.4814$, $\alpha_{r1} = 30/s$ and $\alpha_{r2} = 40/s$ are presented in Figures 2.2 and 2.3. The remaining pertinent data necessary for the computation of evolutionary spectral and cross-spectral densities are included in Table 2.1. It should be noted that the phase plots of the evolutionary cross-spectral densities have not been included in Figure 2.3 since they are very complex and difficult to interpret. More computed results were presented in Ref. [14]. With reference to the presented plots, the responses are due to the first two modes. This implies that the random excitation applied at the base

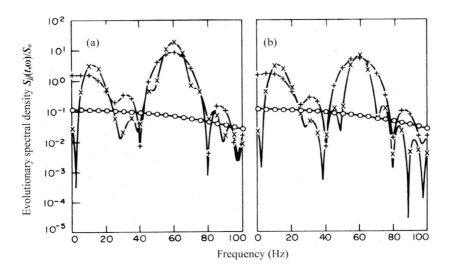

Figure 2.2 Evolutionary spectral densities of displacement at Node 15,
$j = 41$; $\circ$, $t = 0.01$ s; $+$, $t = 0.06$ s; $\times$, $t = 0.11$ s :
(a) $\zeta_r = 0$; (b) $\zeta_r = 0.005$ (to be continued).

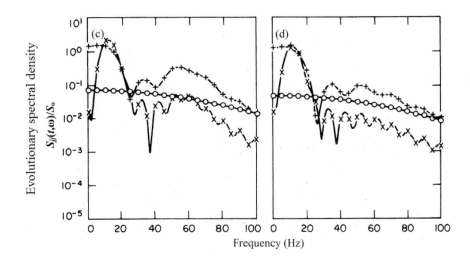

Figure 2.2 Evolutionary spectral densities of displacement at Node 15, $j = 41$; $\circ$, $t = 0.01$ s; $+$, $t = 0.06$ s; $\times$, $t = 0.11$ s : (c) $\zeta_r = 0.05$; (d) $\zeta_r = 0.10$.

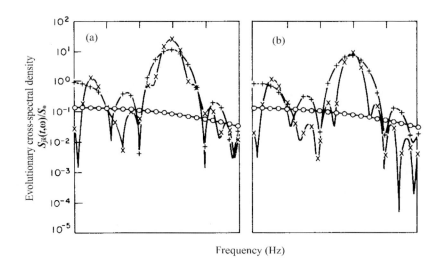

Figure 2.3 Evolutionary cross-spectral densities of displacements at Nodes 15 and 11, $j = 41$, $k = 29$; key as for Figure 2.2(a)-(d) (to be continued).

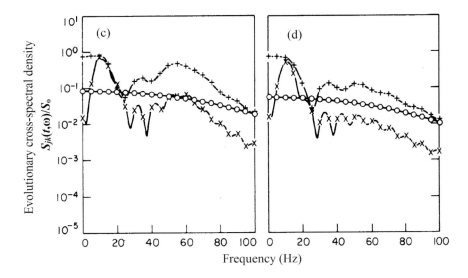

Figure 2.3 Evolutionary cross-spectral densities of displacements at Nodes 15 and 11, $j = 41$, $k = 29$; key as for Figure 2.2(a)-(d).

of the antenna structure had excited only the first two modes. However, the digital computer program written in Fortran for the investigation can be applied to situations involving any number of modes excited.

2.3.2   Evolutionary spectra of cantilever beam structure

In this application a beam structure is considered. It is clamped at one end and free at the other. It is represented by 10 two-dimensional beam bending elements. Every element has two nodes. Each node has 3 dof. These nodal dof are u, v, θ_z where u is the axial displacement, v the transversal displacement, and θ_z the angular displacement about z-axis which is perpendicular to the plane containing u and v. The present two-node beam element is considered as a combination of a two-node bar element and a two-node bending beam element. It is applied here due to the fact that it is simple and in this case the applied external nonstationary random excitation is at the free-end of the cantilever structure. This is in contrast to the mast antenna structure in Sub-section 2.3.1 where the external random excitation is applied at the base or clamped end. For simple direct reference, the consistent element mass and stiffness matrices are presented here. The consistent element mass matrix is

$$[m] = \frac{\rho A \ell}{420} \begin{bmatrix} 140 & 0 & 0 & 70 & 0 & 0 \\ 0 & 156 & 22\ell & 0 & 54 & -13\ell \\ 0 & 22\ell & 4\ell^2 & 0 & 13\ell & -3\ell^2 \\ 70 & 0 & 0 & 140 & 0 & 0 \\ 0 & 54 & 13\ell & 0 & 156 & -22\ell \\ 0 & -13\ell & -3\ell^2 & 0 & -22\ell & 4\ell^2 \end{bmatrix} \quad (2.35)$$

where ρ is the density, A the cross-section area, and ℓ the length of the element. The consistent element stiffness matrix is defined by

$$[k] = \begin{bmatrix} a & 0 & 0 & -a & 0 & 0 \\ 0 & 12b & 6\ell b & 0 & -12b & 6\ell b \\ 0 & 6\ell b & 4\ell^2 b & 0 & -6\ell b & 2\ell^2 b \\ -a & 0 & 0 & a & 0 & 0 \\ 0 & -12b & -6\ell b & 0 & 12b & -6\ell b \\ 0 & 6\ell b & 2\ell^2 b & 0 & -6\ell b & 4\ell^2 b \end{bmatrix} \quad (2.36)$$

in which a and b are given by $a = EA/\ell$, $b = EI/\ell^3$, E is the Young's modulus of elasticity, and I the moment of inertia of the cross-section of the beam.

The beam structure is 1.0 m in length and 0.01 m in diameter. The material of the beam structure is concrete with the following properties: Young's modulus $E = 27.67$ GPa, density $\rho = 2308$ kg/m^3. It is assumed that the cantilever beam is clamped at Node 1 and Node 11 is the free end so that after the application of boundary conditions the beam structure has 30 dof or the order of each of the constrained assembled mass and stiffness matrices of the structure is 30. Therefore, the 29'th dof (or the 32'th dof of the unconstrained structure) is the transverse displacement at the tip or free end of the beam structure. It is also assumed that the random excitation is applied at this nodal dof.

Only the first three modes are incorporated in the response computation. Thus, in Eq. (2.18) the number of modes, $L = 3$. The responses due to the transverse random forces applied at the free end of the beam structure are obtained subsequently for two types of random forces. One is a stationary white noise and the other is an amplitude modulated white noise process.

The governing equation of motion is given by Eq. (2.14) for the constrained structure and the applied stationary random excitation defined by Eq. (2.15) in which $i = 29$ becomes

$$F_{29} = w_{29} , \tag{2.37}$$

whereas the amplitude modulated white noise excitation is given by

$$F_{29} = 3.06\left(e^{-2.51t} - e^{-6.28t}\right)w_{29} . \tag{2.38}$$

This means that the envelope modulating function defined by Eq. (2.15) is

$$e_{29} = 3.06\left(e^{-2.51t} - e^{-6.28t}\right) .$$

By applying Eq. (2.28), computed evolutionary spectral densities of the transverse displacement at the tip of the cantilever beam structure under a stationary white noise excitation at the tip are included in Figure 2.4. Those of the cantilever beam structure under an amplitude modulated white noise excitation are shown in Figure 2.5. In these three-dimensional (3D) plots $S_{29,29}(t,\omega)$ for two different damping ratios, $\zeta_r = 0.01$ and $\zeta_r = 0.10$, where $r = 1$, 2, and 3 were considered.

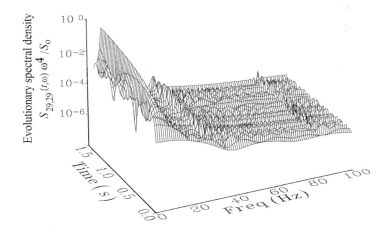

Figure 2.4(a) Evolutionary spectral density of transverse displacement of cantilever beam under stationary random excitation: $\zeta_r = 0.01$.

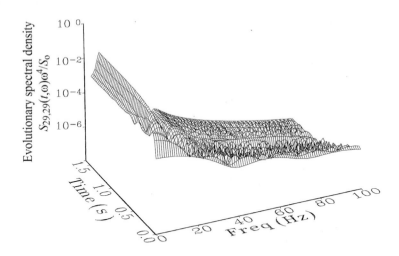

Figure 2.4(b)   Evolutionary spectral density of transverse displacement of cantilever beam under stationary random excitation: $\zeta_r = 0.10$.

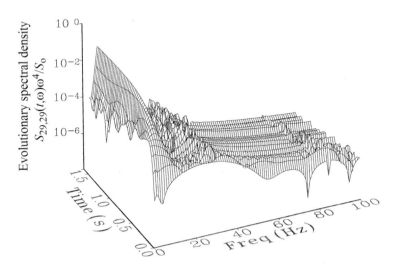

Figure 2.5(a)   Evolutionary spectral density of transverse displacement of cantilever beam under nonstationary random excitation: $\zeta_r = 0.01$.

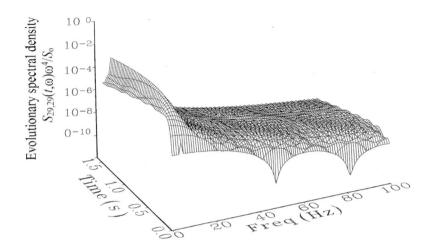

Figure 2.5(b) Evolutionary spectral density of transverse displacement of cantilever beam under nonstationary random excitation: $\zeta_r = 0.10$.

2.3.3 Evolutionary spectra of plate structure

The evolutionary spectra are applied to study the dynamic responses of a typical thin plate structure. The finite element employed in this sub-section is one of the earliest to appear in the literature that was originally and independently developed by Melosh [20], and Zienkiewicz and Cheung [21]. It is the non-conforming plate bending rectangular finite element and is referred to as the MZC [22] rectangular thin plate element. It has four nodes and every node has three dof. The latter are the transverse displacement $w_i^e(x,y)$ at node i or simply written as w_i^e, slope $\partial w_i^e/\partial x$, and slope $\partial w_i^e/\partial y$, in which x and y are the orthogonal co-ordinates on the plane of the element.

The plate dimensions are 1.0 m by 1.0 m, and thickness 5 mm. It is approximated by 16 MZC elements. This finite element representation is included in Figure 4.1. The plate structure is simply-supported at all sides. These boundary conditions are denoted as S4 and the nodal data are listed in Table 2.2. Note that Node 13 in Figure 4.1 is at the center of the plate structure. The material of the plate structure is steel with the following material properties: Young's modulus $E = 200$ GPa, density $\rho = 7.83 \times 10^3$ kg/m^3, and Poisson's ratio $\nu = 0.3$.

The eigenvalue problem is solved to determine the natural frequencies and mode shapes. However, only the first 8 modes are considered in the computation. These natural frequencies and mode-shapes have been obtained by Wang [23]. For brevity, only the first 8 natural frequencies with their corresponding analytical values for comparison are presented in Table 4.1. Clearly, the 16 element representation of the S4 plate structure is adequate as these natural frequencies agree very well with the analytical results presented in Refs. [24, 25].

By applying Eq. (2.28), the evolutionary spectral and cross-spectral densities of transverse displacements at Nodes 7 and 13, are evaluated for the two cases of random excitations applied at Node 13 and along the transverse direction. These two cases of random excitations are identical to Eqs. (2.37) and (2.38) except that now the dof, $i = 19$, assuming Eq. (2.14) is employed for the constrained system. Note that if Eq. (2.14) is applied to the plate structure before the application of constraints or boundary conditions, the dof is $i = 37$. The 3D plots of $S_{5,5}(t,\omega)$, $S_{19,19}(t,\omega)$, and $S_{5,19}(t,\omega)$ presented in Figures 2.6 through 2.11 are for the lightly damped case in which $\zeta_r = 0.01$. The dof, i refers to in the figures is that of the constrained structure. From the plots it is observed that the main contribution to the magnitude of the evolutionary spectrum is due to the first three modes. This is consistent with the fact that only the first three modes of the structure have been included in the computation. Note that the evolutionary cross-spectral densities have been evaluated in the course of computation. They are not included in this sub-section because the phase information is extremely difficult to interpret.

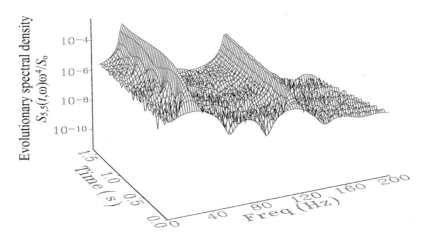

Figure 2.6 Evolutionary spectral density of transverse displacement at Node 7: white noise excitation case.

Table 2.2   Nodal data for S4 plate structure.

Node number	Nodal co-ordinates (m)		
	X	Y	Z
1	0.0000	0.0000	0.0000
2	0.2500	0.0000	0.0000
3	0.5000	0.0000	0.0000
4	0.7500	0.0000	0.0000
5	1.0000	0.0000	0.0000
6	0.0000	0.2500	0.0000
7	0.2500	0.2500	0.0000
8	0.5000	0.2500	0.0000
9	0.7500	0.2500	0.0000
10	1.0000	0.2500	0.0000
11	0.0000	0.5000	0.0000
12	0.2500	0.5000	0.0000
13	0.5000	0.5000	0.0000
14	0.7500	0.5000	0.0000
15	1.0000	0.5000	0.0000
16	0.0000	0.7500	0.0000
17	0.2500	0.7500	0.0000
18	0.5000	0.7500	0.0000
19	0.7500	0.7500	0.0000
20	1.0000	0.7500	0.0000
21	0.0000	1.0000	0.0000
22	0.2500	1.0000	0.0000
23	0.5000	1.0000	0.0000
24	0.7500	1.0000	0.0000
25	1.0000	1.0000	0.0000

2.3.4   Remarks

It should be mentioned that in the foregoing three sub-sections the results presented are two-sided evolutionary spectral and cross-spectral densities. All these quantities along the ordinates of the plots are divided by the spectral density of the applied white noise excitation. The phase components of the evolutionary cross-spectral densities are difficult to interpret and therefore they are not included in this section. In practice, if these phase components are retained, extreme care has to be taken in their interpretation.

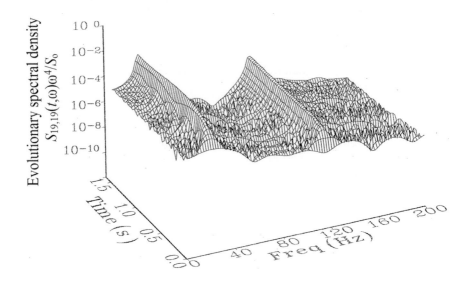

Figure 2.7   Evolutionary spectral density of transverse displacement
at Node 13: white noise excitation case.

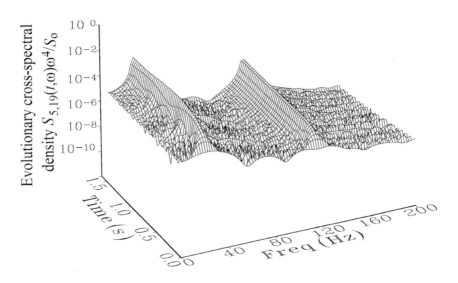

Figure 2.8   Evolutionary cross-spectral density of transverse displacements
at Nodes 7 and 13: white noise excitation case.

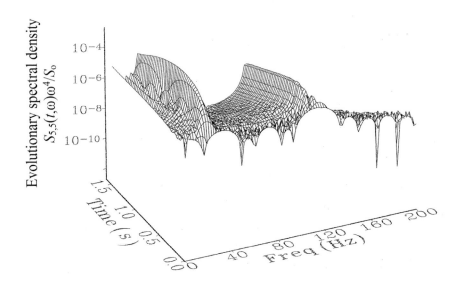

Figure 2.9   Evolutionary spectral density of transverse displacement
at Node 7: nonstationary random excitation case.

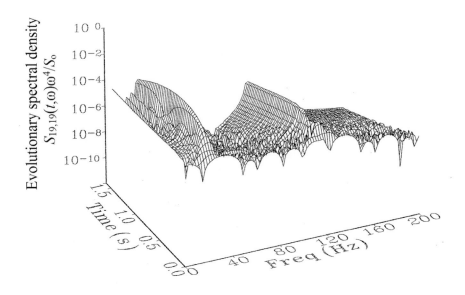

Figure 2.10   Evolutionary spectral density of transverse displacement
at Node 13: nonstationary random excitation case.

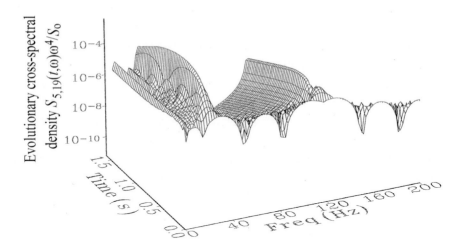

Figure 2.11 Evolutionary cross-spectral density of transverse displacements at Nodes 7 and 13: nonstationary random excitation case.

2.4 Modal Analysis and Time-Dependent Response Statistics

The response of a nuclear power plant containment shell structure excited by earthquake ground motion was analysed by Yousafzai and Ahmadi [26]. The nonstationary earthquake motion was modelled as a product of a deterministic envelope function and the Gaussian white noise process, while the structure was approximated by the FE of the computer package, SAP IV.

Extension of the evolutionary spectral analysis [14] to the time-dependent response analysis of discretized structures under a wide class of nonstationary random disturbances was reported by the author. Explicit expressions for time-dependent variances of displacements, variances of velocities, covariances of displacements, and covariances of displacements and velocities were derived and presented by the author [15, 27]. The analysis of the distribution of first passage time of mast antenna structures represented by the FEM and under amplitude modulated zero mean Gaussian white noise excitations was also investigated [28]. Applying the derived explicit expressions in Ref. [15], To and Wang [29] investigated the time-dependent response statistics of axisymmetric shell structures. The latter were idealized by the truncated conical shell finite element, the explicit expressions of consistent element mass and stiffness matrices of which were obtained earlier in Ref. [30]. For completeness and possible application by the reader explicit expressions of the special case of the cylindrical shell element are included in Appendix 2F.

Application of the FEM and normal mode approach for structural reliability analysis was made by Shinozuka, Kako and Tsurui [31]. An extensive review of random vibration analysis by the normal mode approach was also presented.

The nonstationary random response analysis of laminated composite structures, which were approximated by the hybrid strain-based or mixed formulation laminated flat triangular shell finite element, was performed and reported by To and Wang [32]. In this latter reference normal mode method was applied. The three-node laminated composite shell finite elements developed by To and Wang [33] were employed. Every one of which has 18 dof. Thus, there are three translational and three rotational dof at every node where the important drilling dof (ddof) is included.

2.4.1 Time-dependent covariances of displacements

Applying complex contour integration, the closed form time-dependent covariances of displacements of discretized structural systems under nonstationary random excitations defined in Eq. (2.26) have been obtained by the author [15] and they are included in this sub-section.

Starting from Eq. (2.27) and going through some lengthy algebraic manipulation and simplifying, one can show that

$$\left\langle x_j(t)x_k(t)\right\rangle = \sum_{i=1}^{14} N_i^{(1)}\left(x_j x_k\right), \tag{2.39}$$

where the terms on the rhs are defined in Appendix 2C. The first eight terms on the rhs of Eq. (2.39) are the contribution due to the evolutionary coincident spectral density functions whereas the remaining six terms on the rhs of Eq. (2.39) are due to the evolutionary quadrature spectral density functions.

Note that if the subscripts $j = k$, Eq. (2.39) gives the variances of the generalized displacements.

2.4.2 Time-dependent covariances of displacements and velocities

Similar to the procedure used in the last sub-section, the closed form time-dependent covariances of displacements of discretized structural systems under nonstationary random excitations defined in Eq. (2.26) have been obtained by the author [15]. They are included in the following

$$\left\langle x_j(t)\dot{x}_k(t)\right\rangle = i \int_{-\infty}^{\infty} \omega S_{jk}(t,\omega)d\omega, \tag{2.40}$$

where the overdot on the lhs designates the derivative with respect to time t while on the rhs i is the imaginary number and not to be confused with the subscript i which is an integer denoting the nodal dof.

After some lengthy algebraic manipulation the results can be expressed as

$$\left\langle x_j(t)\dot{x}_k(t)\right\rangle = \sum_{i=1}^{14} N_i^{(12)}\left(x_j \dot{x}_k\right), \tag{2.41}$$

where the terms on the rhs are defined in Appendix 2D. Similar to Eq. (2.39) one can identify the first eight terms on the rhs of Eq. (2.41) as contribution due to the evolutionary coincident spectral density functions and the remaining six terms are due to the evolutionary quadrature spectral density functions.

2.4.3   Time-dependent covariances of velocities
Similar to the derivation of the results in Sub-section 2.4.1, the closed form time-dependent covariances of velocities y_j and y_k can be expressed as

$$\left\langle \dot{x}_j(t)\dot{x}_k(t)\right\rangle = \int_{-\infty}^{\infty} \omega^2 S_{jk}(t,\omega)\,d\omega . \tag{2.42}$$

Operating on the integral, one can show that

$$\left\langle \dot{x}_j(t)\dot{x}_k(t)\right\rangle = \sum_{i=1}^{14} N_i^{(2)}\left(\dot{x}_j \dot{x}_k\right), \tag{2.43}$$

where the terms on the rhs are now defined in Appendix 2E. Note that the first eight terms on the rhs of Eq. (2.43) are the contribution due to the evolutionary coincident spectral density functions. The remaining six terms are contributions from the evolutionary quadrature spectral density functions.

2.4.4 Remarks
From the last three sub-sections and Appendices 2B through 2E, one may appreciate the amount of algebraic manipulation required in the derivation of individual covariance relation. To evaluate the derived closed form results a digital computer program is necessary. To this end, therefore, a digital computer program written in Fortran has been developed. The number of modes that can be taken into account during the computation has no limit in theory although in practice it is restricted by the memory of the computing machine employed.

Before leaving this section it should be noted that the variances and covariances of responses derived in the foregoing are based on the condition that $0 < \xi_r < 1$ where ξ_r is defined by Eq. (2B.15) in Appendix 2B. That is, ξ_r has to be inside the contour of integration on the upper half of the complex plane in Figure 2B.1. This, in turn, requires that the damping ratio $\zeta_r < 1$, which is consistent with the assumption that the system is lightly damped.

2.5 Response Statistics of Engineering Structures

The explicit expressions derived in Section 2.4, Appendices 2B through 2E, and the associated digital computer program developed can now be applied to evaluate the variances and covariances of displacements, and velocities of structures under stationary and nonstationary random excitations. To limit the scope only three representative cases were investigated and included in the following sub-sections.

2.5.1 Mast antenna structure

The physical model of the mast antenna structure studied in Sub-section 2.3.1 is considered in this sub-section. Thus, the governing matrix equation of motion and relationships in this latter sub-section are applied. The FE model is identical to that in Sub-section 2.3.1 and therefore the pertinent input data in Table 2.1 are employed. The digital computer program based on the explicit expressions derived in Section 2.4 and Appendices 2B through 2E is applied. The obtained time-dependent variances of displacement responses at Node 15 (see, Figure 2.1) for various damping ratios, $\zeta_r = \zeta = 0.01, 0.025$, and 0.04 with $r = 1, 2$ are presented in Figure 2.12 in which variances of displacements are normalized by the spectral density of white noise, S_o. The results without the coupled terms were superimposed on those with the coupled terms. Graphically, there is no observable difference. The time-dependent covariances of displacement responses at Nodes 15 and 11 for $\zeta_r = \zeta = 0.01, 0.025$, and 0.04 with $r = 1, 2$ are presented in Figure 2.13. The results in Figure 2.13 without the coupled terms were superimposed on those with the coupled terms. Graphically, there is no observable difference.

It may be appropriate to mention that the derived explicit expressions in Eqs. (2.37) through (2.39) are hinged on the assumption that $0 \le \zeta_r \le 1.0$ although these explicit expressions can be modified for cases where $\zeta_r > 1$. However, it is very likely that a mode for which $\zeta_r > 1$ has an insignificant effect on the overall responses. Therefore, the case for $\zeta_r > 1$ has not been pursued in this study.

Before leaving this sub-section it should be noted that more computed results including covariances of velocities have been obtained and representative ones have been reported in Refs. [15, 27]. They are not included here for brevity.

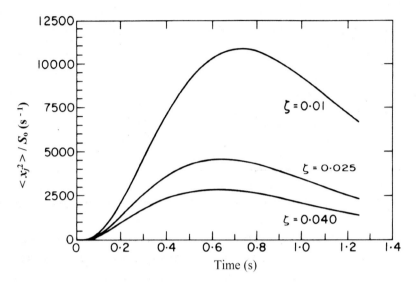

Figure 2.12 Variance of displacement response $x_j(t)$ with $j = 29$, $E_r = 9.4814$, $\alpha_{r1} = 1.5/s$ and $\alpha_{r2} = 2.0/s$.

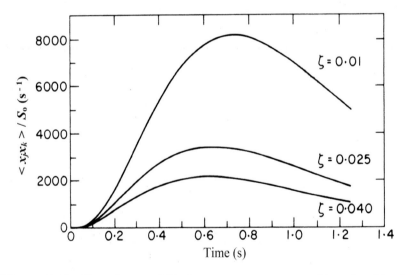

Figure 2.13 Covariance of displacement responses $x_j(t)$ and $x_k(t)$ with $j = 29$, $k = 41$, $E_r = 9.4814$, $\alpha_{r1} = 1.5/s$ and $\alpha_{r2} = 2.0/s$.

2.5.2   Truncated conical shell structures

One of the truncated shell structures studied [29] and included here is shown in Figure 2.14. It is represented by the axisymmetrical conical thin shell finite element developed by To and Wang [30]. The explicit expressions for the consistent element mass and stiffness matrices of this displacement formulation-based truncated shell finite element and its corresponding cylindrical counterparts have been presented in Ref. [30]. For brevity and direct reference, only the explicit expressions for the consistent element mass and stiffness matrices of the cylindrical shell finite element are included in Appendix 2F. For computational efficiency, the so-called equivalent cylindrical shell element method (ECSEM) reported in [29] is adopted in the evaluation of variances of displacements in the following. In the ECSEM every conical shell element is represented by an equivalent cylindrical shell (ECS) element, the length and radius of which are, respectively, the slant length and average radius of the truncated conical shell element. The thickness and material properties for both the truncated conical shell elements and the ECS elements are identical. Each of the truncated conical shell element and the ECS element has two nodes. Each node has four dof.

As shown in Figure 2.12 the truncated shell structure has a vertical length H and semi-vertex angle φ. It is clamped at the base. While more finite element meshes were studied in Refs. [29, 30], two are employed in this sub-section for the computation of response statistics. These are the 20 element (20E) and 11 element (11E) approximations. Its geometrical properties are: small end radius $R_s = 0.3048$ m (12 in), base radius $R_b = 0.3358$ m (13.22 in), vertical length or height $H = 0.5950$ m (23.43 in), thickness $h = 0.8128$ mm (0.032 in), and semi-vertex angle $\varphi = 3°$. The material is aluminum whose Young's modulus of elasticity $E = 68.95$ GPa, shear modulus $G_e = 26.0$ GPa, Poisson's ratio $v = 0.3$, and density $\rho = 2.823 \times 10^3$ kg/m^3. In the figure, the encircled numbers refer to the element numbers while those without circles are nodal numbers.

The discretized cantilever truncated conical shell structure is excited by a nonstationary random disturbance at the base and its matrix equation of motion is similar to Eq. (2.29) in which the deterministic modulating function is defined by Eq. (2.26). After the application of boundary conditions, Eq. (2.29) reduces to Eq. (2.30). Disregarding the damping term, the eigenvalue solution of Eq. (2.30) is performed before the computation of the response statistics. For brevity, only the first five natural frequencies $f_r^c = 2\pi\omega_r$, where the subscript $r = 1, 2, ... , 5$ denote the modal numbers, are presented in Table 2.3. In the latter n_c is the circumferential wave number and the experimental results from Ref. [34] for the cylindrical shell of radius

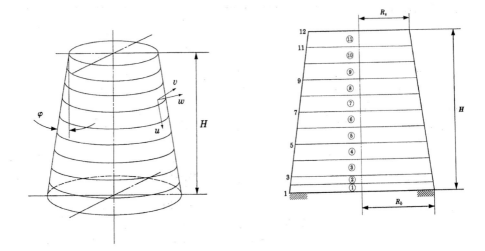

Figure 2.14   Truncated conical shell: (a) ring finite element model with deformation directions shown, and (b) 11-element model.

$R = R_s = 0.3048$ m (12 in) are also included. Note that the other geometrical and material properties of this cylindrical shell are the same as the truncated conical shell. In the table, 6ECS20E designates the computed results for the 6° apex angle shell using the ECSEM with 20 shell elements. These results has been verified and reported in Ref. [29]. The results for the 11-element model, shown in Figure 2.14(b), are denoted by 6ECS11E. The 20-element model is presented in Figure 2.15.

As the shell element has two nodes, each of which has four dof, namely, u_i^e, v_i^e, w_i^e, and $\beta_i = \partial w_i^e/\partial s$ with the slant length s along the u-axis in Figure 2.14(a). The subscript i designates the node number. Therefore, there are 80 equations in total for the truncated conical shell structure represented by 20 elements. For brevity, two cases are considered in this sub-section. One has included the first two modes and the other has the first five modes during the computation of response statistics. Thus, the parameters in Eq. (2.26) are: $E_r = 9.4814$, $\alpha_{r1} = 30/s$ and $\alpha_{r2} = 40/s$ where the subscript $r = 1, 2$ for the first case and $r = 1, 2, ... , 5$ for the second case. Other pertinent input data for the 20 element-case are included in Table 2.4.

With reference to Table 2.3 one can see that the lowest natural frequency for the truncated conical aluminum shell appears at the circumferential number $n_c = 6$ and therefore, this circumferential number is applied in the computation of variances of displacements w_i^e. Damping ratios $\zeta_r = \zeta = 0.01, 0.025$, and 0.050 are considered.

Table 2.3 Natural frequencies of clamped-free (CF) truncated conical shell.

		Natural frequencies (Hz)				
n_c	Source	$f_1^{\,c}$	$f_2^{\,c}$	$f_3^{\,c}$	$f_4^{\,c}$	$f_5^{\,c}$
3	Experiment	210.8/206.2	N/A	N/A	N/A	N/A
	6ECS20E	225.9	845.8	1525.4	1901.5	2106.6
4	Experiment	131.7	N/A	N/A	N/A	N/A
	6ECS20E	143.2	599.6	1195.1	1623.5	1899.9
5	Experiment	100.8	429.1	N/A	N/A	N/A
	6ECS20E	105.9	444.9	944.3	1371.7	1688.4
6	Experiment	96.9	326.3/334.4	N/A	N/A	N/A
	6ECS20E	97.3	345.3	759.8	1159.5	1489.9
	6ECS11E	97.9	352.7	783.5	1208.6	1556.3
7	Experiment	113.0	273.5	N/A	N/A	N/A
	6ECS20E	108.6	283.1	625.3	987.2	1315.5
8	Experiment	140.4	247.4	N/A	N/A	N/A
	6ECS20E	132.0	249.3	528.8	850.9	1162.3
9	Experiment	174.3	244.2	N/A	N/A	N/A
	6ECS20E	162.8	239.0	462.7	745.7	1036.5

Representative computed results for the truncated conical shell with 6° apex angle by applying Eq. (2.39) are presented in Figures 2.16 through 2.19. More computed results for other geometrical properties can be found in Refs. [29, 30]. In these figures the responses are the displacements perpendicular to the surface of the shell and the variances are normalized by the spectral density $S_{rr} = S_o$ of the base random displacement treated as a zero mean Gaussian white noise process.

Four observations should be made at this stage. First, with reference to Table 2.3 the natural frequencies obtained by using the 11-element and 20-element models are in excellent agreement. These FE results compare very well with those obtained experimentally. Second, very significant reductions in the amplitudes of variances of displacements are achieved with increasing damping ratios. Third, the amplitude of the largest variance of displacement for the two mode case is about 10% lower than that of the five mode case. Fourth, with reference to Figure 2.17 the difference between amplitudes of the variances of the 20-element and 11-element models is less than 2%. Thus, for computational efficiency, the 11-element model can be adopted in order to provide sufficient accurate results.

Table 2.4   Data for variances of displacements for 6ECS20E model.

| Node j | Elements of modal matrix | | | | |
	$\psi_{j,1}$	$\psi_{j,2}$	$\psi_{j,3}$	$\psi_{j,4}$	$\psi_{j,5}$
1	1.0000	1.0000	1.0000	1.0000	1.0000
2	0.5945	1.9400	4.8700	18.3100	-106.7000
3	-7.3520	-15.1700	-34.4300	-135.1000	909.9000
4	-211.8000	-487.1000	-1167.0000	-4721.0000	31920.000
79	-459.6000	139.1000	-130.7000	332.2000	1718.0000

Material and structural properties:

$m_{11} = 73710.0,$ $m_{22} = 7827.9,$ $m_{33} = 7628.4,$ $m_{44} = 53676.0$
$m_{55} = 1523500.0,$ $k_{53} = 2.9021 \times 10^{7},$ $k_{63} = 1.5473 \times 10^{7},$
$k_{73} = 5.0264 \times 10^{5},$ $k_{53} = 2.9021 \times 10^{7},$

$f_1^{c} = 97.3$ Hz, $f_2^{c} = 345.3$ Hz, $f_3^{c} = 759.7$ Hz,
$f_4^{c} = 1159.5$ Hz, $f_5^{c} = 1489.9$ Hz

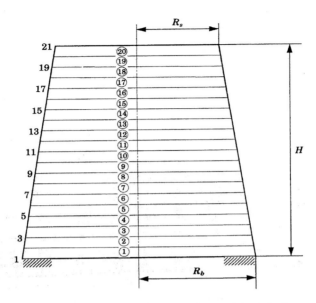

Figure 2.15   Truncated conical shell approximated by 20 elements.

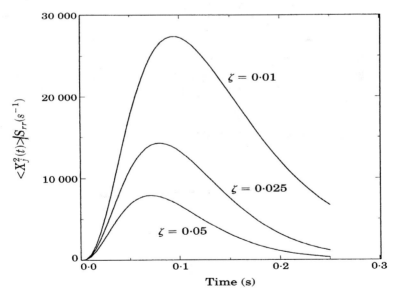

Figure 2.16   Variances of responses at Node 21 of 20-element model: $j = 79$, five modes considered.

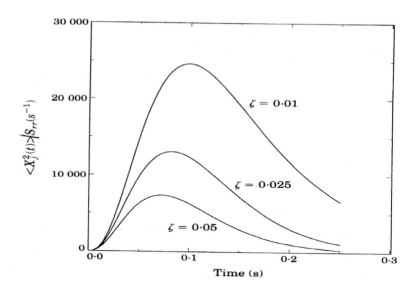

Figure 2.17   Variances of responses at Node 21 of 20-element model:
$j = 79$, two modes considered.

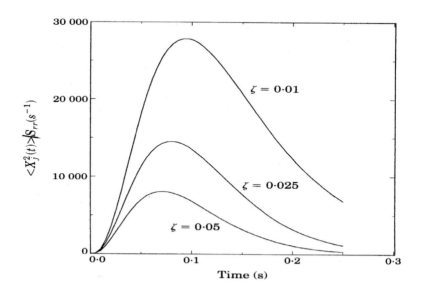

Figure 2.18   Variances of responses at Node 12 of 11-element model:
$j = 43$, five modes considered.

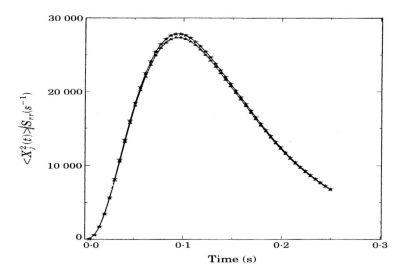

Figure 2.19   Variances of responses: ▲, $j = 79$ for 6ECS20E; and ★, $j = 43$ for 6ECS11E; $\zeta = 0.01$ and five modes considered.

2.5.3   Laminated composite plate and shell structures

The prediction, analysis, and design of laminated composite structures under nonstationary random excitations are of considerable interest to design engineers in the aerospace and automobile engineering fields. The prediction and analysis of response statistics are central to the design process. Thus, the response statistics of a nine-layer cross-ply square plate and a four-layer cross-ply cylindrical panel are studied. They are both symmetrically laminated and simply-supported at all sides.

It may be appropriate to mention that while application is made of the Fortran digital computer program based on the modal analysis and bi-modal approach to be introduced in Sub-section 4.1.1 which, in turn, is based on Cumming's approach [35], the computed results in this sub-section are concerned with linear structures.

The structures are approximated by the laminated composite flat triangular shell finite element which is identified as HLCTS^{qd} [33]. It is based on the mixed formulation and has many good features compared with others available in the literature. Explicit expressions for the consistent element mass and stiffness matrices were derived but are not included here for the fact that they require over a hundred printed pages. The HLCTS^{qd} element has three nodes, every one of which has six dof. The latter include three translational and three rotational dof. The element is capable of providing the six rigid-body modes correctly. The finite element mesh type employed in both cases is shown at the corner point B of the cylindrical panel in Figure 2.20. The nonstationary random excitation is applied at the center of each structure. More coarse and refined meshes have been applied in the studies but for brevity, computational efficiency, and the fact that results for the finite element representation of the whole structure of 6×6 mesh are accurate compared with those of refined mesh cases, the 6×6 mesh representation is adopted. With this mesh the finite element model for both the plate and shell structures has 85 nodes and 144 elements, respectively.

2.5.3.1   *Responses of simply-supported nine-layer square plate*

As mentioned in the foregoing the plate has nine layers. Its fibre orientation is (0/90/0/90/0/90/0/90/0). The side length of the plate, $L = 1.0$ m and total thickness $h = 0.01$ m. The thickness of each $0°$ and $90°$ layer are, respectively $h/10$ and $h/8$. The material considered in the present studies is the high modulus graphite/epoxy

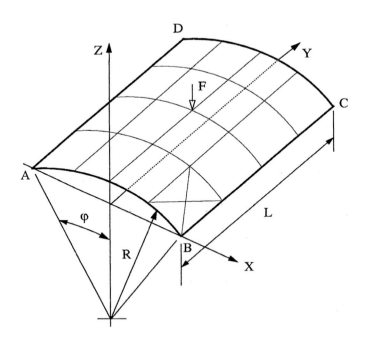

Figure 2.20   Cylindrical panel with D mesh indicated near point B.

composite with $E_1/E_2 = 40$, $G_{12}/E_2 = 0.60$, and $G_{13}/E_2 = G_{23}/E_2 = 0.50$ in which the modulus of elasticity $E_1 = 206.85$ GPa. The Poisson's ratio is $v_{12} = 0.25$. The shear correction factors are $\kappa_4 = \kappa_5 = (5/6)^{1/2}$. The density $\rho = 1.605 \times 10^3$ kg/m^3.

The nonstationary random excitation in this case is defined by Eq. (4.2b) and is included here for direct reference

$$F_i = 9.48148\left(e^{-900t} - e^{-1200t}\right)w(t) ,$$

where the subscript i denotes the dof at the center of the structure and $w(t)$ the zero mean Gaussian white noise process with autocorrection, $2\pi\delta(\tau)$ in which $\delta(\tau)$ is the Dirac delta function.

Since it is a plate problem, therefore all in-plane dof and drilling dof (ddof) are constrained. Consequently, there are 203 unknown equations of motion to be solved. In the free vibration analysis, for example, the first three natural frequencies are obtained as 344.4, 970.7, and 1053.8 rad/s. More natural frequencies and mode-shapes were obtained but not included here fore brevity. It may be appropriate to note that the analytical solution for the fundamental frequency of the same problem reported in Ref. [36] is 338.5 rad/s. In other words, the finite element result is 1.7% higher with respect to the analytical one. This upper bound is consistent with the analytical prediction and is very accurate considering the mesh applied.

Representative computed variance of transverse displacement, variance of transverse velocity, and covariance of transverse displacement and transverse velocity at the central node are presented in Figures 2.21 through 2.23, respectively. In these figures the results are of 2% damping for every mode considered. In the following, for conciseness, these results will be collectively referred to as response statistics or individually as variance of displacement and so on. Note that in the bi-modal approach presented in Chapter 4 the fourth-order Runge-Kutta (RK4) algorithm is employed to solve the first-order differential equations of statistical moments. In the present studies the time step size employed in the RK4 algorithm is $\Delta t = 0.02$ ms. With reference to the plots in the figures, it is clear that the results with the first 25 modes considered in the computation are in excellent agreement with those including the first 30 modes. In Figure 2.21, the maximum variance of displacement at the center of the plate is 5.75×10^{-9} m^2. One can conclude that in this particular case the first 25 modes have to be considered in the computation if accurate response statistics are required. It may be noted that in the foregoing and subsequent computations the number of modes to be included in the variances of

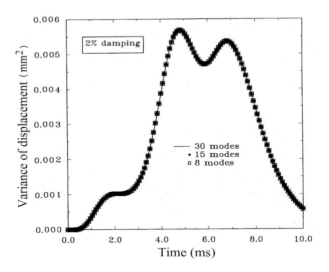

Figure 2.21   Transverse displacement at the center of the nine-layer plate.

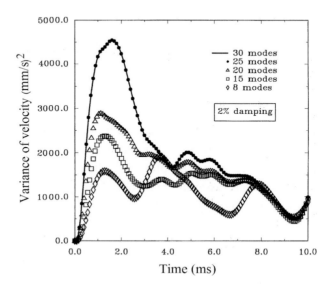

Figure 2.22   Transverse velocity at the center of the nine-layer plate.

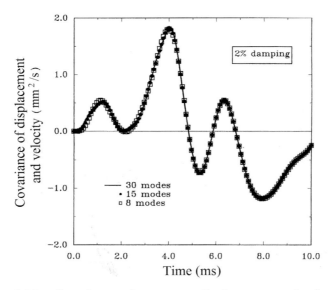

Figure 2.23 Covariance of transverse displacement and velocity at the center of the nine-layer plate.

velocities has to be decided first by performing several computations of different numbers of modes. As mentioned in Ref. [32] if the number of modes included in the computation is decided on the basis of variances of displacements, incorrect numbers of modes to be included in the computation may result. This is due to the fact that the order of accuracy of the RK4 algorithm for velocity responses is one order lower than those for the displacement responses.

2.5.3.2 Responses of simply-supported four-layer cylindrical panel

Now, consider the cross-ply cylindrical panel shown in Figure 2.20 having four layers. Its fiber orientation is (0/90/90/0). In this case the finite element model has 426 unknowns after the application of simply-supported boundary conditions.

The geometrical properties of the shell panel are: length of the sides of the cylindrical shell panel is $L = 0.254$ m (10.0 in), radius $R = 1.270$ m (50.0 in), thickness $h = 0.0254$ m (1.0 in). The angle $\varphi = 5.7392°$. The material properties are: $E_1/E_2 = 25$, $G_{12}/E_2 = G_{13}/E_2 = 0.50$, and $G_{23}/E_2 = 0.20$ in which the modulus of elasticity $E_1 = 51.675$ GPa (7.50×10^6 psi), Poisson's ratio is $v_{12} = v_{13} = 0.25$, and density $\rho = 2.768 \times 10^4$ kg/m^3 (1.0 lb/in^3). The computed first natural frequency is

64.3 rad/s compared with that obtained by Reddy and Liu [37] of 64.8 rad/s.

The nonstationary random excitation in this case is defined by Eq. (4.2b) but different from that for the nine-layer laminated composite square plate above

$$F_i = 9.48148 \left(e^{-45t} - e^{-60t} \right) w(t) ,$$

in which the auto-correlation of the zero mean Gaussian white noise process $w(t)$ is the same as that for the nine-layer composite plate structure studied in the foregoing. In applying the RK4 algorithm, the time step size $\Delta t = 1.25$ ms is employed for this case. Representative computed response statistics are presented in Figures 2.24 through 2.26. With reference to the results presented in the figures, one can observe that the computed results of the variance of velocity with the first 15 modes included in the computation are identical, graphically speaking, to those considering the first 20 modes. Therefore, the first 15 modes of the panel have to be included in the computations in order to provide accurate results of variances of displacements. The computed results are of 1% damping for every mode considered.

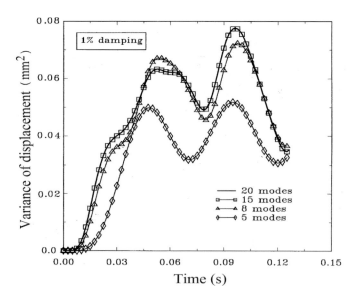

Figure 2.24   Variance of transverse displacement at the center
of the four-layer panel.

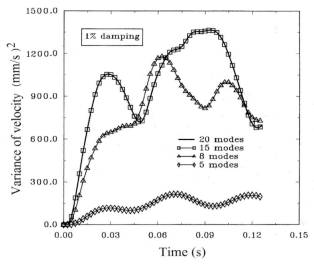

Figure 2.25   Variance of transverse velocity at the center of the four-layer panel.

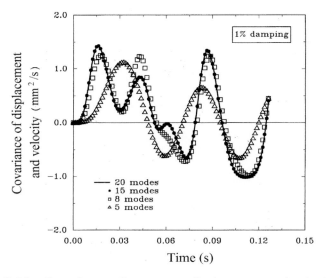

Figure 2.26   Covariance of transverse displacement and velocity
at the center of the four-layer panel.

References

[1] Lindberg, G.M. and Olson, M.D. (1967). Vibration modes and random response of a multi-bay panel system using finite elements, National Research Council of Canada, Report LR-492.

[2] Olson, M.D. (1970). Free vibrations and random response of an integrally stiffened panel, National Research Council of Canada, Report LR-544492.

[3] Olson, M.D. (1972). A consistent finite element method for random response problems, *Computers and Structures*, **2**, 163-180.

[4] Newsom, C.D., Fuller, J.R. and Sherrer, R.E. (1967). A finite element approach for the analysis of randomly excited complex elastic structures, *A.I.A.A./A.S.M.E. 8th Structures, Structural Dynamics and Material Conf.*, Palm Springs, California.

[5] Jacobs, L.D. and Lagerquist, D.R. (1968). Finite element analysis of complex panel to random loads, Wright-Patterson Air Force Base, Ohio, AFFDL-TR-68-44.

[6] Handa, K.N. (1970). *The Response of Tall Structures to Atmospheric Turbulence*, Ph.D. Thesis, University of Southampton, Southampton, England.

[7] Jones, A. and Beadle, C. (1970). Random vibration response of cantilever plates using the finite element method, *American Institute of Aeronautics and Astronautics*, **8**, 1905-1907.

[8] Orris, R.M. and Petyt, M. (1975). Random response of periodic structures by a finite element technique, *Journal of Sound and Vibration*, **43**, 1-8.

[9] Dey, S.S. (1979). Finite element method for random response of structures due to stochastic excitation, *Comput. Meth. Appl.Mech. Eng.*, **20**, 173-194.

[10] Weeks, E. and Cost, T. (1980). Complex stress response and reliability analysis of a composite elastic-viscoelastic missile configuration using finite elements, *Mechanics Research Communications*, **7(2)**, 59-63.

[11] Pfaffinger D. (1981). Probabilistic dynamic analysis with ADINA, *Computers and Structures*, **13**, 637-646.

[12] Yang, T. and Kapania, R. (1984). Finite element random response analysis of cooling tower, *ASCE J. of Engineering Mechanics*, **110(4)**, 589-609.

[13] Elishakoff, I. and Zhu, L. (1993). Random vibration of structures by the finite element method, *Computer Methods in Applied Mechanics and Engineering*, **105**, 359-373.

[14] To, C.W.S. (1982). Non-stationary random responses of a multi-degree-of-freedom system by the theory of evolutionary spectra, *Journal of Sound and Vibration*, **83(2)**, 273-291.

[15] To, C.W.S. (1984). Time-dependent variance and covariance of responses of structures to non-stationary random excitations, *J. Sound and Vibr.*, **93**, 135-156.

[16] Priestley, M.B. (1965). Evolutionary spectra and nonstationary processes, *Journal of the Royal Statistical Society B*, **27**, 204-237.

[17] Priestley, M.B. (1967). Power spectra analysis of nonstationary random processes, *Journal of Sound and Vibration*, **6**, 86-97.

[18] Hung, D. (1992). Unpublished report, Department of Mechanical Engineering, University of Western Ontario, Canada.

[19] To, C.W.S. (1979). Higher order tapered beam finite elements for vibration analysis, *Journal of Sound and Vibration*, **63**, 33-50.

[20] Melosh, R.J. (1963). Basis of derivation of matrices for the direct stiffness method, *American Institute of Aeronautics and Astronautics Journal*, **1(7)**, 1631-1637.

[21] Zienkiewicz, O.C. and Cheung, Y.K. (1964). The finite element method for analysis of elastic isotropic and orthotropic slabs, *Proceedings of the Institution of Civil Engineers*, **28**, 471-488.

[22] Weaver, Jr., W. and Johnson, P.R. (1984). *Finite Elements for Structural Analysis*, Prentice-Hall, Englewood Cliffs, New Jersey.

[23] Wang, B. (1991). Non-stationary random response of plates with geometric non-linearity, M.E.S. Thesis, University of Western Ontario, Canada.

[24] Blevins, R.D. (1979). *Formulas for Natural Frequency and Mode Shape*, Van Nostrand Reinhold, New York.

[25] Cowper, G.R., Kosko, E., Lindberg, G.M., and Olson, M.D. (1969). Static and dynamic applications of a high-precision triangular plate bending element, *American Institute of Aeronautics and Astronautics Journal*, **7**, 1957-1965.

[26] Yousafzai, A.H. and Ahmadi, G. (1982). Deterministic and stochastic earthquake response analysis of the containment shell of a nuclear power plant, *Nuclear Engineering Design*, **72**, 309-320.

[27] To, C.W.S. (1986). Response statistics of discretized structures to non-stationary random excitation, *Journal of Sound and Vibration*, **105(2)**, 217-231.

[28] To, C.W.S. (1986). Distribution of the first-passage time of mast antenna structures to non-stationary random excitation, *J. of Sound and Vibr.*, **108(1)**, 11-23.

[29] To, C.W.S. and Wang, B. (1993). Time-dependent response statistics of axisymmetric shell structures, *J. of Sound and Vibration*, **164(3)**, 554-564.

[30] To, C.W.S. and Wang, B. (1991). An axisymmetric thin shell finite element for vibration analysis, *Computers and Structures*, **40(3)**, 555-568.

[31] Shinozuka, M., Kako, T. and Tsurui, A. (1986). A random vibration analysis in finite element formulation, in *Random Vibration-Status and Recent Developement, Stephen H. Crandall Anniversary Volume*, Elishakoff, I. and Lyon, R., Eds., Elsevier, New York, 415-450.

[32] To, C.W.S. and Wang, B. (1996). Nonstationary random response of laminated composite structures by a hybrid strain-based laminated flat triangular shell finite element, *Finite Elements in Analysis and Design*, **23(1)**, 23-35.

[33] To, C.W.S. and Wang, B. (1998). Hybrid strain-based three-node flat triangular laminated composite shell elements, *Fin. Elem. in Anal. and Des.*, **28**, 177-207.

[34] Sewall, J.L. and Pusey, C.G. (1971). Vibration study of clamped-free elliptical cylindrical shells, *A.I.A.A. Journal*, **9**, 1004-1011.

[35] Cumming, I.G. (1967). Derivation of the moments of a continuous stochastic system, *International Journal of Control*, **5**, 85-90.

[36] Noor, A.K. and Mathers, M.D. (1975). Shear-flexible finite element models of laminated composite plates and shells, NASA TN D-8044.

[37] Reddy, J.N. and Liu, C.F. (1985). A higher order shear deformation theory of laminated elastic shells, *Int. Journal of Engineering Science*, **23**, 319-330.

3
Direct Integration Methods for Linear Structural Systems

The normal mode method and complex modal analysis are powerful when the structures have well-spaced modes. However, in large-scale FEA such an assumption is frequently not satisfied. Furthermore, complex modal analysis subroutines are not commonly available in commercial FE packages. Consequently, an alternative procedure, the stochastic central difference (SCD) method, based on the established central difference method was proposed by the author [1] and subsequently developed for the determination of response statistics of beams and plates [2] as well as plates and shells [3].

In this chapter the SCD method for response statistics of discretized linear structural systems under wide-band stationary and nonstationary random excitations, the time co-ordinate transformation (TCT) for the SCD method, the extended SCD method for response statistics of discretized systems under narrow-band random excitations, and the Newmark family of algorithms as well as their stochastic counterparts are presented. Selected computed results applying these methods to linear systems idealized by the finite elements are included for illustration.

3.1 Stochastic Central Difference Method

The original SCD method has been presented by To [1] for the analysis of discretized dynamic systems whose matrix equation of motion is

$$M\ddot{x} + C\dot{x} + Kx = F(t) = F \tag{3.1}$$

and the time domain discretized version of Eq. (3.1) is

Stochastic Structural Dynamics: Application of Finite Element Methods, First Edition. C.W.S. To.
© 2014 John Wiley & Sons, Ltd. Published 2014 by John Wiley & Sons, Ltd.

$$M\ddot{x}_s + C\dot{x}_s + Kx_s = F_s \qquad (3.2)$$

where the subscript s denotes the time step, for instance, x_s is the value of x at time step t_s such that the time step size $\Delta t = t_{s+1} - t_s$ and $t_0 = 0$.

Assuming x and its derivatives are single-valued, finite, and continuous functions of t, then by Taylor theorem,

$$x_{s+1} = x(t + \Delta t) = x_s + (\Delta t)\dot{x}_s + \frac{1}{2}(\Delta t)^2 \ddot{x}_s + \cdots ,$$

$$x_{s-1} = x(t - \Delta t) = x_s - (\Delta t)\dot{x}_s + \frac{1}{2}(\Delta t)^2 \ddot{x}_s - \cdots .$$

By adding and subtracting the above two equations, one has, respectively

$$x_{s+1} + x_{s-1} = 2x_s + \frac{1}{2}(\Delta t)^2 \ddot{x}_s ,$$

$$x_{s+1} - x_{s-1} = 2(\Delta t)\dot{x}_s ,$$

with a leading error of order $(\Delta t)^2$.

By substituting the last two equations into Eq. (3.2), one obtains

$$x_{s+1} = (\Delta t)^2 N_1 F_s + N_2 x_s + N_3 x_{s-1} \qquad (3.3)$$

where

$$N_1 = \left[M + \frac{1}{2}(\Delta t)C \right]^{-1} , \qquad N_2 = N_1 \left[2M - (\Delta t)^2 K \right] ,$$

$$N_3 = N_1 \left[\frac{1}{2}(\Delta t)C - M \right] .$$

Equation (3.3) is the central difference method that expresses the displacement vector at the next time step in terms of the displacement vectors at the current and last time steps. It is a member of the Newmark algorithms presented in Appendix 3A.

Before proceeding further, one assumes

$$F_s = A_s w_s , \qquad (3.4)$$

where A_s is a vector of discretized deterministic amplitude modulating functions, w_s

is the zero mean discrete Gaussian white noise (DGWN) process.

Taking the transpose of Eq. (3.3) and making use of Eq. (3.4) one has

$$
\begin{aligned}
x_{s+1}^{T} &= \left[(\Delta t)^2 N_1 F_s + N_2 x_s + N_3 x_{s-1} \right]^{T} \\
&= (\Delta t)^2 F_s^{T} N_1^{T} + x_s^{T} N_2^{T} + x_{s-1}^{T} N_3^{T} .
\end{aligned}
\tag{3.5}
$$

Multiplying Eqs. (3.3) and (3.5), taking ensemble average and rearranging gives

$$
\begin{aligned}
R_{s+1} &= N_2 R_s N_2^{T} + N_3 R_{s-1} N_3^{T} + (\Delta t)^4 N_1 B_s N_1^{T} \\
&\quad + N_2 D_s N_3^{T} + N_3 D_s^{T} N_2^{T} + (\Delta t)^2 N_2 \left\langle x_s F_s^{T} \right\rangle N_1^{T} \\
&\quad + (\Delta t)^2 N_1 \left\langle F_s x_s^{T} \right\rangle N_2^{T} + (\Delta t)^2 N_3 \left\langle x_{s-1} F_s^{T} \right\rangle N_1^{T} \\
&\quad + (\Delta t)^2 N_1 \left\langle F_s x_{s-1}^{T} \right\rangle N_3^{T}
\end{aligned}
\tag{3.6}
$$

in which

$$
R_s = \left\langle x_s x_s^{T} \right\rangle, \quad B_s = \left\langle F_s F_s^{T} \right\rangle ,
$$

$$
\begin{aligned}
D_s &= \left\langle x_s x_{s-1}^{T} \right\rangle \\
&= N_2 R_{s-1} + N_3 D_{s-1}^{T} + (\Delta t)^2 N_1 \left\langle F_{s-1} x_{s-1}^{T} \right\rangle .
\end{aligned}
\tag{3.7a,b,c}
$$

For excitations modeled as Gaussian white noise or shot noise or Wiener process or their modulated forms the terms associated with $\langle F_s x_s^{T} \rangle$ and $\langle F_s x_{s-1}^{T} \rangle$ become zero as

$$
\left\langle F_s x_s^{T} \right\rangle = 0 \ , \quad \left\langle F_s x_{s-1}^{T} \right\rangle = 0 .
\tag{3.8a,b}
$$

Applying Eq. (3.8) to (3.6), it reduces to

$$
\begin{aligned}
R_{s+1} &= N_2 R_s N_2^{T} + N_3 R_{s-1} N_3^{T} + (\Delta t)^4 N_1 B_s N_1^{T} \\
&\quad + N_2 D_s N_3^{T} + N_3 D_s^{T} N_2^{T}
\end{aligned}
\tag{3.9}
$$

where now

$$D_s = N_2 R_{s-1} + N_3 D_{s-1}^T .$$

Equation (3.9) is the recursive covariance matrix expression for displacement responses of mdof systems under external nonstationary random excitations. Of course, when the discretized deterministic modulating functions are of unity, then the excitations are stationary random. It should be emphasized that because Eq. (3.9) is discretized in the time domain, the class of nonstationary random excitations is large.

3.2 Stochastic Central Difference Method with Time Co-ordinate Transformation

For stiff system or when the number of finite elements is large, and therefore the highest natural frequency of the discretized structural system is very large, the time step size for the SCD method is very small. A TCT was presented by To [4] to deal with this difficulty. The SCD method with TCT was applied to the analysis of discretized beams and plates [2], and plates and shells [3].

For simplicity and illustration of the strategy, consider Eq. (3.1) again. Assuming the stiff system governed by Eq. (3.1) has its highest natural frequency Ω. Divide both sides of Eq. (3.1) by the square of Ω and transform the resulting equation of motion in the t-domain to the dimensionless τ-domain such that

$$M\frac{d^2x}{d\tau^2} + \frac{1}{\Omega}C\frac{dx}{d\tau} + \frac{1}{\Omega^2}Kx = \frac{1}{\Omega^2}F \tag{3.10}$$

where the response vector x and the forcing vector F are functions of the dimensionless time τ which is being chosen as

$$\tau = \Omega t . \tag{3.11}$$

The variance or covariance of responses for the system described by Eq. (3.10) can then be evaluated with Eq. (3.9) in which the assembled mass matrix M is identical to that in Eq. (3.1), the damping matrix C is equal to the original damping matrix C in Eq. (3.1) divided by the highest natural frequency Ω, the stiffness matrix K is equal to the original stiffness matrix K in Eq. (3.1) divided by Ω^2, and the excitation vector $F(t)$ is equal to $F(\tau)/\Omega^2$. In other words, if one writes the matrix equation of motion in the τ-domain as

$$M^\tau \frac{d^2x}{d\tau^2} + C^\tau \frac{dx}{d\tau} + K^\tau x = F^\tau , \tag{3.12}$$

where

$$M^\tau = M \quad , \quad C^\tau = \frac{1}{\Omega} C \quad , \quad K^\tau = \frac{1}{\Omega^2} K \quad , \quad F^\tau = \frac{1}{\Omega^2} F(\tau)$$

so that Eq. (3.9) can be applied to evaluate the variance or covariance of responses in the τ-domain. In applying Eq. (3.9) to obtain the variance or covariance of responses in the τ-domain M, C, K, and F_s in Eq. (3.9) have to be replaced, respectively with M^τ, C^τ, K^τ, and $F_s^{\ \tau}$ defined by Eq. (3.12), where $F_s^{\ \tau}$ is F^τ at the dimensionless time step τ_s while Δt in Eq. (3.9) is replaced by $\Delta\tau$.

Once the variance or covariance of responses in the τ-domain is determined by applying Eq. (3.9), it is converted back to the t-domain. They are related by the following expression,

$$R_s = \Omega R_{\tau s} , \tag{3.13}$$

where R_s is the recursive covariance matrix of displacement vector x_s in the t-domain and $R_{\tau s}$ is that in the τ-domain. Equation (3.13) can be easily proved if one considers the stationary variance of displacement of a single degree-of-freedom (sdof) linear system excited by a zero mean Gaussian white noise [5].

Remark 3.2.2.1
Applying the above steps, the computational instability can be circumvented. In addition, this strategy can reduce drastically the computational time for stiff systems because of the fact that the time step in the τ-domain is relatively much larger than that in the t-domain.

Remark 3.2.2.2
Further, applying the SCD method without the TCT the time step size for very stiff systems can often render determination of the recursive responses impossible.

Remark 3.2.2.3
The relations of time step size Δt and the natural frequency ω for systems under wide-band stationary and nonstationary random excitations have been investigated by To and Liu [2]. It has been found that as the natural frequency reduces to a small value, the time step size Δt approaches unity. The relation may be written as

$$\Delta t = 0.83 - 0.72 \log_{10} \omega , \quad 1.0 \le \omega < 5.0$$
$$\tag{3.14a,b}$$
$$\Delta t = 1.0 - 0.053 \omega - 0.12 \omega^2 , \quad \omega \le 1.0$$

where ω is the highest angular natural frequency of the system in radian per second while Δt is the time step size in second.

Equation (3.14) does not include a formula for $\omega \geq 5.0$ rad/s because if the highest angular natural frequency is higher than 5.0 rad/s, the time step size becomes small. In this case, the TCT introduced in the foregoing should be used. Equation (3.14) also applies to dimensionless systems.

3.3 Applications

In this section applications of the techniques presented in the last two sections are made on the computation of response statistics such as the variances and covariances of beam and plate structures. It may be appropriate to mention that while dampings considered in this section are proportional, the techniques in the last two sections can include non-proportional dampings.

3.3.1 Beam structures under base random excitations

The beam structure considered in this sub-section is subjected to a nonstationary random excitation applied at the base or clamped end. Thus, the governing matrix equation of motion is similar to that for the mast antenna structure dealt with in Sub-section 2.3.1. That is, the governing equation of motion for the present structural system is defined by Eq. (2.30). Equations (2.32) and (2.33) also apply here. However, the cantilever beam structure is approximated by a two-node beam element whose consistent element mass and stiffness matrices are given by Eqs. (2.35) and (2.36). This gives the displacement vector y as $y = [\, 0 \;\; \hat{y}(t) \;\; 0 \,]^T$ and the excitation vector defined by Eq. (2.34) as

$$\underset{n \times 1}{F_x} = - \underset{1 \times n}{\left[k_{42} \;\; k_{52} \;\; k_{62} \;\; 0 \;\; \cdots \;\; 0 \right]^T} \hat{y}(t) , \tag{3.15}$$

where $\hat{y}(t) = a_i(t) w(t)$ in which $a_i(t)$ is defined as in Eq. (2.26) and $w(t)$ is the zero mean Gaussian white noise. Note that the subscript i here is the dof.

In applying the SCD method together with the TCT technique $w(t)$ is understood to be the discrete Gaussian white noise (DGWN) excitation. Note that in Eq. (3.15) k_{ij} denotes the ij'th entry or element of assembled stiffness matrix K in Eq. (2.30). The main difference between the beam elements employed in Sub-section 2.3.1 for the mast antenna structure and the structural system considered in this sub-section is that in the present problem the axial displacement dof u_i at every node is included. Thus, here, $k_{42} = 0$ since no axial random excitation is applied at the base.

For comparison to the exact solution in Chapter 2 in which damping is assumed to be proportional, one considers

$$2\zeta_r\omega_r = \lambda_m + \lambda_k\omega_r^2, \qquad C_{xx} = \lambda_m M_{xx} + \lambda_k K_{xx}, \qquad (3.16a,b)$$

where the Rayleigh damping coefficients λ_m and λ_k are determined by the given damping ratios ζ_r and natural frequencies ω_r of the first two modes. Equation (3.16) is simple and damping of the higher modes depends on the first two damping ratios.

In the present studies the beam structure has the following properties: density ρ = 7860 kg/m^3, Young's modulus of elasticity E = 207 GPa, cross-section area A = 6.25×10^{-4} m^2, second moment of cross-section area I = 3.26×10^{-8} m^4, length of beam structure L = 1.0 m. Four element representations or meshes are included in Table 3.1 while Table 3.2 includes computed natural frequencies for the four finite element models. The 11-element model with $L = (2\ell_1 + 9\ell_2)$ is illustrated in Figure 3.1. The deterministic modulating function is

$$a_r(t) = 9.4815\left(e^{-45t} - e^{-60t}\right), \qquad (3.17)$$

such that at time $t = t_p = 0.15$ s the excitation can be considered insignificantly small since $a_r(t_p) = 0.01$, where t_p is the duration of the nonstationary random excitation.

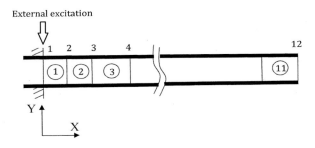

Figure 3.1   Discretized cantilever beam with base excitation.

Note that with reference to Table 3.2 the fundamental period, $T_1 = 2\pi/\omega_1$, is equal to 0.04819 s and therefore the duration of excitation selected above is slightly more than three times the fundamental period. The first two modal damping ratios are: $\zeta_1 = \zeta_2 = 0.05$ or 5% damping. Additional pertinent data for the computation are included in Tables 3.3 through 3.6.

Table 3.1   Finite element discretizations of cantilever beam.

Number of elements	Total dof	Effective dof	ℓ_2 (m)	ℓ_1 (m)
2	9	6	N/A	0.500
11	36	33	0.100	0.050
15	48	45	0.070	0.045
21	66	63	0.050	0.025

Table 3.2   Natural frequencies of cantilever beam (rad/s).

Natural frequency	Effective dof of model			
	6	33	45	63
ω_1	130.3777	130.3138	130.3147	130.3147
ω_2	823.5985	816.6954	816.6749	816.6701
ω_3	2785.5606	2287.2760	2286.8373	2286.7325
ω_4	8084.8891	4485.2487	4482.0623	4481.2894
ω_5	8269.5425	7425.7774	7412.0839	7408.6839
$\vdots$	These natural frequencies are not required in the computation.			
Ω	28,888.7357	404,056.5917	543,912.8794	1,616,226.3648

Table 3.3   Modal damping coefficients of cantilever beam.

Modal damping ratio	Effective dof of discretization			
	6	33	45	63
ζ_1	0.05	0.05	0.05	0.05
ζ_2	0.05	0.05	0.05	0.05
ζ_3	0.1413	0.1231	0.1231	0.1232
ζ_4	0.4049	0.2381	0.2379	0.2379
ζ_5	0.4142	0.3938	0.3921	0.3919

Table 3.4   Rayleigh damping coefficients of cantilever beam.

Rayleigh damping coefficients	Effective dof of model			
	6	33	45	63
λ_m	11.3379	11.3333	11.2382	11.2382
λ_k	0.0001	0.0001	0.0001	0.0001

Table 3.5   Parameters for cantilever beam variances by EQ75.

Elements of stiffness matrix ($\times 10^6$)	Effective dof of discretization			
	6	33	45	63
k_{52}	- 0.64783	- 647.83	- 888.65	- 5,182.6
k_{62}	0.16196	16.196	19.995	64.783

To provide a comparison the execution times are presented in Table 3.7. The digital computer program written in Fortran based on the SCD method provides all the variances and covariances in the covariance matrix. This is why the execution time increases rapidly as the number of effective dof of the discretized system is increased. For brevity, only the variances of displacement and rotation at the free end of the cantilever beam structure are included in Figures 3.2 through 3.6. In the figures the computed results obtained by the SCD method with the TCT are denoted by SCD while those by the exact solution in Chapter 2 are designated as EQ75.

With reference to the figures presented in this sub-section, two observations should be made. Firstly, the computed results obtained by the SCD method and EQ75 by including the first five modes of vibration are in excellent agreement. The agreement for the two element model in which no damping was considered and their results are presented in Figure 3.2 is also excellent when the first five modes are included in the computation by applying EQ75. When only the first three modes were included in using the EQ75 significant differences appear. This indicates that the influence of every mode on the response of the undamped beam is important. This modal influence on the accuracy of the computed responses is evident particularly in the plots for rotations at the free-end of the cantilever beam. Secondly, the computed results for the eleven, fifteen, and twenty-one element models indicated that the eleven element model can give very accurate results.

Table 3.6   Modal data for cantilever beam variances by EQ75.

Two-element model				
i	2	3	5	6
ψ_{i1}	- 0.3067	- 1.0505	- 0.9032	- 1.2433
ψ_{i2}	- 0.6578	0.3959	0.9112	4.3872
ψ_{i3}	0.1031	- 7.7504	1.0135	9.7742
ψ_{i4}	0.4309	8.8571	1.7019	33.8970
ψ_{i5}	$- 0.7823 \times 10^{-16}$		$- 0.4154 \times 10^{-15}$	
		$- 0.1850 \times 10^{-14}$		$- 0.2144 \times 10^{-14}$

Eleven-element model				
i	2	3	32	33
ψ_{i1}	- 0.0039	- 0.1532	- 0.9024	- 1.2420
ψ_{i2}	- 0.0229	- 0.8754	0.9024	4.3140
ψ_{i3}	- 0.0605	- 2.2390	- 0.0928	- 7.0860
ψ_{i4}	- 0.1116	- 3.9670	0.9040	9.9420
ψ_{i5}	- 0.1734	- 5.8770	- 0.9066	- 1.2820

Fifteen-element model				
i	2	3	44	45
ψ_{i1}	- 0.0031	0.1384	0.9024	1.2420
ψ_{i2}	- 0.0187	- 0.7985	0.9024	4.3140
ψ_{i3}	- 0.0497	2.0630	0.9025	7.0830
ψ_{i4}	- 0.0923	- 3.6990	0.9028	9.9270
ψ_{i5}	0.1441	5.5510	0.9035	12.7700

Table 3.6   Modal data for cantilever beam variances by EQ75 (continued).

Twenty-one-element model				
i	2	3	62	63
ψ_{i1}	0.0010	0.0780	0.9024	1.2420
ψ_{i2}	0.0060	0.4674	- 0.9024	- 4.3140
ψ_{i3}	0.0163	1.2550	0.9024	7.0830
ψ_{i4}	- 0.0310	- 2.3530	0.9025	9.9240
ψ_{i5}	- 0.1734	- 5.8770	- 0.9066	- 12.7600

Table 3.7   Time step sizes and execution times by the SCD method.

Effective dof of beam models	$\Delta\tau$	Execution time (Hour)	Error (%) with respect to EQ75	
			Tip displacement	Tip rotation
6	0.95	0.15@	4.02	2.22
33	1.00	11.50@	2.43	- 6.67
45	1.00	50.00@	2.62	- 6.51
63	1.00	89.50*	2.57	- 7.12

@   Denotes computation performed with SUN 4.0 workstation.
 *   Denotes computation performed with Silicon Graphics machine.

Before leaving this sub-section it may be appropriate to point out that since the above results are based on linear random vibration theory, the following condition must be satisfied:

$$\max\left\{\left\langle x_L^2(t)\right\rangle\right\} \le \left(\frac{5L}{100}\right)^2 , \tag{3.18}$$

where $\langle x_L^2 \rangle$ is the variance of displacement at the tip. Thus,

$$0.25 \times 10^5 \, S_0 \le 0.05^2 , \quad \text{or} \quad S_0 \le 0.9804 \times 10^{-7} \, \text{m}^2/\text{rad} . \tag{3.19}$$

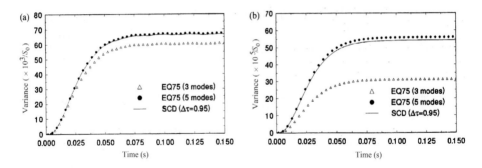

Figure 3.2   Variances at free-end for undamped two-element model:
(a) displacement, and (b) rotation.

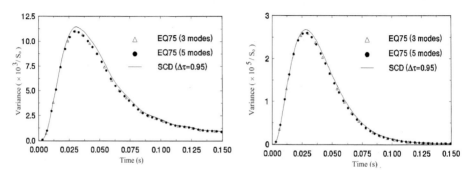

Figure 3.3   Variances at free-end for damped two-element model:
(a) displacement, and (b) rotation.

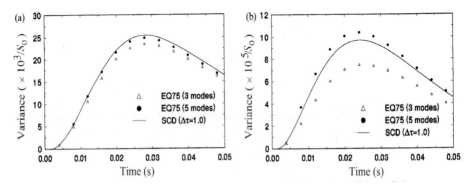

Figure 3.4   Variances at free-end for damped eleven-element model:
(a) displacement, and (b) rotation.

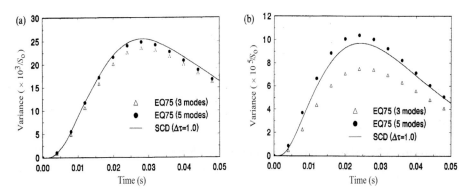

Figure 3.5   Variances at free-end for damped fifteen-element model:
(a) displacement, and (b) rotation.

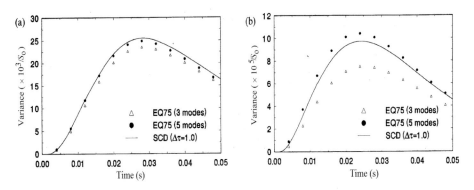

Figure 3.6   Variances at free-end for damped twenty one-element model:
(a) displacement, and (b) rotation.

3.3.2   Plate structures

In this sub-section Eq. (3.9) and the TCT are employed to obtain the variances and covariances of displacements of plate structures idealized by the FEM [2]. The square plate considered has geometrical dimensions $1.0 \times 1.0 \times 0.05 \, \text{m}^3$ while its material properties are: Young's modulus of elasticity $E = 200 \, \text{GPa}$, Poisson's ratio $\nu = 0.3$, and density $\rho = 7830 \, \text{kg/m}^3$. Two cases are studied in this sub-section. The first case is clamped at one side and free at the remaining three sides and is denoted as CF3.

The second case is clamped at all four sides and designated as C4. These discretizations are shown in Figure 3.7 where the triangular finite elements are based on the high precision triangular bending finite element identified as TBH6 [6]. It has 3 nodes and every node has six dof. The latter are: $w_i^e, \partial w_i^e/\partial\xi, \partial w_i^e/\partial\eta, \partial^2 w_i^e/\partial\xi^2$, $\partial^2 w_i^e/\partial\xi\partial\eta$, and $\partial^2 w_i^e/\partial\eta^2$, where w_i^e is the transversal displacement at Node i on the middle plane of the plate while ξ and η are two perpendicular axes on the middle plane of the triangular element. The displacement field $w^e(\xi,\eta)$ or simply w^e within the element is defined by the quintic polynomial. As illustrated in Figure 3.7 the CF3 plate is represented by 8 elements. It has 27 dof after application of boundary conditions. The C4 plate has 32 elements and 25 nodes such that it has 66 dof after application of boundary conditions. Additional information about the finite element representations and computed natural frequencies are included, respectively, in Tables 3.8 and 3.9.

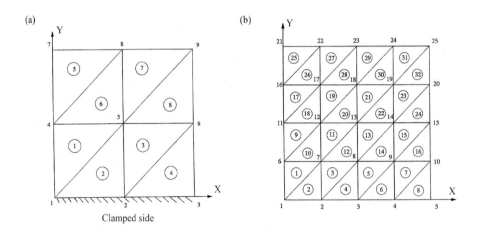

Figure 3.7 Discretized plate structures: (a) CF3 plate; and (b) C4 plate.

The governing equation of motion for the plate structure can be expressed as Eq. (3.1) in which the assembled damping matrix is given by

$$C = M\left(\sum_i 2\zeta_i\omega_i[\psi_i][\psi_i]^T \right) M,$$

where ψ_i is the i'th mode shape vector which is orthonormal relative to the mass matrix M and the summation may be performed over several modes, the modal damping ratios of which are specified. Those unspecified modes are in fact undamped modes.

Table 3.8   Finite element representations of square plate.

Elements in model	Boundary conditions	Total dof of model	Effective dof of model
8	CF3	54	27
32	C4	150	66

For the CF3 case the nodal random force, applied at Node 8 and for the C4 plate at Node 13, and both forces are in the direction perpendicular to the plane of the plate, is modeled as a product of a modulating function $a_i(t)$ and the zero mean Gaussian white noise $w(t)$ the spectral intensity S_0 of which is unity. That is, $S_0 = 1.0$ was applied and the modulating function $a_i(t)$ is defined by

$$a_i(t) = 9.4815 \left(e^{-45t} - e^{-60t} \right). \tag{3.20}$$

It is understood that in applying the SCD method with TCT for this plate structure the DGWN process is employed, while for direct comparison to the closed form solution by using EQ75 in which the continuous Gaussian white noise process is assumed. In applying EQ75, additional data are required and they are provided in Table 3.10 in which w_j^e is the j'th nodal displacement and not to be confused with the white noise $w(t)$ mentioned above. Representative computed results are included in Figures 3.8 through 3.10 in which SCD denotes results obtained by the SCD method whereas EQ75 means those employing the EQ75 computer program. Note that $\Delta\tau$ for the SCD method are included in the legends of the figures.

The execution time by using the SCD method for the CF3 plate is 0.2 hour whereas that for the C4 plate structure is 13.5 hrs. Both computations have been performed with a SUN 4.0 workstation. The relatively long execution time for the C4 plate has to do with the fact that it has 66 dof compared with 27 dof for the CF3 structure. In each structure all elements of the covariance matrix have been evaluated. This explains the considerable amount of computational time difference between the two discretized structures.

Table 3.9   Natural frequencies (rad/s) and damping ratios.

Modal number i	CF3, 27 dof		C4, 66 dof	
	ω_i	ζ_i	ω_i	ζ_i
1	26.7193	0.05	275.2883	0.05
2	67.5943	0.05	562.0487	0.05
3	164.0702	0.05	562.9375	0.05
4	213.2028	0.05	833.8993	0.05
5	244.4920	0.05	1011.3255	0.05
6	425.9661	0.05	1018.6845	0.05
7	These data are not required		1277.8441	0.05
8	in the computation.		1290.5347	0.05
.			These data are not required	
.			in the computation.	
Ω	4959.0863		21,641.7317	

Table 3.10   Data for variances of plates by EQ75.

	CF3 plate, w_j^e		C4 plate
	w_8^e	w_5^e	w_{13}^e
ψ_{j1}	0.3230	0.1100	-0.3935
ψ_{j2}	0.1590×10^{-2}	0.2130×10^{-2}	0.1496×10^{-13}
ψ_{j3}	0.3844	-0.1718	0.1705×10^{-13}
ψ_{j4}	-0.2533	-0.2326	-0.2281×10^{-2}
ψ_{j5}	0.7412×10^{-2}	0.1059×10^{-1}	0.1204×10^{-12}
ψ_{j6}	0.4643	-0.2652	0.4873
ψ_{j7}	These data are not required for		-0.4910×10^{-14}
ψ_{j8}	computation.		0.9097×10^{-14}

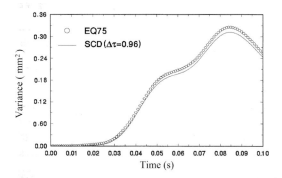

Figure 3.8 Variance of displacement at Node 5 for CF3 plate.

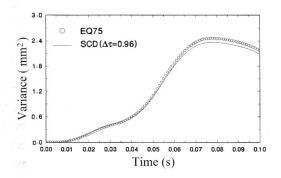

Figure 3.9   Variance of displacement at Node 8 for CF3 plate.

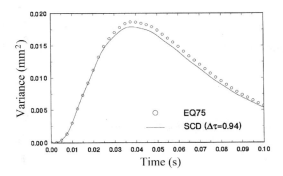

Figure 3.10   Variance of displacement at Node 13 for C4 plate.

3.3.3   Remarks

In the foregoing sub-sections, the computed results indicate that the SCD method is very efficient and accurate when the dimensionless time step size $\Delta\tau = 1.0$ is applied in the computation. It is interesting to note that $\Delta\tau$ is not affected by the number of dof in a particular structural system. This agrees with the theoretical $\Delta\tau$ obtained in [5] and included in Eq. (3.43).

3.4 Extended Stochastic Central Difference Method and Narrow-band Force Vector

The SCD method introduced in Sections 3.1 and 3.2 was extended by Chen and To [7] to cases with narrow-band stationary and nonstationary random excitations. This so-called extended stochastic central difference (ESCD) method and selected computed results are included in this section.

3.4.1   Extended stochastic central difference method

Consider a mdof system under narrow-band random excitations, which are obtained from the outputs of filters perturbed by modulated Gaussian white noise excitations. The governing matrix equations of motion for the filter and system are, respectively

$$M_f \ddot{f} + C_f \dot{f} + K_f f = F(t) = e(t)w(t) ,$$
$$M\ddot{Y} + C\dot{Y} + KY = f ,$$
(3.21a,b)

where M_f, C_f, and K_f are the assembled mass, damping, and stiffness matrices of the filters while M, C, and K are the assembled mass, damping, and stiffness matrices of the approximated structural system, respectively; $\ddot{f}, \dot{f}$, and f are the random acceleration, velocity, and displacement vectors of the filters while $\ddot{Y}$, $\dot{Y}$ and Y are the acceleration, velocity and displacement vectors of the system; and $e(t)$ is a time-dependent deterministic modulating function vector. The zero-mean stationary Gaussian white noise process $w(t)$ has the spectral density S_0. By changing the natural frequencies and damping ratios of the filters and the spectral density of the Gaussian white noise excitation, a series of narrow-band random processes, in the time t-domain can be provided.

By employing the same procedure introduced in Section 3.1 for the SCD method and after some algebraic manipulation, one can show that the recursive covariance matrix expression of the system is

$$R_{s+1} = N_2 R_s N_2^T + N_3 R_{s-1} N_3^T + (\Delta t)^4 N_1 R_s^f N_1^T$$
$$+ N_2 D_s N_3^T + N_3 (D_s)^T N_2^T + (\Delta t)^2 N_2 G_s N_1^T$$
$$+ (\Delta t)^2 N_1 (G_s)^T N_2^T + (\Delta t)^2 N_3 H_s N_1^T$$
$$+ (\Delta t)^2 N_1 (H_s)^T N_3^T , \tag{3.22}$$

in which now the stationary Gaussian white noise process $w(t)$ is understood to be the DGWN process w_s but $G_s = \langle Y_s f_s^T \rangle$, $H_s = \langle Y_{s-1} f_s^T \rangle$, $R_s^f = \langle f_s f_s^T \rangle$, $R_s = \langle Y_s Y_s^T \rangle$, $Y_s = Y(t_s) = Y(t_0 + \Delta t)$, and $f_s = f(t_s)$;

$$D_s = \langle Y_s Y_{s-1}^T \rangle = N_2 R_{s-1} + N_3 (D_{s-1})^T + (\Delta t)^2 N_1 (G_{s-1})^T ;$$

and N_1, N_2, and N_3 have already been defined in Eq. (3.3).

The recursive covariance expression for the filters can be obtained by applying Eq. (3.9) to give

$$R_{s+1}^f = N_{2f} R_s^f N_{2f}^T + N_{3f} R_{s-1}^f N_{3f}^T + (\Delta t_f)^4 N_{1f} B_s N_{1f}^T$$
$$+ N_{2f} D_s^f N_{3f}^T + N_{3f} (D_s^f)^T N_{2f}^T , \tag{3.23}$$

where now $B_s = \langle F_s F_s^T \rangle$,

$$D_s^f = \langle f_s f_{s-1}^T \rangle = N_{2f} R_{s-1}^f + N_{3f} (D_{s-1}^f)^T ;$$

$$N_{1f} = \left[M_f + \frac{1}{2} (\Delta t_f) C_f \right]^{-1} , \qquad N_{2f} = N_{1f} \left[2 M_f - (\Delta t_f)^2 K_f \right] ,$$

$$N_{3f} = N_{1f} \left[\frac{1}{2} (\Delta t_f) C_f - M_f \right] .$$

In the foregoing, the subscript or superscript f designates the filter. Thus, in general the time step size for the filter system Δt_f is very much different from that for the structural system Δt.

The major differences between the ESCD method for narrow-band random excitations and SCD method are the addition of G_s and H_s terms in the former method. These two additional terms are the vehicles carrying frequencies and bandwidth contents from the filters to the system.

Before applying Eq. (3.22), recursive relations of G_s and H_s have first to be derived. To this end, one substitutes the following central difference scheme for the system

$$Y_s = (\Delta t)^2 N_1 f_{s-1} + N_2 Y_{s-1} + N_3 Y_{s-2}$$

into the expression for G_s. Thus,

$$G_s = \left\langle Y_s f_s^T \right\rangle = \left\langle \left[(\Delta t)^2 N_1 f_{s-1} + N_2 Y_{s-1} + N_3 Y_{s-2} \right] f_s^T \right\rangle$$

$$= (\Delta t)^2 N_1 \left(D_s^f \right)^T + N_2 H_s + N_3 \left\langle Y_{s-2} f_s^T \right\rangle .$$

$$(3.24)$$

Similarly, one can substitute the recursive expression for f_s^T into the last term of the above equation such that

$$\left\langle Y_{s-2} f_s^T \right\rangle = \left\langle Y_{s-2} \left[(\Delta t_f)^2 F_{s-1}^T N_{1f}^T + f_{s-1}^T N_{2f}^T + f_{s-2}^T N_{3f}^T \right] \right\rangle$$

$$= H_{s-1} N_{2f}^T + G_{s-2} N_{3f}^T$$

$$(3.25)$$

in which the term associated with $\left\langle Y_{s-2} F_{s-1}^T \right\rangle$ is zero has been used.

After combining the last two equations and re-arranging, it leads to

$$G_s = (\Delta t)^2 N_1 \left(D_s^f \right)^T + N_2 H_s + N_3 H_{s-1} N_{2f}^T$$

$$+ N_3 G_{s-2} N_{3f}^T .$$

$$(3.26)$$

A similar operation can be performed on the term H_s. Thus, one can show that

$$H_s = \left\langle Y_{s-1} f_s^T \right\rangle$$

$$= \left\langle Y_{s-1} \left[(\Delta t_f)^2 F_{s-1}^T N_{1f}^T + f_{s-1}^T N_{2f}^T + f_{s-2}^T N_{3f}^T \right] \right\rangle$$

$$(3.27)$$

$$= G_{s-1} N_{2f}^T + \left\langle Y_{s-1} f_{s-2}^T \right\rangle N_{3f}^T .$$

Upon further substitution in the last term on the rhs of Eq. (3.27), one has

$$\left\langle Y_{s-1} f_{s-2}^T \right\rangle = \left\langle \left[(\Delta t)^2 N_1 f_{s-2} + N_2 Y_{s-2} + N_3 Y_{s-3} \right] f_{s-2}^T \right\rangle$$

$$= (\Delta t)^2 N_1 R_{s-2}^f + N_2 G_{s-2} + N_3 H_{s-2} .$$

Thus, Eq. (3.27) reduces to

$$H_s = G_{s-1} N_{2f}^T + (\Delta t)^2 N_1 R_{s-2}^f N_{3f}^T + N_2 G_{s-2} N_{3f}^T$$
$$+ N_3 H_{s-2} N_{3f}^T . \tag{3.28}$$

Remark 3.4.1.1
Equations (3.22), (3.23), (3.26) and (3.28) constitute the ESCD method for discretized structural systems under narrow-band random excitations.

In the above operations, the ensemble average of the narrow-band random force vector has been assumed to be zero. If the input does not have a zero ensemble average one can follow the procedure provided by To and Liu [8, 9].

Remark 3.4.1.2
The TCT introduced above and Eq. (3.14) for the relationship between natural frequency and time step size have been found to be applicable to systems under narrow-band stationary and nonstationary random excitations [7].

Remark 3.4.1.3
It was shown numerically [7] that the natural frequency of the filter does not affect the time step size of the discretized system.

Remark 3.4.1.4
Although the center frequency and bandwidth of the filter response can be adjusted by changing the natural frequency and damping ratio of the filter, the amplitude and the shape of amplitude for nonstationary random response cannot be conveniently and simply controlled without further modification. A technique of providing narrow-band random forces which can be conveniently controlled was introduced by Chen and To [7]. In the latter reference it was shown that the recursive relations representing the narrow-band random vector process are

$$R_{s+1}^f = \left(e_{s+1} e_{s+1}^T \right) I_f , \quad D_s^f = N_{3f} D_{s-1}^f + N_{2f} R_{s-1}^f , \tag{3.29a,b}$$

where I_f is a constant. By applying different envelope functions for the filters, e_s, constant I_f, the natural frequencies of the filters, and the ratios of damping to mass, a series of different shapes, spectral densities, center frequencies, and bandwidths of the narrow-band random processes from the filter can be obtained. This is a unique feature of the ESCD method that the Monte Carlo simulation (MCS) techniques do not possess.

Remark 3.4.1.5

The time step size for the computation of response statistics of a discretized structural system is governed by its highest natural frequency. However, the natural frequencies of filter and system are generally different. Therefore, in general, the time step size of the system Δt is different from that of the filter Δt_f. In other words, an output from the filter cannot be directly used as an input to the system since the time steps of the filter and the system will not match. To resolve this problem, one can adopt the interpolation or extrapolation to adjust the time step of the filter so that it can be applied to the computation of the response of the system. In the investigation by Chen and To [7] an interpolation strategy was found to give accurate results. This approach is adopted in the following computations.

3.4.2   Beam structure under narrow-band excitations

The beam structure considered in this sub-section is modeled by two two-node beam finite elements whose consistent element mass and stiffness matrices are defined by Eqs. (2.35) and (2.36), respectively. This model is presented in Figure 3.11. Of course, more refined finite element representations of the beam structure may be considered. However, the present main objectives are: (i) to implement the proposed ESCD method in a digital computer program that employs the finite elements, (ii) to provide a relatively simple example for comparison to the MCS data, and (iii) the finite element model is included here to illustrate the simplicity of the ESCD method

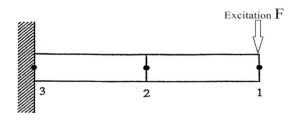

Figure 3.11 Two-element representation of cantilever beam.

and the difference between responses of the beam structure under broad-band and narrow-band random excitations. Therefore, it is believed that the two-element idealization is appropriate to demonstrate all the features of the ESCD method, while it maintains a very modest computational time for the MCS. The beam structure has the following properties: density $\rho = 7860.0\,\text{kg/m}^3$, Young's modulus of elasticity $E = 207\,\text{GPa}$, cross-section area $A = 6.25 \times 10^{-4}\,\text{m}^2$, second moment of area of cross-section $I_b = 3.26 \times 10^{-8}\,\text{m}^4$, and length of beam $L = 1.0$ m.

For this simple cantilever beam model, there are 6 dof after the application of boundary conditions and the natural frequencies are found to be: $\omega_1 = 130.3777$ rad/s, $\omega_2 = 823.5985$ rad/s, $\omega_3 = 2785.5606$ rad/s, $\omega_4 = 8084.8891$ rad/s, $\omega_5 = 8269.5425$ rad/s, and $\Omega = \omega_6 = 28888.7357$ rad/s.

The excitation vector in Eq. (3.21b) is defined by

$$f = [0 \quad -e(t)\sqrt{I_f} \,\ldots\, 0]^T, \tag{3.30}$$

and the discretized modulating function according to Eq. (3.30) is given by

$$e(t_s) = 9.4815\left(e^{-45\,t_s} - e^{-60\,t_s}\right).$$

The duration of the random excitation, $t_p = 0.15$ s. Note that the fundamental period, $T_1 = 2\pi/\omega_1$, is 0.04819 s. Therefore, the duration of random excitation is slightly more than 3 times the fundamental period of the structure. Damping in the beam is of the proportional type and for simplicity only the first two modal damping ratios of the beam structure are assumed. They are: $\zeta_1 = \zeta_2 = 0.05$ such that the Rayleigh damping coefficients are found to be: $\lambda_m = 11.3379$ and $\lambda_k = 0.0001$.

As the present formulation is based on linear theory, the following condition must be satisfied:

$$\max\left\{\langle Y_L^2 \rangle\right\} \leq \left(\frac{5L}{100}\right)^2,$$

where $\langle Y_L^2 \rangle$ is the variance of displacement at the free end of the beam.

3.4.2.1 Nonstationary random responses

A number of numerical tests is designed to examine the efficiency and accuracy of the ESCD method. For simplicity and illustration, a single dof filter is employed in the computation. The center frequency ω_f and damping ratio ζ_f of the filter are 130.0 rad/s and 0.05, respectively. The intensity of the excitation process in Eq. (3.30) is $I_f = 2\pi \times 10^4$ units. For brevity, only representative computed results by the ESCD method are presented in Figures 3.12 and 3.13. The dimensionless time step size employed was $\Delta\tau = 1.0$ after the application of the TCT. Computed results obtained by the MCS are also included in the figures for comparison. It is noted that the ratio of computation time for the MCS results to that of the ESCD method is approximately 89. With reference to the results in Figures 3.12 and 3.13 and the ratio of computation times, one can conclude that the ESCD method is very accurate and efficient compared with the MCS.

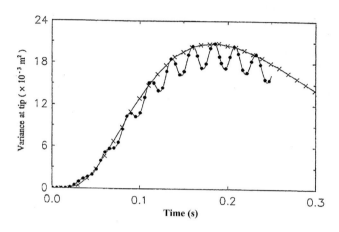

Figure 3.12 Variance of displacement at tip of cantilever: × ESCD; • MCS.

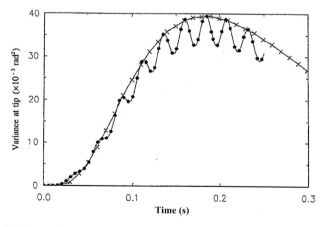

Figure 3.13 Variance of rotation at tip of cantilever: × ESCD; • MCS.

3.4.2.2 *Effect of bandwidths of excitation process*

The effect of bandwidths of the excitation process from the single dof filter on the responses of the discretized beam structure is the logical next step in the present investigation. All the filter and system parameters remain the same as in the foregoing except that the filter damping ratios, $\zeta_f = 0.05$ and 0.10 are considered here. Again, for brevity, only the displacement and rotation variances at the tip of the cantilever by using the ESCD method are provided in Figures 3.14 and 3.15. It is observed that the smaller the bandwidth of the filter, the higher the variance responses at the tip of the cantilever beam structure.

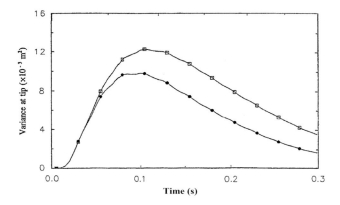

Figure 3.14   Effect of bandwidth on tip displacement: □, $\zeta_f = 0.05$; •,$\zeta_f = 0.10$.

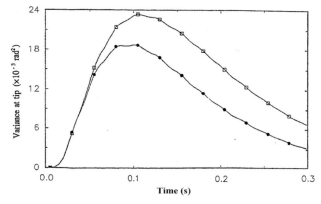

Figure 3.15   Effect of bandwidth on tip rotation: □, $\zeta_f = 0.05$; •,$\zeta_f = 0.10$.

3.4.3 Concluding remarks

In the foregoing the SCD method has been extended to systems under narrow-band random excitations. Two important points should be noted. First, the TCT and Eq. (3.14) for the relationship between natural frequencies and time step size are applicable to systems under narrow-band stationary and nonstationary random excitations. Second, the relations representing the narrow-band random vector process defined by Eq. (3.29) constitute a unique feature of the ESCD method.

3.5 Stochastic Newmark Family of Algorithms

For completeness this section includes the deterministic and stochastic versions of the Newmark family of algorithms [10]. The deterministic Newmark algorithms are commonly employed in the computation of responses of discretized dynamic systems and crucial in the derivation of their stochastic counterparts. In this presentation, responses of mdof systems disturbed by nonstationary random excitations are considered.

In the following sub-section the deterministic Newmark family of algorithms is outlined. They are applied in Sub-section 3.5.2 to the derivation of their stochastic counterparts. The detailed derivation of the deterministic Newmark family of algorithms is presented in Appendix 3A. Sub-section 3.5.3 is concerned with applications of the latter algorithms to plate structures idealized by the FEM.

3.5.1   Deterministic Newmark family of algorithms

The original Newmark method [10] for the numerical integration of the following matrix equation of motion of a spatially discretized structural dynamic system

$$Ma + Cv + Kd = P \tag{3.31}$$

has the relations

$$d_{s+1} = d_s + (\Delta t)v_s + \left(\frac{1}{2} - \beta \right)(\Delta t)^2 a_s + \beta(\Delta t)^2 a_{s+1},$$

$$v_{s+1} = v_s + (\Delta t)\left[(1 - \gamma)a_s + \gamma a_{s+1} \right] \tag{3.32, 3.33}$$

with P in Eq. (3.31) being the deterministic excitation vecctor; a, v, and d are the acceleration, velocity, and displacement vectors, respectively; β and γ are real positive constants; and the remaining symbols have already been defined in previous sections. Note that a here is not to be confused with the deterministic modulating function defined in Eq. (3.17) or (3.20) or elsewhere. Applying Eq. (3.31) for three consecutive time steps, t_{s-1}, t_s, and t_{s+1}, one has

$$Ma_{s-1} + Cv_{s-1} + Kd_{s-1} = P_{s-1} \, ,$$

$$Ma_s + Cv_s + Kd_s = P_s \, , \qquad (3.34\text{-}3.36)$$

$$Ma_{s+1} + Cv_{s+1} + kd_{s+1} = P_{s+1} \, .$$

Multiply the sum of Eqs. (3.34) and (3.36) by $(\Delta t)^2 \beta$, Eq. (3.35) by $(\Delta t)^2 (1 - 2\beta)$, the difference between Eqs. (3.35) and (3.34) by $(\Delta t)^2 (\gamma - 1/2)$, and add all these resulting equations. That is, symbolically $(\Delta t)^2 \beta [(3.34) + (3.36)] + (\Delta t)^2 (1 - 2\beta) [(3.35)] + (\Delta t)^2 (\gamma - 1/2) [(3.35) - (3.34)]$. With this relation and going through the detailed steps in Appendix 3A, one can obtain the following equation,

$$M_1 d_{s+1} = M_2 d_s - M_3 d_{s-1} + (\Delta t)^2 F_{es} \, , \qquad (3.37)$$

where the coefficient matrices M_i, in which $i = 1, 2, 3$, and the force vector F_{es} are given by

$$M_1 = M + \gamma(\Delta t)C + \beta(\Delta t)^2 K \, ,$$

$$M_2 = 2M - (1 - 2\gamma)(\Delta t)C - \left(\gamma + \frac{1}{2} - 2\beta \right)(\Delta t)^2 K \, ,$$

$$M_3 = M - (1 - \gamma)(\Delta t)C + \left(\beta + \frac{1}{2} - \gamma \right)(\Delta t)^2 K \, ,$$

$$F_{es} = \beta P_{s+1} + \left(\frac{1}{2} + \gamma - 2\beta \right) P_s + \left(\frac{1}{2} + \beta - \gamma \right) P_{s-1} \, .$$

Equation (3.37) gives the displacement vector in the next time step, d_{s+1} in terms of the displacement vector of the current time step, d_s, the displacement vector of the past time step, d_{s-1}, and forces in three consecutive time steps. It agrees with that reported in [11]. It was pointed out in [11] and [12] that a similar relation to Eq. (3.37) has been derived by Chaix and Leleux [13].

Before leaving this sub-section it suffices to mention that for a sdof system with small damping ratio ζ the algorithms with the stability condition of $2\beta \geq \gamma \geq 1/2$ are unconditional while that of $\gamma \geq 1/2$ and $\beta < \gamma/2$ are conditional [14]. When $\beta = 0$ and $\gamma = 1/2$ it is the central difference method whose order of accuracy is 2. In this case the critical time step size as shown in Appendix 3A is

$$(\Delta t)_c = \frac{2}{\omega_1} \, ,$$

where ω_1 is the undamped natural frequency of the sdof system. More discussion on the critical time step size can be found in Chapter 9 of [14].

3.5.2   Stochastic version of Newmark algorithms

Now, it is assumed that P in Eq. (3.31) is a vector of Gaussian white noises or modulated Gaussian white noise excitations so that upon application of Eq. (3.37), and remembering that $x_{s+1} = d_{s+1}$, $x_s = d_s$, $x_{s-1} = d_{s-1}$, one obtains

$$x_{s+1} = (\Delta t)^2 N_1 F_s + N_2 x_s + N_3 x_{s-1} \, , \tag{3.38}$$

where the coefficient matrices are defined as

$$N_1 = \left[M + \gamma(\Delta t)C + \beta(\Delta t)^2 K \right]^{-1},$$

$$N_2 = N_1 \left[2M - (1 - 2\gamma)(\Delta t)C - \left(\gamma + \frac{1}{2} - 2\beta \right)(\Delta t)^2 K \right],$$

$$N_3 = N_1 \left[(1 - \gamma)(\Delta t)C - M - \left(\beta + \frac{1}{2} - \gamma \right)(\Delta t)^2 K \right],$$

$$F_s = \beta P_{s+1} + \left(\gamma + \frac{1}{2} - 2\beta \right) P_s + \left(\beta + \frac{1}{2} - \gamma \right) P_{s-1} \, .$$

It should be noted that the temporally discretized random force vector P_s contains DGWN process whose properties are very much different from those of the continuous Gaussian white noise process.

Following the steps in Section 3.1 the recursive covariance matrix expression of the generalized displacement vector can be shown to be

$$R_{s+1} = N_2 R_s N_2^T + N_3 R_{s-1} N_3^T + (\Delta t)^4 N_1 B_s N_1^T + R_s^{(c)} \, , \tag{3.39}$$

in which

$$
R_s^{(c)} = N_2 D_s N_3^T + N_3 D_s^T N_2^T + (\Delta t)^2 N_2 \left\langle x_s F_s^T \right\rangle N_1^T
$$

$$
+ (\Delta t)^2 N_1 \left\langle F_s x_s^T \right\rangle N_2^T + (\Delta t)^2 N_3 \left\langle x_{s-1} F_s^T \right\rangle N_1^T
$$

$$
+ (\Delta t)^2 N_1 \left\langle F_s x_{s-1}^T \right\rangle N_3^T ,
$$

$$
R_s = \left\langle x_s x_s^T \right\rangle , \quad B_s = \left\langle F_s F_s^T \right\rangle
$$

$$
D_s = \left\langle x_s x_{s-1}^T \right\rangle = N_2 R_{s-1} + N_3 D_{s-1}^T + (\Delta t)^2 N_1 \left\langle F_{s-1} x_{s-1}^T \right\rangle .
$$

Note that Eq. (3.39) is similar in form to the corresponding covariance matrix relation of the SCD method in Section 3.1. The main differences are in the definitions of the parameter matrices N_1 , N_2 , N_3, and the covariance matrix of forces B_s or simply called the force matrix at time step t_s. For excitations modeled as white noise or shot noise or Wiener processes or their modulated counterparts, the terms

$$
\left\langle F_s x_s^T \right\rangle = 0 , \quad \left\langle F_s x_{s-1}^T \right\rangle = 0 . \tag{3.40a,b}
$$

With similar reasons the force matrix becomes

$$
B_s = \beta^2 B^{(1)} + \left(\gamma + \frac{1}{2} - 2\beta \right)^2 B^{(2)} + \left(\beta + \frac{1}{2} - \gamma \right)^2 B^{(3)} , \tag{3.41}
$$

in which

$$
B^{(1)} = \left\langle P_{s+1} P_{s+1}^T \right\rangle , \quad B^{(2)} = \left\langle P_s P_s^T \right\rangle , \quad B^{(3)} = \left\langle P_{s-1} P_{s-1}^T \right\rangle .
$$

Applying Eqs. (3.40) and (3.41) to Eq. (3.39), one can obtain

$$
R_{s+1} = N_2 R_s N_2^T + N_3 R_{s-1} N_3^T + (\Delta t)^4 N_1 B_s N_1^T
$$

$$
+ N_2 D_s N_3^T + N_3 D_s^T N_2^T . \tag{3.42}
$$

Equation (3.42) is the recursive covariance matrix expression for mdof systems. For identification purposes, it is called the stochastic Newmark (SN) method.

It may be appropriate to mention that the dimensionless time step size $\Delta\tau$ for the above SN method has been shown in page 136 of Ref. [5] as

$$\Delta\tau = \left[\beta^2 + \left(\gamma + \frac{1}{2} - 2\beta \right)^2 + \left(\beta + \frac{1}{2} - \gamma \right)^2 \right]^{-1}. \qquad (3.43)$$

When $\beta = 0$ and $\gamma = 1/2$, which is the SCD method, the dimensionless time step size becomes unity. More discussion on the SN method will follow in Chapter 6. It is noted, however, that this dimensionless time step size does not change with the increase in the number of dof. It will be shown in Chapter 6 that the SCD method for nonlinear mdof systems is symplectic.

3.5.3   Responses of square plates under transverse random forces

Two square plates studied in Sub-section 3.3.2 are applied here. The same high precision triangular bending finite element, TBH6 is employed to idealize the plate structures. The two FE models considered have been presented in Figure 3.7. The applied random point forces and their locations are identical to those in Sub-section 3.3.2. Transverse response statistics by applying Eq. (3.42) together with the TCT for the C4 plate are determined for $\gamma = 1/2$ and $\beta = 0, 3/10, 1/3$ [15]. In this case, the damping matrix for the plate structure is assumed to be proportional and only the first five modes are included in the computation. The damping ratio for these five modes is 5%. Representative results for the variances of displacements at Node 13 (center of C4 plate) and Node 18 are plotted in Figures 3.16 and 3.17.

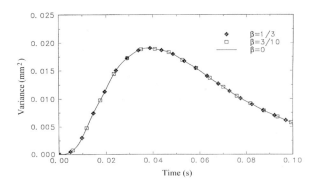

Figure 3.16   Variance of displacement at Node 13 of C4 plate: $\gamma = 1/2$.

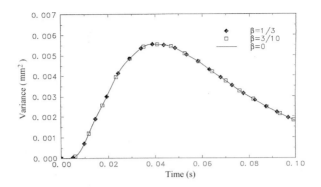

Figure 3.17   Variance of displacement at Node 18 of C4 plate: $\gamma = 1/2$.

Note that results for the other members of the SN method considered in the present studies are in excellent agreement with those by applying the SCD method. The latter algorithm is when $\gamma = 1/2$ and $\beta = 0$.

For the CF3 plate, transverse response statistics are similarly obtained for $\gamma = 1/2$ and $\beta = 0, 1/4, 1/3$ [15]. The damping matrix is similarly formed by including the first five models as for the C4 case. Selected computed results are presented in Figures 3.18 through 3.20, for brevity. Once again, results obtained for the other members of the SN algorithms are in excellent agreement with those by the SCD method. Graphically, there is no observable difference.

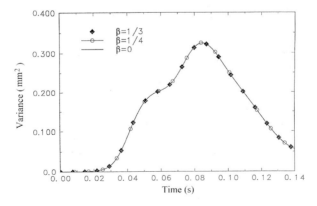

Figure 3.18   Variance of displacement at Node 5 of CF3 plate: $\gamma = 1/2$.

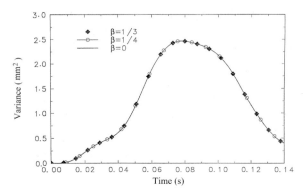

Figure 3.19   Variance of displacement at Node 8 of CF3 plate: $\gamma = 1/2$.

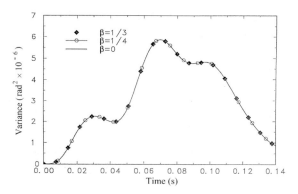

Figure 3.20   Variance of rotation at Node 8 of CF3 plate: $\gamma = 1/2$.

References

[1] To, C.W.S. (1986). The stochastic central difference method in structural dynamics, *Computers and Structures*, **23(6)**, 813-818.

[2] To, C.W.S. and Liu, M.L. (1994). Random responses of discretized beams and plates by the stochastic central difference method with time co-ordinate transformation, *Computers and Structures*, **53**, 727-738.

[3] Liu, M.L. and To, C.W.S. (1994). Random responses of discretized plates and shells by the stochastic central difference method. In *Proc. of 12th U.S. National Congress of Appl. Mech.*, pp. 10, June 26-July 1, Seattle, Washington.

[4] To, C.W.S. (1992). Response of multi-degree-of-freedom systems with geometrical nonlinearity under random excitations by the stochastic central difference method. In *Proc. of 4th Conf. on Nonl. Vibr., Stab., and Dyn. of Struct. and Mech.*, June 7-11, V. P. I. and State Univ., Blacksburg, Virginia.

[5] To, C.W.S. (2010). *Nonlinear Random Vibration: Computational Methods*, Zip Publishing, Columbus, Ohio.

[6] To, C.W.S. and Wang, B. (1993). Response analysis of discretized plates under in-plane and external nonstationary random excitations, in *Proceedings of the 14th ASME Biennial Conference on Mechanical Vibration and Noise*, DE-Vol. 56, pp. 51-62, Sept. 19-22, Albuquerque, New Mexico.

[7] Chen, Z. and To, C.W.S. (2005). Responses of discretized systems under narrow band nonstationary random excitations, Part 1: linear problems, *Journal of Sound and Vibration*, **287**, 433-458.

[8] Liu, M.L. and To, C.W.S. (1994). Adaptive time schemes for responses of non-linear multi-degree-of-freedom systems under random excitations, *Computers and Structures*, **52(3)**, 563-571.

[9] To, C.W.S. and Liu, M.L. (2000). Large nonstationary random responses of shell structures with geometrical and material nonlinearities, *Finite Elements in Analysis and Design*, **35**, 59-77.

[10] To, C.W.S. (1992). A stochastic version of the Newmark family of algorithms for discretized dynamic systems, *Computers and Structures,* **44(3)**, 667-673.

[11] Wood, W.L. (1984). A further look at Newmark, Houbolt, etc., time-stepping formulae, *Int. J. for Numerical Methods in Engineering*, **20**, 1009-1017.

[12] Zienkiewicz, O.C. (1977). A new look at the Newmark, Houbolt and other time stepping formulas: A weighted residual approach, *Int. J. for Numerical Methods in Engineering*, **7**, 413-418.

[13] Chaix, M. and Leleux, F. (1975). Private communication to O.C. Zienkiewicz, CETIM Senlis, France.

[14] Hughes, T.J.R. (1987). *The Finite Element Method: Linear Static and Dynamic Finite Element Analysis*, Prentice-Hall, New Jersey.

[15] Liu, M.L. and To, C.W.S. (2000). Linear and nonlinear random responses of discretized systems by the stochastic Newmark method, TR-001, Department of Mechanical Engineering, University of Nebraska, Lincoln, Nebraska.

4

Modal Analysis and Response Statistics of Quasi-linear Structural Systems

The structural systems considered in this chapter are governed by the second order quasi-linear differential equations of motion. In the language of FEM the systems have geometrical nonlinearities of small deformations. The SFEA for linear systems is included in this class which consists of temporally stochastic quasi-linear systems, and temporally and spatially stochastic quasi-linear systems.

The temporally stochastic quasi-linear systems include structures with geometrical nonlinearities due to axial and in-plane random disturbances. The modal analysis and response statistics by Cumming's approach are considered in Section 4.1 while its application to plates under external and parametric random excitations and approximated by the rectangular plate finite element is presented in Section 4.2. Plate structures approximated by the triangular plate finite element are studied in Section 4.3. Concluding remarks are included in Section 4.4.

4.1 Modal Analysis of Temporally Stochastic Quasi-linear Systems

To and Orisamolu [1, 2] applied the Melosh-Zienkiewicz-Cheung (MZC) [3] rectangular bending element in dealing with response analysis of plate structures under external and in-plane nonstationary random excitations. Normal mode analysis was made and response moments were obtained in closed form. Explicit expressions for the MZC element stiffness, mass and initial stress stiffness matrices were employed, resulting in considerable saving in computational time. The associated first passage time problem was considered [4].

Differential moment equations of FE modeled structures with geometrical nonlinearities were presented by Di Paola and Muscolino [5]. Their starting equations of motion, Eq. (20) in the latter reference, treated the nonlinearity term as a pseudo-loading vector. A more general formulation with thermal effect included in the pseudo-loading vector was presented by the author [6]. In the publication of Di Paola and Muscolino modal matrix was required in addition to application of a truncation closure. The latter was necessary because of the infinite hierarchy of moments inherent in the analysis. No computed results were included in [5].

In order to provide a means for more accurate random response analysis of plate structures with relatively complicated geometries and subjected to external and in-plane nonstationary random excitations, To and Wang [7] employed a displacement formulation-based three-node higher order triangular bending finite element denoted as TBH6. Explicit expressions of the element mass, stiffness and initial stress stiffness matrices were obtained by Wang [8].

4.1.1   Modal analysis and bi-modal approach

For systems with both external and parametric or in-plane random excitations the matrix equation of motion becomes

$$M\ddot{x} + C\dot{x} + \left[K - p(t)K_G\right]x = F \tag{4.1}$$

where $p(t)$ or written simply as p is the scalar parametric or in-plane random excitation whereas K_G is the assembled stability or initial stress stiffness matrix, K is the linear assembled stiffness matrix, and the remaining symbols have already been defined in previous chapters.

Assuming that the random excitations are all nonstationary random, and they are modeled as products of uniformly modulated functions and zero mean Gaussian white noise processes

$$p = e_p w_p, \quad F_i = e_i w_i, \qquad i = 1, 2, \cdots, n \tag{4.2a,b}$$

where e_p is the envelope modulating function for the parametric excitation with w_p being the zero mean Gaussian white noise process, while e_i is the envelope modulating function for the i'th excitation with w_i being the corresponding zero mean Gaussian white noise process.

By following the usual modal analysis and normalizing the mode shapes as in Sub-section 2.2.2 and [1, 2], a complete set of n normalized mode shapes Ψ_m is obtained. The i'th column of the normal mode matrix Ψ_m is defined by

$$\Psi_m^{(i)} = \frac{1}{\sqrt{m_{1i}}} \Psi^{(i)}, \qquad i = 1, 2, \cdots, n \tag{4.3}$$

where $\Psi^{(i)}$ is the i'th column of the modal matrix, Ψ whereas m_{ii} is the diagonal element of the matrix $\Psi^T M \Psi$ associated with mode shape $\Psi^{(i)}$. In other words, $(\Psi^{(i)})^T M \Psi^{(i)} = M_{ii}$. Note that because of the above definition $\Psi_m{}^T M \Psi_m$ is an identity matrix. That is, symbolically, $\Psi_m{}^T M \Psi_m = I$.

Introducing the transformation

$$\{x\}_{n\times 1} = [\Psi_m]_{n\times L} \{q\}_{L\times 1}$$

or simply written as

$$x = \Psi_m q \qquad (4.4)$$

to Eq. (4.1) with q being the vector of normal co-ordinates, and premultiplying Eq. (4.1) throughout by $\Psi_m{}^T$, it leads to the uncoupled equation,

$$\ddot{q}_j + 2\zeta_j \omega_j \dot{q}_j + \left[\omega_j^2 - p(t)\omega_{jg}^2 \right] q_j = f_j(t) , \qquad j = 1,2,\cdots,L \qquad (4.5)$$

where

$$2\zeta_j \omega_j = \left(\Psi_m^{(j)} \right)^T C \Psi_m^{(j)} , \qquad \omega_j^2 = \left(\Psi_m^{(j)} \right)^T K \Psi_m^{(j)} ,$$

$$\omega_{jg}^2 = \left(\Psi_m^{(j)} \right)^T K_G \Psi_m^{(j)} , \qquad f_j(t) = \left(\Psi_m^{(j)} \right)^T F$$

and here the integer L is the total number of modes considered in the analysis. In general, L is less than n. This has to do with the fact that in structural dynamic analysis of continuum structures discretized by the FEM the response is usually reasonably accurately represented by the first few modes even though the discretized structure may have a large number of dof.

In the foregoing normal mode analysis it is assumed that the damping matrix is of the proportional type. It may also be appropriate to mention that the approach introduced by Lin and Shih [9] can be employed to solve for the response of the system. However, such an approach would require a relatively much larger computer storage and computation time. In the present work therefore, a so-called "bi-modal" interaction procedure [1, 2] is applied so that the off-diagonal elements of the covariance matrix can be accounted for in addition to the diagonal elements which are the variances. It is believed that the "bi-modal" approach is more flexible computationally than that given in Ref. [9].

The equations of motion for any two modes, j and k can be written in normal co-ordinates as

$$\ddot{q}_j + 2\zeta_j \omega_j \dot{q}_j + \omega_j^2 q_j - p(t)\omega_{jg}^2 q_j = f_j(t) ,$$

$$\ddot{q}_k + 2\zeta_k \omega_k \dot{q}_k + \omega_k^2 q_k - p(t)\omega_{kg}^2 q_k = f_k(t) ,$$

(4.6a,b)

or in matrix form

$$\begin{bmatrix} 1 & 0 \\ 0 & 1 \end{bmatrix}\begin{Bmatrix} \ddot{q}_j \\ \ddot{q}_k \end{Bmatrix} + \begin{bmatrix} 2\zeta_j \omega_j & 0 \\ 0 & 2\zeta_k \omega_k \end{bmatrix}\begin{Bmatrix} \dot{q}_j \\ \dot{q}_k \end{Bmatrix} + \begin{bmatrix} K_{jk} \end{bmatrix}\begin{Bmatrix} q_j \\ q_k \end{Bmatrix} = \begin{Bmatrix} f_j \\ f_k \end{Bmatrix} ,$$

where

$$\begin{bmatrix} K_{jk} \end{bmatrix} = \begin{bmatrix} \omega_j^2 & 0 \\ 0 & \omega_k^2 \end{bmatrix} - p \begin{bmatrix} \omega_{jg}^2 & 0 \\ 0 & \omega_{kg}^2 \end{bmatrix} .$$

This equation can be solved by the "bi-modal" approach. For completeness, it is to be introduced in the following. But first, comments on the modal random excitations f_j and f_k are in order. Recall that the original external random excitations were given as F_i, $i = 1, 2, ..., n$. The latter is the total number of dof.

Let the elements of the transformation matrix, Ψ_m be denoted as ψ_{ij}, where $i = 1, 2, ... , n$ and $j = 1, 2, ..., L$. In details, the transformation matrix is

$$\Psi_m = \begin{bmatrix} \psi_{11} & \psi_{12} & \cdots & \psi_{1L} & 0 & \cdots & 0 \\ \psi_{21} & \psi_{22} & \cdots & \psi_{2L} & 0 & \cdots & 0 \\ \cdots & \cdots & \cdots & \cdots & 0 & \cdots & 0 \\ \cdots & \cdots & \cdots & \cdots & 0 & \cdots & 0 \\ \psi_{n1} & \psi_{n2} & \cdots & \psi_{nL} & 0 & \cdots & 0 \end{bmatrix} .$$

Also, the externally applied random excitation vector is

$$F = \begin{bmatrix} F_1 & F_2 & \cdots & F_n \end{bmatrix}^T , \qquad f = \Psi_m^T F , \qquad \text{as} \qquad f_j(t) = \left(\Psi_m^{(j)}\right)^T F$$

defined in Eq. (4.5) in the foregoing.

To provide a clearer understanding of the relationship between the applied random excitations and the corresponding random modal excitations, it is necessary to expand on the definition of f. To this end, one has

$$f = \Psi_m^T F = \begin{Bmatrix} \psi_{11}F_1 + \psi_{21}F_2 + \psi_{31}F_3 + \cdots + \psi_{n1}F_n \\ \psi_{12}F_1 + \psi_{22}F_2 + \psi_{32}F_3 + \cdots + \psi_{n3}F_n \\ \cdots \quad\quad \cdots \quad\quad \cdots \quad\quad \cdots \quad\quad \cdots \\ \psi_{1L}F_1 + \psi_{2L}F_2 + \psi_{3L}F_3 + \cdots + \psi_{nL}F_n \\ 0 \;+\; 0 \;+\; 0 \;+\cdots+\; 0 \\ 0 \;+\; 0 \;+\; 0 \;+\cdots+\; 0 \\ 0 \;+\; 0 \;+\; 0 \;+\cdots+\; 0 \end{Bmatrix} \tag{4.7}$$

or

$$f = \Psi_m^T F = \begin{Bmatrix} \psi_{11} \\ \psi_{12} \\ \cdots \\ \psi_{1L} \\ 0 \\ \cdots \\ 0 \end{Bmatrix} F_1 + \begin{Bmatrix} \psi_{21} \\ \psi_{22} \\ \cdots \\ \psi_{2L} \\ 0 \\ \cdots \\ 0 \end{Bmatrix} F_2 + \cdots + \begin{Bmatrix} \psi_{n1} \\ \psi_{n2} \\ \cdots \\ \psi_{nL} \\ 0 \\ \cdots \\ 0 \end{Bmatrix} F_n . \tag{4.8}$$

With the form of the vector of modal random forces in Eq. (4.8), it is easy to see the effect of every external random excitation. This form is better than that in Eq. (4.7) in indicating the effect of the externally applied random forces because in practice it is the physical excitation that is directly modeled and not the modal random excitations, which are the intermediate results of the mathematical analysis. With reference to Eq. (4.8), it is clear that the total response is calculated by adding up all the responses due to every random excitation applied at every nodal dof. It should be noted that the correlation between any two forces F_j and F_k is not accounted for. In other words, it is assumed that

$$\langle F_j F_k \rangle = 0 , \qquad j \neq k . \tag{4.9}$$

On the other hand,

$$\langle F_j p \rangle \neq 0 . \tag{4.10}$$

That is, the ensemble average of the product of the externally applied and parametric or in-plane random excitations is not independent in general.

Now, consider the "bi-modal" approach. The first step is to translate Eq. (4.6) into four first order differential equations. For direct identification of the four-dimensional first order differential equations, one replaces the subscripts j and k respectively with 1 and 2 so that Eq. (4.6) becomes

$$\ddot{q}_1 + 2\zeta_1 \omega_1 \dot{q}_1 + \omega_1^2 q_1 - p(t)\omega_{1g}^2 q_1 = f_1(t) ,$$
$$\ddot{q}_2 + 2\zeta_2 \omega_2 \dot{q}_2 + \omega_2^2 q_2 - p(t)\omega_{2g}^2 q_2 = f_2(t) . \tag{4.11}$$

Let $q_1 = Z_1$, $\dot{q}_1 = Z_2$, $q_2 = Z_3$, $\dot{q}_2 = Z_4$ and $Z = [Z_1 \ Z_2 \ Z_3 \ Z_4]^T$. Then Eq. (4.11) may be written as

$$\dot{Z}_2 = \alpha_1 Z_1 + \alpha_2 Z_2 + \alpha_5 Z_1 p + f_1$$
$$\dot{Z}_4 = \beta_3 Z_3 + \beta_4 Z_4 + \beta_6 Z_3 p + f_2 \tag{4.12}$$

where

$$\alpha_1 = - \omega_1^2 , \qquad \alpha_2 = - 2\zeta_1\omega_1 , \qquad \alpha_5 = - \omega_{1g}^2 ,$$
$$\beta_3 = - \omega_2^2 , \qquad \beta_4 = - 2\zeta_2\omega_2 , \qquad \beta_6 = - \omega_{2g}^2 ,$$

and in the general case considered here

$$p = e_p w_p , \qquad f_1 = e_1 w_1 , \qquad f_2 = e_2 w_2 .$$

By making use of the above definitions, Eq. (4.11) can be rewritten as a four-dimensional Itô stochastic differential equation,

$$dZ = \xi(Z,t)dt + G(Z,t)dB ,\qquad(4.13)$$

where the symbols are individually defined in the following

$$\xi(Z,t) = \begin{Bmatrix} Z_2 \\ \alpha_1 Z_1 + \alpha_2 Z_2 \\ Z_4 \\ \beta_3 Z_3 + \beta_4 Z_4 \end{Bmatrix}, \quad dB = \begin{Bmatrix} w_p \\ w_p \\ w_1 \\ w_2 \end{Bmatrix} dt ,$$

and

$$G(Z,t) = \begin{bmatrix} 0 & 0 & 0 & 0 \\ e_p \alpha_5 Z_1 & 0 & e_1 & 0 \\ 0 & 0 & 0 & 0 \\ 0 & \beta_6 e_p Z_3 & 0 & e_2 \end{bmatrix} .$$

In what follows the solution of Eq. (4.13) will be considered.

4.1.2   Response statistics by Cumming's approach

Applying the approach presented by Cumming[10] to the above two dof system, the statistical moment equations can be obtained from

$$\frac{d\langle h \rangle}{dt} = \left\langle \frac{\partial h}{\partial t} \right\rangle + \sum_{j=1}^{4} \left\langle \xi_j \frac{\partial h}{\partial Z_j} \right\rangle$$

$$+ \sum_{i=1}^{4} \sum_{j=1}^{4} \left\langle (GDG^T)_{ij} \frac{\partial^2 h}{\partial Z_i \partial Z_j} \right\rangle ,\qquad(4.14)$$

where $h = h(Z,t)$ is an arbitrary function whose partial derivatives are jointly continuous and bounded over any finite interval of Z and t, whereas D is the matrix of white noise intensities with elements or entries D_{ij}. Thus,

$$\langle dB_i \, dB_j \rangle = 2 D_{ij} \, dt \ .$$

By applying Cumming's approach, one can show that

$$\frac{d\langle h \rangle}{dt} = \left\langle \frac{\partial h}{\partial t} \right\rangle + \sum_{j=1}^{4} \left\langle \xi_j \frac{\partial h}{\partial Z_j} \right\rangle + \left\langle \sigma_{22} \frac{\partial^2 h}{\partial Z_2} \right\rangle$$

$$+ \left\langle \sigma_{44} \frac{\partial^2 h}{\partial Z_4^2} \right\rangle + 2 \left\langle \sigma_{24} \frac{\partial^2 h}{\partial Z_2 \partial Z_4} \right\rangle , \tag{4.15}$$

in which

$$\sigma_{22} = D_{11} e_p^2 \alpha_5^2 Z_1^2 + 2 D_{13} e_p e_1 \alpha_5 + D_{33} e_2 \ ,$$

$$\sigma_{44} = D_{22} e_p^2 \beta_6^2 Z_3^2 + 2 D_{24} e_p e_2 \beta_6 Z_3 + D_{44} e_2^2 \ ,$$

$$\sigma_{24} = \sigma_{42} = D_{12} e_p^2 \alpha_5 \beta_6 Z_1 Z_3 + D_{14} e_p e_2 \alpha_5 Z_1$$

$$+ D_{32} e_p e_1 \beta_6 Z_3 + D_{34} e_1 e_2 \ .$$

By selecting a suitable function h, the statistical moment equations may be generated from Eq. (4.15). For example, if one chooses h successively as Z_1, Z_2, Z_3, and Z_4, respectively the first statistical moment equation can be obtained as

$$\{\dot{m}^q\} = [a]\{m^q\} \ , \tag{4.16}$$

where the moment vector and coefficient matrix are defined as

$$\{m^q\} = \left[m_1^q \, m_2^q \, m_3^q \, m_4^q \right]^T , \quad m_i^q = \langle Z_i \rangle , \quad [a] = \begin{bmatrix} (a_{ij})_{11} & (a_{ij})_{12} \\ (a_{ij})_{21} & (a_{ij})_{22} \end{bmatrix} ,$$

in which the submatrices are defined as

$$\left(a_{ij}\right)_{11} = \begin{bmatrix} 0 & 1 \\ \alpha_1 & \alpha_2 \end{bmatrix}, \quad \left(a_{ij}\right)_{12} = \left(a_{ij}\right)_{21} = \begin{bmatrix} 0 & 0 \\ 0 & 0 \end{bmatrix}, \quad \left(a_{ij}\right)_{22} = \begin{bmatrix} 0 & 1 \\ \beta_3 & \beta_4 \end{bmatrix}.$$

If the initial conditions of the system are given, the exact solution of Eq. (4.16) can be easily determined. In a similar manner, the second moment equations becomes

$$\{\dot{N}^q\} = [A]\{N^q\} + \{C\} \tag{4.17}$$

in which the vector of second order statistical moments is defined by

$$\{N^q\} = \begin{bmatrix} N_1^q & N_2^q & N_3^q & N_4^q & N_5^q & N_6^q & N_7^q & N_8^q & N_9^q & N_{10}^q \end{bmatrix}^T,$$

where

$$N_i^q = \left\langle Z_i^2 \right\rangle, \quad i = 1,2,3,4, \quad N_5^q = \left\langle Z_1 Z_2 \right\rangle, \quad N_6^q = \left\langle Z_1 Z_3 \right\rangle,$$

$$N_7^q = \left\langle Z_1 Z_4 \right\rangle, \quad N_8^q = \left\langle Z_2 Z_3 \right\rangle, \quad N_9^q = \left\langle Z_2 Z_4 \right\rangle, \quad N_{10}^q = \left\langle Z_3 Z_4 \right\rangle,$$

while the coefficient or amplification matrix is

$$[A] = \begin{bmatrix} \left(A_{ij}\right)_{11} & \left(A_{ij}\right)_{12} \\ \left(A_{ij}\right)_{21} & \left(A_{ij}\right)_{22} \end{bmatrix},$$

$$\left(A_{ij}\right)_{11} = \begin{bmatrix} 0 & 0 & 0 & 0 & 2 \\ A_{21} & 2\alpha_2 & 0 & 0 & 2\alpha_1 \\ 0 & 0 & 0 & 0 & 0 \\ 0 & 0 & A_{43} & 2\beta_4 & 0 \\ \alpha_1 & 1 & 0 & 0 & \alpha_2 4 \end{bmatrix}, \quad \left(A_{ij}\right)_{12} = \begin{bmatrix} 0 & 0 & 0 & 0 & 0 \\ 0 & 0 & 0 & 0 & 0 \\ 0 & 0 & 0 & 0 & 2 \\ 0 & 0 & 0 & 0 & 2\beta_3 \\ 0 & 0 & 0 & 0 & 0 \end{bmatrix},$$

$$\left(A_{ij}\right)_{22} = \begin{bmatrix} 0 & 1 & 1 & 0 & 0 \\ \beta_3 & \beta_4 & 0 & 1 & 0 \\ \alpha_1 & 0 & \alpha_2 & 1 & 0 \\ A_{96} & \alpha_1 & \beta_3 & A_{99} & 0 \\ 0 & 0 & 0 & 0 & \beta_4 \end{bmatrix}, \quad \left(A_{ij}\right)_{21} = \begin{bmatrix} 0 & 0 & 0 & 0 & 0 \\ 0 & 0 & 0 & 0 & 0 \\ 0 & 0 & 0 & 0 & 0 \\ 0 & 0 & 0 & 0 & 0 \\ 0 & 0 & \beta_3 & 1 & 0 \end{bmatrix},$$

in which

$$A_{21} = 2D_{11}e_p^2\,\alpha_5^2\,, \quad A_{43} = 2D_{22}\,e_p^2\,\beta_6^2\,,$$

$$A_{96} = 2D_{12}\,e_p^2\,\alpha_5\,\beta_6\,, \quad A_{99} = \alpha_2 + \beta_4\,.$$

The remaining vector on the rhs of Eq. (4.17) consists of elements or entries which contain the deterministic modulating functions of random excitations, white noise intensities, first order statistical moments, and other system parameters. In this particular problem all the elements are zero except

$$C_2 = 2\left(2D_{13}e_p e_1 \alpha_5 m_1^q + D_{33}e_1^2\right), \quad C_4 = 2\left(2D_{24}e_1 e_2 \beta_6 m_3^q + D_{44}e_2^2\right),$$

$$C_9 = 2\left(D_{14}e_p e_2 \alpha_5 m_1^q + D_{32}e_1 e_p \beta_6 m_3^q + D_{34}e_1 e_2\right).$$

With some algebraic manipulation Eqs. (4.16) and (4.17) can be solved exactly. However, it is more convenient to evaluate Eq. (4.17) numerically. To this end, the Runge-Kutta fourth order (RK4) algorithm is employed. Thus, the second order statistical moments of Eq. (4.6) can be determined numerically. That is,

$$\left\langle q_j^2\right\rangle,\ \left\langle \dot{q}_j^2\right\rangle,\ \left\langle q_k^2\right\rangle,\ \left\langle \dot{q}_k^2\right\rangle,\ \left\langle q_j\dot{q}_j\right\rangle,\ \left\langle q_j q_k\right\rangle,$$

$$\left\langle q_j\dot{q}_k\right\rangle,\ \left\langle \dot{q}_j q_k\right\rangle,\ \left\langle \dot{q}_j\dot{q}_k\right\rangle,\ \left\langle q_k\dot{q}_k\right\rangle,$$

can be obtained by applying the RK4 algorithm to Eq. (4.17).

The matrices $\langle q\,q^T \rangle$ and $\langle \dot{q}\,\dot{q}^T \rangle$ are symmetric while $\langle q\,\dot{q}^T \rangle$ is non-symmetric because, in general,

$$\langle q_j\,\dot{q}_k \rangle \neq \langle \dot{q}_j\,q_k \rangle .$$

It is also noted that unlike the diagonal elements in $\langle q\,q^T \rangle$ and $\langle \dot{q}\,\dot{q}^T \rangle$, the non-diagonal elements of $\langle q\,\dot{q}^T \rangle$ are not variances of displacements or velocities and therefore, they are not positive definite in general.

In the digital computer program developed for the above computation, steps have been taken to ensure the correct number of pairs of modal responses were evaluated. Also, in the computation a provision was made to sort out the responses into the matrices in which they belong.

After evaluating the second order statistical moments for the modal co-ordinates, one is required to express the statistical moments in the original co-ordinate system. Thus, the following relationships are employed

$$\langle xx^T \rangle = \Psi_m \langle q\,q^T \rangle \Psi_m^T , \qquad (4.18a)$$

$$\langle x\dot{x}^T \rangle = \Psi_m \langle q\,\dot{q}^T \rangle \Psi_m^T , \qquad \langle \dot{x}\dot{x}^T \rangle = \Psi_m \langle \dot{q}\,\dot{q}^T \rangle \Psi_m^T . \qquad (4.18b,c)$$

4.2 Response Analysis Based on the Melosh-Zienkiewicz-Cheung Bending Plate Finite Element

One of the early displacement formulation-based plate bending finite elements was originally developed by Melosh [11], and Zienkiewicz and Cheung [12]. It is generally referred to as the MZC [3] bending element. It is known as the nonconforming plate bending element since it does not possess normal-slope compatibility at the edges. It has four nodes and every one of which has three dof. The latter are w_i^e, $\partial w_i^e/\partial y$, and $\partial w_i^e/\partial x$, where w_i^e is the transversal displacement at node i on the middle plane of the plate while x and y are two perpendicular axes on the middle plane of the plate. The assumed displacement function for the element is

$$w^e = b_1 + b_2\xi + b_3\eta + b_4\xi^2 + b_5\xi\eta + b_6\eta^2 + b_7\xi^3 + b_8\xi^2\eta$$
$$+ b_9\xi\eta^2 + b_{10}\eta^3 + b_{11}\xi^3\eta + b_{12}\xi\eta^3 , \qquad (4.19)$$

in which $\xi = x/a$, $\eta = y/b$, a and b are the half side length of the element parallel to, respectively the x and y axes of the middle plane of the plate, and the coefficients b's in the polynomial are arbitrary constants. The consistent linear element mass and stiffness matrices, m and k, as well as the element stability matrix k_G have been included in [3]. These element matrices have been applied to the studies of response statistics of plate structures under external and in-plane or parametric nonstationary random excitations in Refs. [1, 2].

In the following two sub-sections simply-supported plate and plate fixed at all sides under external or non-parametric and in-plane or parametric nonstationary random excitations are considered. Responses statistics are computed and compared with available exact solutions. The final sub-section deals with remarks on the obtained results and their implications.

4.2.1   Simply-supported plate structure

The steel square plate structure studied in this sub-section is simply-supported on all four edges. The geometrical properties of this plate structure are: side length of plate $a = b = 1.0$ m, and thickness $t = 5$ mm. Material properties are: Young's modulus of elasticity $E = 200$ GPa, density $\rho = 7.83 \times 10^3$ kg/m^3, and Poisson's ratio $v = 0.3$. The plate is represented by 16, 32, and 64 finite elements. Only the 16 and 64 elements models are shown in Figure 4.1 since the 32-element model is just like the 16-element model in which every square element is divided into two equal rectangular elements.

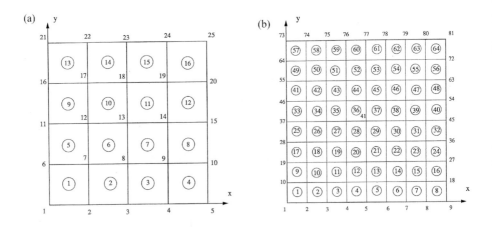

Figure 4.1   Finite element models of simply-supported square plate: (a) 16-element model, and (b) 64-element model.

The nonstationary random excitation applied at the center of the plate and along the axis perpendicular to the plane of the plate is given by

$$F(t) = 9.48148\left(e^{-45t} - e^{-60t}\right)w(t) , \qquad (4.20)$$

where $w(t)$ is a zero-mean Gaussian white noise process with auto-correlation equal to $2\pi\delta(\tau)$, in which $\delta(\tau)$ is the Dirac delta function.

In addition to excitation applied at the center of the plate a compressive stochastic inplane load $p(t)$ applied in the horizontal direction is similar to Eq. (4.20) except that the zero mean Gaussian white noise $w(t)$ is replaced with a zero mean Gaussian white $w_p(t)$ with auto-correlation different from that in Eq. (4.20). The root mean square values of the auto-correlation of the white noise part of the parametric random excitation are chosen to be smaller values of their corresponding static critical buckling stress σ_{cr}.

For the present studies the static critical buckling stress for the simply-supported plate is given by

$$\sigma_{cr} = \frac{\pi^2 E}{3\left(1 - v^2\right)}\left(\frac{t}{b}\right)^2 = 18.0762 \text{ MPa} . \qquad (4.21)$$

Thus, for example, with a 5% static critical buckling stress as the magnitude of the rms of the white noise part of the parametric excitation the white noise intensities

$$D_{11} = D_{22} = \left(5\% \times \sigma_{cr}\right)^2 = 8.168725 \times 10^{11} , \qquad (4.22)$$

where the white noise intensities D_{11} and D_{22} are those in Eq. (4.14).

The parametric random excitation levels considered in the present investigation are denoted as p_2 and p_5. Thus, p_2 means that the parametric random excitation level is such that its rms value of the Gaussian white noise part of the nonstationary random excitation is 2% of σ_{cr} while p_5 means 5% of σ_{cr}.

Modal analysis was performed on the FE models by using SPADAS [13] to obtain the natural frequencies and mode shapes. Only the first eight modes have been applied to the computation of the response statistics. These frequencies with their corresponding exact solutions derived from [14] are included in Table 4.1. In the latter, 16RE, for instance, denotes the plate represented by 16 MZC bending elements. Computed results of the response statistics for the square plate simply-supported at all edges and approximated by the 64RE model are presented in Figures 4.2 through 4.4

Table 4.1   Natural frequencies of S4 plate using different FE models.

| Mode number | Exact | Frequencies (Hz) | | |
		16RE	32RE	64RE
1	24.02	23.30	23.55	23.83
2	60.06	57.62	57.86	59.32
3	60.06	57.62	58.80	59.32
4	96.09	87.28	89.60	93.22
5	120.11	116.60	115.05	118.63
6	120.11	116.65	119.18	118.63
7	156.15	139.39	139.82	150.11
8	156.15	139.39	145.76	150.11

for damping ratio $\zeta_r = 0.01, 0.05$, and 0.10, where $r = 1, 2, 3, ..., 8$. Note that these results are transversal displacements and velocities at the center of the plate. For identification purpose this plate is denoted by S4. For comparison, similar results for the 16RE and 32RE models were obtained and included in Figure 4.5. It is observed that as the structure is approximated by the refined FE model the responses become better estimates in the sense that the results converge to the exact solution with finer meshes. It is also noted the coarser the mesh, the higher the response level obtained.

As the structure is refined the number of dof becomes higher and consequently the cost of computation rises. The computation time rises very fast since the interaction of two modes at a time are considered and 10 simultaneous differential equations must be solved for each available non-parametric excitation, usually corresponding to each available dof in the most general case. The number of time, η_s, the IMSL differential equation subroutine (DVERK) is called at every time step, Δt is given by

$$\eta_s = L(L - 1)(N - e_o) / 2 , \tag{4.23}$$

where e_o, not to be confused with the deterministic modulating function, is the total number of zero force elements or entries in the external random excitation vector, L has been previously defined as the total number of modes considered in the response computation, and N the total number of dof, respectively. Note when only parametric

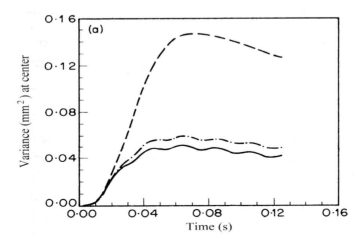

Figure 4.2(a)   Displacement responses of S4 plate to non-parametric and parametric random excitations for 1% damping: ____ , no parametric excitation; – · – · –, parametric excitation p_2 ; - - -, parametric excitation p_5.

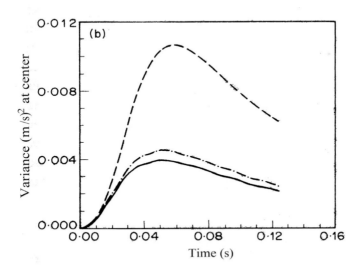

Figure 4.2(b)   Velocity responses of S4 plate to non-parametric and parametric random excitations for 1% damping: ____ , no parametric excitation; – · – · –, parametric excitation p_2 ; - - -, parametric excitation p_5.

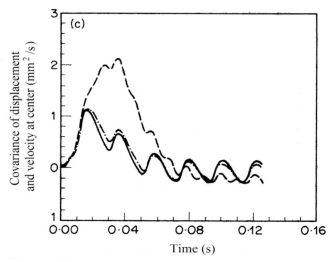

Figure 4.2(c)   Coupled responses of S4 plate to non-parametric and parametric random excitations for 1% damping: ＿＿ , no parametric excitation; – · – · –, parametric excitation p_2 ; - - -, parametric excitation p_5.

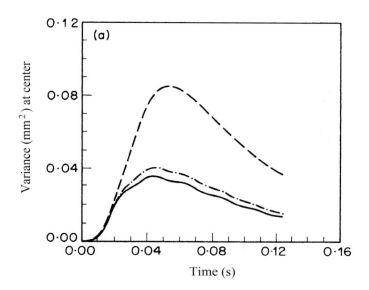

Figure 4.3(a)   Displacement responses of S4 plate to non-parametric and parametric random excitations for 5% damping: ＿＿ , no parametric excitation; – · – · –, parametric excitation p_2 ; - - -, parametric excitation p_5.

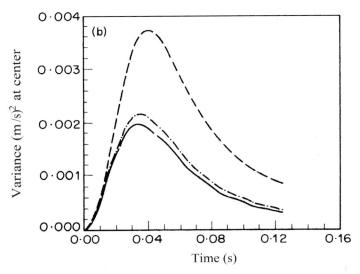

Figure 4.3(b)   Velocity responses of S4 plate to non-parametric and parametric random excitations for 5% damping: ——— , no parametric excitation; – · – · – , parametric excitation p_2 ; - - -, parametric excitation p_5.

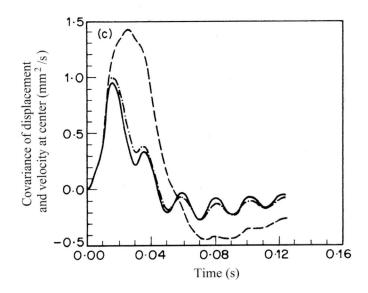

Figure 4.3(c)   Coupled responses of S4 plate to non-parametric and parametric random excitations for 5% damping: ——— , no parametric excitation; – · – · – , parametric excitation p_2 ; - - -, parametric excitation p_5.

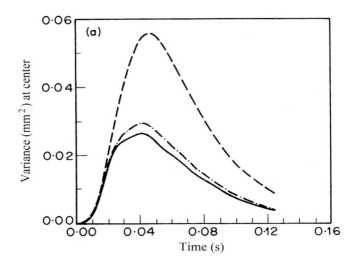

Figure 4.4(a) Displacement responses of S4 plate to non-parametric and parametric random excitations for 10% damping: ——— , no parametric excitation; – · —· –, parametric excitation p_2 ; - - -, parametric excitation p_s.

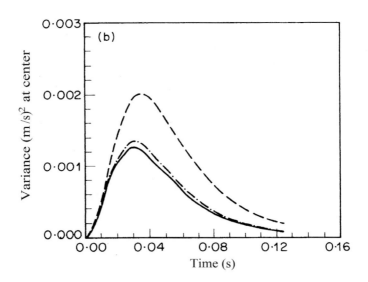

Figure 4.4(b) Velocity responses of S4 plate to non-parametric and parametric random excitations for 10% damping: ——— , no parametric excitation; – · —· –, parametric excitation p_2 ; - - -, parametric excitation p_s.

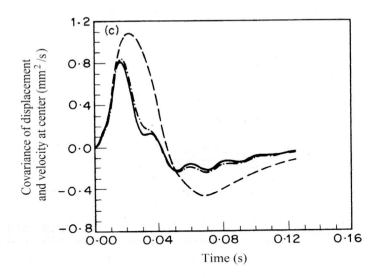

Figure 4.4(c) Coupled responses of S4 plate to non-parametric and parametric random excitations for 10% damping: ——— , no parametric excitation; – · – · – , parametric excitation p_2 ; - - -, parametric excitation p_5.

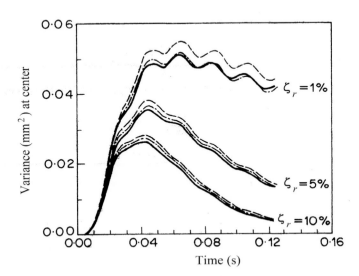

Figure 4.5 Comparison of displacement responses at center of S4 plate: ——— , 64RE; – · – · – , 32RE ; - - -, 16RE.

random excitations are present, e_o is zero. This explains why in cases with parametric random excitations the difference in times of computation is very considerable. For example, for the S4 plate, typical computation times for parametric and non-parametric random excitations for the 16RE, 32RE, and 64RE models are, respectively 5525.615, 10451.926, and 12044.421 cpu seconds when using a CDC Cyber 175 machine with NOS 2.3 operating system. Thus, to save cost, a judicious choice of an appropriate FE model that would take less computation time and yet provide the needed degree of accuracy must be made.

To address the question of accuracy of the technique introduced in Section 4.1 studies have been conducted on the plate structure subjected to the external transverse non-parametric random excitation at the center of the plate. The other pertinent data are those considered in this sub-section. In this case exact solution of variances and covariances of displacements and velocities are available from Ref. [15] and Chapter 2. For brevity, only selected results for the 64RE model are presented in Figures 4.6 and 4.7. As can be observed in Figure 4.6 the variances of displacements by applying the approach presented in Section 4.1 and the exact solution of [15] coincide. That is, the plots obtained by the approach in Section 4.1 cannot be distinguished from those of the exact solution. A practically insignificant amount of discrepancy is observed in the case of the statistical moments of the velocities presented in Figure 4.7. This small discrepancy is believed to be due to the fact that the approach in Section 4.1 employed the RK4 algorithm such that the order of accuracy for velocities is one order lower than that for the displacements.

4.2.2   Square plate clamped at all sides

The square plate structure clamped at all sides is denoted as C4. The geometrical and material properties of this plate are the same as that of the S4 plate considered in the last sub-section. For this C4 plate case, the theoretical buckling stress due to uniformly distributed forces normal to the axes parallel to x-direction, say, is [1, 2]

$$\sigma_{cr} = \left(\frac{8 \times 13.29}{a^2 t} \right) \frac{E t^3}{12\left(1 - v^2\right)} = 48.681 \text{ MPa.} \qquad (4.24)$$

Similar to Eq. (4.22), for a 2% static critical buckling stress as the magnitude of the rms of the white noise part of the parametric random excitation, the white noise intensities are

$$D_{11} = D_{22} = \left(2\% \times \sigma_{cr}\right)^2 = 9.47936 \times 10^{11}. \qquad (4.25)$$

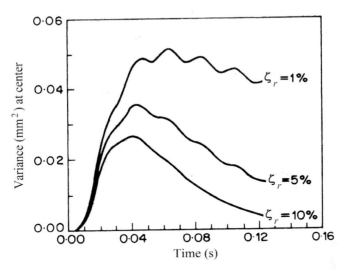

Figure 4.6   Comparison of displacement responses with closed form solution
(note that the two sets of results coincide).

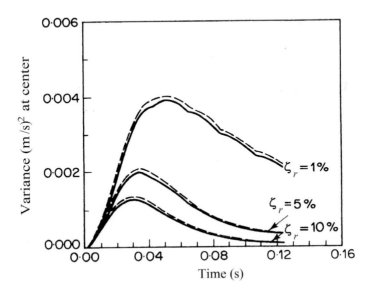

Figure 4.7   Comparison of velocity responses with closed form solution:
_____ , present study; - - -, closed form solution.

This level of parametric random excitation is denoted as p_2 while p_5 means that the parametric random excitation whose white noise intensities are

$$D_{11} = D_{22} = \left(5\% \times \sigma_{cr}\right)^2 = 5.92460 \times 10^{12}. \qquad (4.26)$$

where the white noise intensities D_{11} and D_{22} are those in Eq. (4.14).

Modal analysis was performed by using SPADAS [13] for the 16RE and 64RE models as shown in Figure 4.1. Natural frequencies and their exact solution from [14] are included in Table 4.2 for comparison. In this latter table N/A means the solution is not available. As can be observed in the table the 64RE model provides natural frequencies much closer to the "exact" solution. Response statistics for the 16RE and 64RE models are obtained by the approach in Section 4.1 with the first eight modes included in the computation. This is because the computed results by the approach in Section 4.1 including the first eight modes have excellent agreement with those closed form solutions (see, Figures 4.6 and 4.7). Representative computed results are presented in Figures 4.8 and 4.9 in which computed results by including the first three modes are also included. Note that the 16RE results in Figure 4.8 are concerned with both parametric and transverse random excitations, every one of the latter is defined by Eq. (4.19), applied at Nodes 12, 18, and 19 in Figure 4.1(a). The parametric random excitations are defined by Eqs. (4.20) and (4.25). The 64RE results in Figure 4.9 are based on the parametric excitation defined by Eqs. (4.20) and (4.26), and the transverse excitation defined by Eq. (4.19) applied at Node 41.

4.2.3   Remarks

In the foregoing two sub-sections square plates simply-supported (S4) and clamped at all four sides (C4) have been studied. Natural frequencies and mode shapes have been obtained for the discretized plates. However, for brevity, only the first eight natural frequencies were presented in Tables 4.1 and 4.2.

Response statistics have been computed for the S4 and C4 plates by applying the bi-modal approach introduced in Sub-section 4.1. For the S4 plate with external transverse nonstationary random excitation applied at the center, closed form solutions were also obtained and included for direct comparison in Figures 4.6 and 4.7. The closed form solutions are obtained by the approach in Section 2.4. The agreement as indicated in the latter figures is excellent. This implies that the bi-modal approach is accurate and much simpler than the closed form solution method.

Furthermore, the bi-modal approach introduced above is applicable to systems with and without parametric random excitations.

Table 4.2   Comparison of natural frequencies of C4 plate for FE models.

Mode number	Exact	Frequencies (Hz)	
		16RE	64RE
1	43.8021	41.7565	43.1494
2	89.3446	85.2363	87.6751
3	89.3446	85.2363	87.6751
4	131.8079	119.3510	126.2210
5	160.1654	155.2700	157.5010
6	160.8957	157.7550	158.5600
7	N/A	183.7850	191.0190
8	N/A	183.7850	191.0190

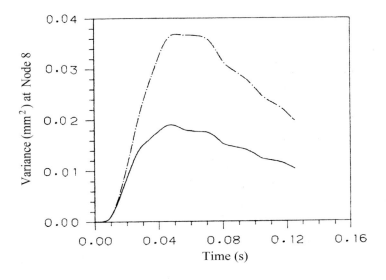

Figure 4.8(a)   Displacement responses at Node 8 of C4 with 1% damping:
_____ , no parametric excitation; − · −· −, parametric excitation p_2.

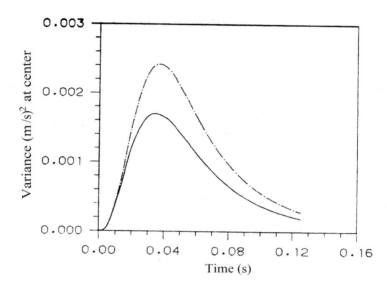

Figure 4.8(b) Velocity responses at Node 13 of C4 with 5% damping: ——— , no parametric excitation; – · – · – , parametric excitation p_2.

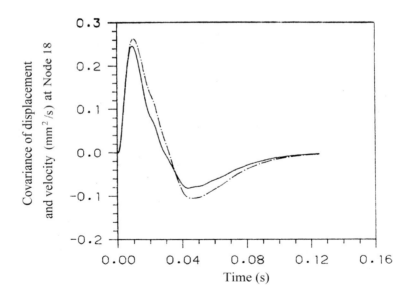

Figure 4.8(c) Coupled responses at Node 18 of C4 with 10% damping: ——— , no parametric excitation; – · – · – , parametric excitation p_2.

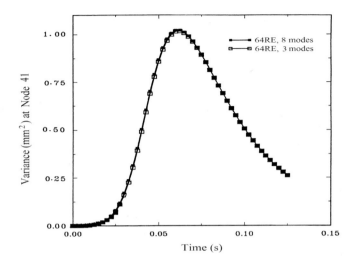

Figure 4.9(a)   Displacements at Node 41 (center) of C4 plate with 5% damping and parametric excitation p_5.

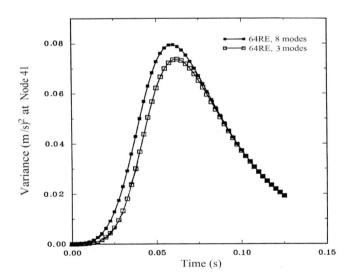

Figure 4.9(b)   Velocities at Node 41 (center) of C4 plate with 5% damping and parametric excitation p_5.

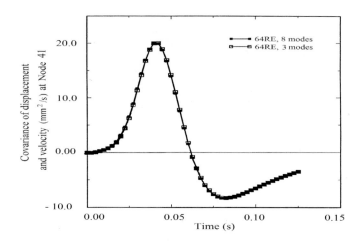

Figure 4.9(c) Coupled responses at Node 41 (center) of C4 plate with 5% damping and parametric excitation p_5.

4.3 Response Analysis Based on High Precision Triangular Plate Finite Element

The high precision triangular plate bending element was developed by the author and his associate and part of the studies was reported in Ref. [7]. This is a displacement formulation-based element. Detailed studies together with response statistics can be found in Wang [8]. It has three nodes every one of which has six dof. These six dof are w_i^e, $\partial w_i^e/\partial \xi$, $\partial w_i^e/\partial \eta$, $\partial^2 w_i^e/\partial \xi^2$, $\partial^2 w_i^e/\partial \xi \partial \eta$, and $\partial^2 w_i^e/\partial \eta^2$, where w_i^e is the transversal displacement at node i on the middle plane of the plate while ξ and η are two perpendicular axes on the middle plane of the triangular element. The displacement field $w^e(\xi, \eta)$ or simply written as w^e within the element is defined by the quintic polynomial as

$$
\begin{aligned}
w^e = a_1 &+ a_2\xi + a_3\eta + a_4\xi^2 + a_5\xi\eta + a_6\eta^2 \\
&+ a_7\xi^3 + a_8\xi^2\eta + a_9\xi\eta^2 + a_{10}\eta^3 + a_{11}\xi^4 \\
&+ a_{12}\xi^3\eta + a_{13}\xi^2\eta^2 + a_{14}\xi\eta^3 + a_{15}\eta^4 + a_{16}\xi^5 \\
&+ a_{17}\xi^3\eta^2 + a_{18}\xi^2\eta^3 + a_{19}\xi\eta^4 + a_{20}\eta^5 .
\end{aligned}
\tag{4.27}
$$

In Eq. (4.27) the term $\xi^4\eta$ is omitted. Thus, the expression satisfies the condition that normal slope be a cubic function in ξ along the edge $\eta = 0$. In order to have cubic variation of the normal slope along the two remaining edges, two additional conditions are required. These conditions and the outline of the derivation as well as explicit expressions for the element matrices were included in [8]. This high precision triangular plate element has been identified as TBH6.

The computed results in this section are essentially selected from [8]. In the latter subroutines for the explicit or closed form expressions for the consistent element stiffness matrix k, stability or initial stress stiffness matrix k_G, and mass matrix m, respectively, were obtained by the application of the symbolic package MACSYMA. These explicit or closed form expressions lead to much less algebraic operations than those of Stavitsky et al. [16], and Jeyachandrabose and Kirkhope [17]. Thus, the computational time is drastically reduced and loss of accuracy due to numerical inversion or transformation and numerical integration of the element matrices is removed. Aside from increasing accuracy this finite element can be applied to deal with cases having boundary conditions that include curvatures. This is different from the MZC bending element which only satisfies boundary conditions involving displacements and slopes.

In the following two sub-sections responses of simply-supported plate and plate clamped at all four sides under external or non-parametric and in-plane or parametric nonstationary random excitations are considered since they are valuable for the modification of design codes for plates in many applications in the aerospace, shipbuilding, and other industries. Further, these two plate structures have been studied in the last section by using the MZC bending element and therefore, results are available for comparison.

4.3.1 Simply-supported plate structures

For comparison the geometrical and material properties of the simply-supported plate structure considered in this sub-section are those presented in Sub-section 4.2.1. The FE models employed in this investigation are 8TE, 32TE, 128TE, 16RE, and 64RE. The 8TE and 32TE meshes have been presented in Figure 3.7 while the 16RE and 64RE meshes are shown in Figure 4.1. Thus, only the 128TE mesh is given in Figure 4.10. Note that as in Section 4.2, 16RE means the square plate being represented by 16 MZC rectangular bending finite element while 8TE means the square plate being idealized by 8 TBH6 high precision triangular plate bending elements.

Before the response statistics are computed the eigenvalue solution of the S4 square plate represented by the TBH6 element is performed applying the FE package,

SPADAS. The natural frequencies for the first eight modes are included in Table 4.3 in which results for the 64RE are also presented.

As can be observed from Table 4.3, the first eight natural frequencies for the 128TE case are extremely close to their analytical counterparts listed as "exact" in the second column of Table 4.3. These exact solutions are derived from Ref. [14]. It is interesting to note that the natural frequencies obtained by using the 32TE and 64RE models are very close to the exact solution. Therefore, response statistics for the 32TE and 64RE representations of the square plate subjected to a nonstationary random excitation applied at the center node and along the axis normal to the plane of the plate are evaluated. In-plane nonstationary random excitation is also included. The external and in-plane nonstationary random excitations applied are the same as those defined by Eqs. (4.20) and (4.21). Response statistics for the foregoing FE models with simultaneous external and in-plane random excitations are computed and presented in Figures 4.11 and 4.12. In Figure 4.11 the computed results by using the bi-modal approach with the first 8 modes included but without in-plane or parametric random excitation are also presented for direct comparison. In the figures the parametric excitation p_s denotes 5% static critical buckling stress as the magnitude of the rms of the white noise part of the parametric random excitation. The details of the latter have been introduced in Sub-section 4.2.1 and will not be repeated here.

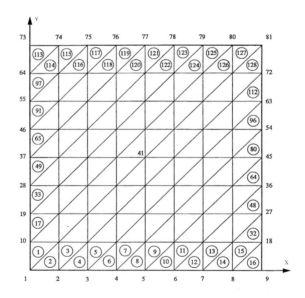

Figure 4.10   Discretized plate structure: 128TE model.

Table 4.3   Natural frequencies of S4 plate using different FE models.

		Frequencies (Hz)			
Mode number	Exact	8TE	32TE	128TE	64RE
1	24.0239	24.0326	24.0240	24.0239	23.8288
2	60.0596	60.4787	60.0704	60.0598	59.3175
3	60.0596	62.0931	60.0792	60.0600	59.3175
4	96.0954	99.3286	96.1319	96.0958	63.2262
5	120.1193	123.5820	120.3260	120.1240	118.6460
6	120.1193	136.7020	120.3590	120.1240	118.6460
7	156.1551	170.5350	156.5310	156.1610	150.1530
8	156.1551	182.7790	156.6320	156.1630	150.1530

To investigate the influence of number of modes on the responses, computed results are provided in Figures 4.13(a-c) in which the legend "8M" denotes the first eight modes included in the computation. Similarly, "3M" indicates that the first three modes are included in the computation. It is noted that the computed responses with the first eight modes included in the computation for the cases of 32TE and 64RE are much closer and consistent with the observation made in Figures 4.6 and 4.7 that inclusion of the first eight modes can lead to accurate response statistics results.

4.3.2   Square plate clamped at all sides

For the same reasons as those of the S4 plate, the TBH6 element is applied in this sub-section for the analysis of the square plate clamped at all sides. All geometrical and material properties as well as nonstationary random excitations are the same as those in Sub-section 4.2.2. The first eight natural frequencies and mode shapes by using SPADAS [13] are obtained for the 8TE, 32TE, 128TE, and 64RE models. These natural frequencies are presented in Table 4.4. Note that the analytical solutions denoted as "exact" in the latter table are derived from Ref. [14].

From Table 4.4 it is apparent that the first six natural frequencies for the 32TE, 128TE, and 64RE FE models are very close to the exact solutions. Note that "exact" solutions for the seventh and eighth modes are not available for comparison. In the present studies response statistics of the C4 plate approximated by the 32TE and 64RE meshes as well as including the first eight modes are evaluated. Selected representative plots are presented in Figure 4.14

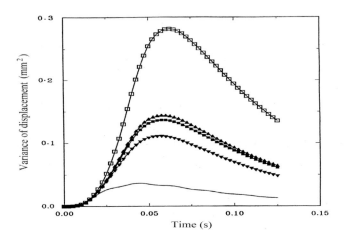

Figure 4.11(a)　Displacement responses at center of S4 with 5% damping and p_5: □---□, 64RE; □--□, 16RE; ▽−▽, 32TE; △−△, 8TE; − , 32TE without p_5.

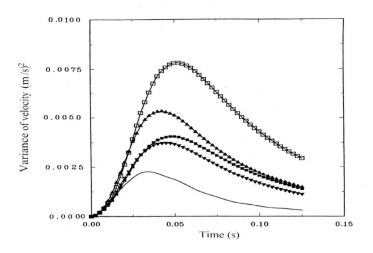

Figure 4.11(b)　Velocity responses at center of S4 with 5% damping and p_5: □---□, 64RE; □--□, 16RE; ▽−▽, 32TE; △−△, 8TE; − , 32TE without p_5.

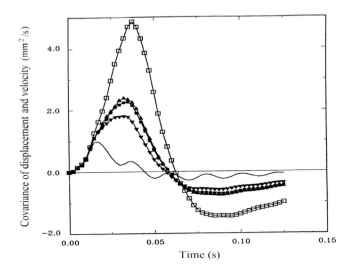

Figure 4.11(c)   Coupled responses at center of S4 with 5% damping and p_5: □---□, 64RE; □--□, 16RE; ▽–▽, 32TE; △–△, 8TE; – , 32TE without p_5.

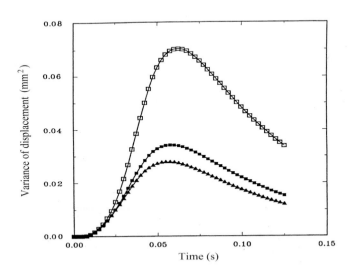

Figure 4.12(a)   Displacement responses of S4 with 5% damping and p_5: □--□, 16RE at Node 7; □---□, 64RE at Node 21; △–△, 32TE at Node 7.

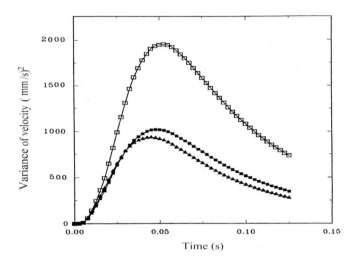

Figure 4.12(b)   Velocity responses of S4 with 5% damping and p_5: □--□ , 16RE at Node 7; □---□, 64RE at Node 21; △−△, 32TE at Node 7.

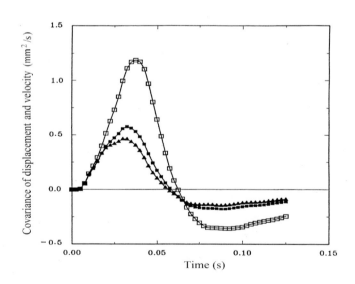

Figure 4.12(c)   Coupled responses of S4 with 5% damping and p_5: □--□ , 16RE at Node 7; □---□, 64RE at Node 21; △−△, 32TE at Node 7.

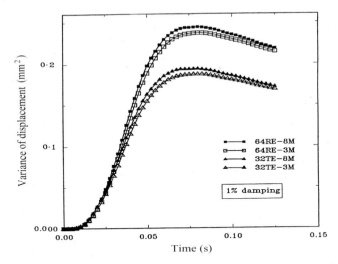

Figure 4.13(a) Displacement responses at center of S4 with 1% damping and p_5: □----□, 64RE 8M; □---□, 64RE 3M; ▲−▲, 32TE 8M; △−△, 32TE 3M.

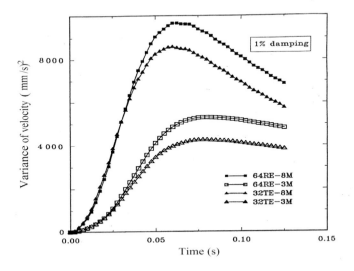

Figure 4.13(b) Velocity responses at center of S4 with 1% damping and p_5: □----□, 64RE 8M; □---□, 64RE 3M; ▲−▲, 32TE 8M; △−△, 32TE 3M.

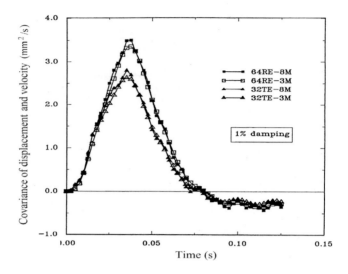

Figure 4.13(c) Coupled responses at center of S4 with 1% damping and p_5: □---□, 64RE 8M; □---□, 64RE 3M; ▲–▲, 32TE 8M; △–△, 32TE 3M.

Table 4.4 Natural frequencies of C4 plate using different FE models.

		Frequencies (Hz)			
Mode number	Exact	8TE	32TE	128TE	64RE
1	43.8021	45.4439	43.8207	43.7970	43.1494
2	89.3446	91.0093	89.4508	89.3277	87.6751
3	89.3446	103.3830	89.5927	89.3297	87.6751
4	131.8079	167.6250	132.7140	131.7230	126.2210
5	160.1654	171.5460	160.9540	160.1600	157.5010
6	160.8957	191.2860	162.1260	160.9230	158.5600
7	N/A	283.1840	203.3680	200.8690	191.0190
8	N/A	414.8180	205.3880	200.8930	191.0190

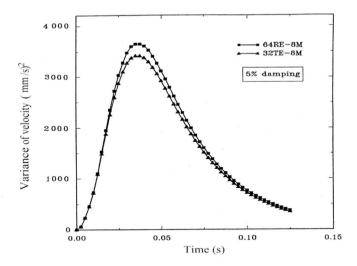

Figure 4.14(a)   Displacement responses at center of C4 with 5% damping and p_2: □----□, 64RE 8M; ▲—▲, 32TE 8M.

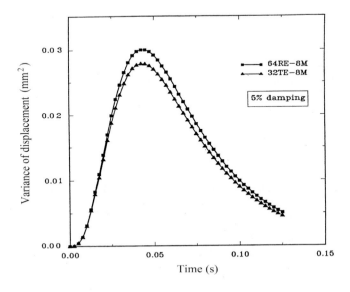

Figure 4.14(b)   Velocity responses at center of C4 with 5% damping and p_2: □----□, 64RE 8M; ▲—▲, 32TE 8M.

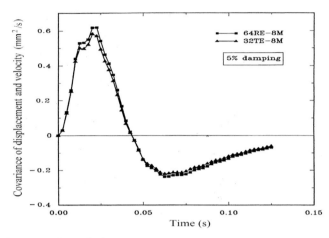

Figure 4.14(c)   Coupled responses at center of C4 with 5% damping and p_2: □---□, 64RE 8M; ▲—▲, 32TE 8M.

4.4 Concluding Remarks

In this chapter the so-called bi-modal analysis based on Cumming's approach has been introduced. Applications have been made to plate structures under both external and parametric nonstationary random excitations. These plate structures have been approximated by the MZC rectangular element and the high precision triangular element TBH6. Comparison to closed form solutions for the cases with external nonstationary random excitations has been made and it is observed that the agreement between the closed form solutions and those evaluated by the bi-modal approach is excellent.

References

[1] To, C.W.S. and Orisamolu, I.R. (1985). Response of mdf systems to parametric and non-parametric nonstationary random excitations, Society of Engineering Science, *Engineering Science Preprint* No. ESP22/85053.

[2] To, C.W.S. and Orisamolu, I.R. (1987). Response of discretized plates to transversal and in-plane non-stationary random excitations, *Journal of Sound and Vibration*, **114(3)**, 481-494.

[3] Weaver, Jr., W. and Johnson, P.R. (1984). *Finite Elements for Structural Analysis*, Prentice-Hall, Englewood Cliffs, New Jersey.

[4] To, C.W.S. (1986). First passage time of discretized plates with geometrical nonlinearity, *Computers and Structures*, **24(6)**, 893-900.

[5] Di Paola, M. and Muscolino, G. (1990). Differential moment equations of FE modelled structures with geometrical non-linearities, *International Journal of Non-Linear Mechanics*, **25(4)**, 363-373.

[6] To, C.W.S. (1986). The stochastic central difference method in structural dynamics, *Computers and Structures*, **23(6)**, 813-818.

[7] To, C.W.S. and Wang, B. (1993). Response analysis of discretized plates under in-plane and external nonstationary random excitations, in *Proceedings of the 14th ASME Biennial Conference on Mechanical Vibration and Noise*, DE-Vol. 56, pp. 51-62, Sept. 19-22, Albuquerque, New Mexico.

[8] Wang, B. (1991). Non-stationary random response of plates with geometric nonlinearity, M.E.S. Thesis, University of Western Ontario, Canada.

[9] Lin, Y.K. and Shih, T.Y. (1982). Vertical seismic load effect on building response, *J. of Engineering Mechanics Division, A.S.C.E.*, **107**, 331-343.

[10] Cumming, I.G. (1967). Derivation of the moments of a continuous stochastic system, *International Journal of Control*, **5**, 85-90.

[11] Melosh, R.J. (1963). Basis of derivation of matrices for the direct stiffness method, *A.I.A.A. Journal*, **1(7)**, 1631-1637.

[12] Zienkiewicz, O.C. and Cheung, Y.K. (1964). The finite element method for analysis of elastic isotropic and orthotropic slabs, *Proceedings of the Institution of Civil Engineers*, **28**, 471-488.

[13] Petyt, M. and Abdel-Rahman, A.H. (1974). *SPADAS I: User's Manual*, Institute of Sound and Vibration Research, University of Southampton.

[14] Leissa, A.W. (1973). The free vibration of rectangular plates, *Journal of Sound and Vibration*, **31**, 257-293.

[15] To, C.W.S. (1986). Response statistics of discretized structures to non-stationary random excitation, *Journal of Sound and Vibration*, **105(2)**, 217-231.

[16] Stavitsky, D., Macagno, E. and Christensen, J. (1981). On the eighteen degrees of freedom triangular element, *Computer Methods in Applied Mechanics and Engineering*, **26**, 265-283.

[17] Jeyachandrabose, C. and Kirkhope, J. (1984). Explicit formulation for the high precision triangular plate-bending element, *Computers and Structures*, **19**, 511-519.

5

Direct Integration Methods for Response Statistics of Quasi-linear Structural Systems

The modal and bi-modal analyses of structural systems under external and in-plane random excitations have been studied in Chapter 4. The two basic assumptions of such approaches are that the dampings in these systems are light and that the modes are well separated. However, in many structural dynamic systems these two basic assumptions cannot be satisfied. For example, in stiffened aerospace and naval structures the modes are not well separated. Further, in flow-induced vibration problems or piping systems containing a moving medium the damping is not light and not of the proportional type.

Thus, in Section 5.1 the SCD method for quasi-linear structural systems is introduced while Section 5.2 deals with applications of the SCD method to piping systems containing turbulent fluids. Response analysis of quasi-linear systems under narrow-band random excitations is included in Section 5.3. Concluding remarks are presented in Section 5.4.

5.1 Stochastic Central Difference Method for Quasi-linear Structural Systems

In this section the recursive covariance matrix of displacements for systems under external and parametric or in-plane random excitations is derived and applied to a simple system. The latter is applied to demonstrate the efficiency and accuracy of the derived expressions.

5.1.1 Derivation of covariance matrix of displacements

For the linear structural system under external and parametric or in-plane random excitations, the spatially discretized matrix equation of motion given by Eq. (4.1) is included in the following for direct use

Stochastic Structural Dynamics: Application of Finite Element Methods, First Edition. C.W.S. To.
© 2014 John Wiley & Sons, Ltd. Published 2014 by John Wiley & Sons, Ltd.

$$M\ddot{x} + C\dot{x} + \left[K - p(t)K_G\right]x = F \tag{5.1}$$

where the symbols have been defined in Sub-section 4.1.1 and are not repeated in this sub-section. The time discretized form corresponding to Eq. (5.1) is

$$M\ddot{x}_s + C\dot{x}_s + \left[K - p_s K_G\right]x_s = F_s \tag{5.2}$$

where the subscript s denotes the time step as defined in Section 3.1.

A displacement relation similar to Eq. (3.3) can be obtained as

$$x_{s+1} = (\Delta t)^2 N_1\left(F_s + p_s K_G x_s\right) + N_2 x_s + N_3 x_{s-1} \tag{5.3}$$

where the coefficient matrices are defined in Section 3.1.

Before proceeding further, it is convenient to write

$$\beta_s = F_s + p_s K_G x_s \ . \tag{5.4}$$

By making use of Eqs. (5.3) and (5.4), and following similar steps in the derivation of Eq. (3.6), one can show that the recursive expression for the covariance matrix of displacements becomes

$$\begin{aligned}
R_{s+1} &= N_2 R_s N_2^T + N_3 R_{s-1} N_3^T + (\Delta t)^4 N_1 B_s N_1^T \\
&\quad + N_2 D_s N_3^T + N_3 D_s^T N_2^T + (\Delta t)^2 N_2 \left\langle x_s \beta_s^T \right\rangle N_1^T \\
&\quad + (\Delta t)^2 N_1 \left\langle \beta_s x_s^T \right\rangle N_2^T + (\Delta t)^2 N_3 \left\langle x_{s-1} \beta_s^T \right\rangle N_1^T \\
&\quad + (\Delta t)^2 N_1 \left\langle \beta_s x_{s-1}^T \right\rangle N_3^T
\end{aligned} \tag{5.5}$$

in which

$$R_s = \left\langle x_s x_s^T \right\rangle, \quad B_s = \left\langle \beta_s \beta_s^T \right\rangle,$$

$$D_s = \left\langle x_s x_{s-1}^T \right\rangle \tag{5.6a,b,c}$$

$$= N_2 R_{s-1} + N_3 D_{s-1}^T + (\Delta t)^2 N_1 \left\langle \beta_{s-1} x_{s-1}^T \right\rangle.$$

Equation (5.5) is similar in form to the corresponding covariance matrix relation of the SCD method in Section 3.1. The only difference is in the definition of random excitation vector β_s. Computationally speaking, Eq. (5.5) provides the response statistics which are much simpler than those derived in Chapter 4.

For excitations modeled as DGWN or discrete shot noise or discrete Wiener process or their modulated forms, the terms associated with $\langle \beta_s x_s^T \rangle$ and $\langle \beta_s x_{s-1}^T \rangle$ become zero as

$$\langle \beta_s x_s^T \rangle = 0 \quad , \quad \langle \beta_s x_{s-1}^T \rangle = 0 \ . \qquad (5.7a,b)$$

Applying Eqs. (5.7) to (3.5), and assuming x_s, F_s, and p_s are jointly Gaussian, one can show that

$$B_s = B_s^{(1)} + B_s^{(2)} \left(K_G R_s K_G^T \right) , \qquad (5.8)$$

and

$$R_{s+1} = N_2 R_s N_2^T + N_3 R_{s-1} N_3^T + (\Delta t)^4 N_1 B_s N_1^T$$
$$+ N_2 D_s N_3^T + N_3 D_s^T N_2^T \qquad (5.9)$$

where

$$D_s = N_2 R_{s-1} + N_3 D_{s-1}^T , \quad B_s^{(1)} = \langle F_s F_s^T \rangle, \quad B_s^{(2)} = \langle p_s p_s^T \rangle .$$

Equation (5.9) is the recursive covariance matrix expression for displacement responses of mdof systems under external and parametric nonstationary random excitations. When the temporally discretized deterministic modulating functions are of unity, then the excitations are stationary random.

Before leaving this sub-section it should be pointed out that when the highest natural frequency of the system is large, the TCT introduced in Section 3.2 should be applied together with Eq. (5.9).

5.1.2   Column under external and parametric random excitations

The recursive expressions for the covariance matrices of displacements and excitations derived in the foregoing are applied to a step column. This is a simple illustration of the use of the derived expressions for a step column of circular cross-section fixed at one end and free at the other end [1]. It should be emphasized that this simple structure is employed in this sub-section for the purpose of illustration and therefore if accurate responses are required, more refined FE models should naturally be applied.

Meanwhile, for simplicity and tractability the step column is represented by two two-node bar finite elements. Each node has one nodal dof. A nonstationary random excitation is applied axially. The distributed longitudinal random loading on the column was approximated as nodal nonstationary random excitations. For simplicity, the lumped mass element matrix is applied.

The material properties for the step column are: Young's modulus of elasticity for the two bar elements is 106 GPa while density of the first bar element which is fixed at one end is 7686 kg/m^3 and the density of the other element is 8540 kg/m^3. The geometrical properties are: each of the two bar elements has a length of 4982.4 m and the radius of the element with the fixed end is 56.42 cm whereas the radius of the other element is 17.84 cm. After assembling, applying the fixed boundary condition, and dividing all the terms on both sides of the matrix equation of motion by $\rho A \ell$ with ρ being the density of the second bar element which has a free end, A being the cross-section of the first bar element, and ℓ being the element length, one can show that the assembled mass, stiffness, damping, and stability or initial stress stiffness matrices are, respectively

$$M = \begin{bmatrix} 1.00 & 0.00 \\ 0.00 & 0.10 \end{bmatrix}, \quad K = \begin{bmatrix} 1.10 & -0.10 \\ -0.10 & 0.10 \end{bmatrix}, \qquad (5.10\text{a,b})$$

$$C = \begin{bmatrix} 0.020738 & -0.000988 \\ -0.000988 & 0.001975 \end{bmatrix}, \quad K_G = \begin{bmatrix} 1.10 & -0.10 \\ -0.10 & 0.10 \end{bmatrix}. \qquad (5.10\text{c,d})$$

The external random excitation vector is defined as

$$F = \begin{bmatrix} f_1 & f_2 \end{bmatrix}^T, \quad f_1 = \left(e^{-0.10t} - e^{-1.50t} \right) w(t), \quad f_2 = 0.10 f_1 .$$

The variance of the zero mean DGWN at t_s is

$$\left\langle w_s^2(0) \right\rangle = 2\pi S_o \delta_K(0) = 1.0 ,$$

where $\delta_K(.)$ is Kronecker delta such that $\delta_K(0) = 1$. The two natural frequencies of the step column are $\omega_1 = 0.854$ rad/s and $\omega_2 = 1.70$ rad/s, and the parametric random excitation is given as $p(t) = 0.5 f_1$. The damping matrix in Eq. (5.10c) corresponds to 1% damping for each mode.

The computed variances and covariances of the displacement responses by applying Eq. (5.5) are presented in Figure 5.1 and the plots are denoted as SCD.

Solutions by the Itô's calculus approach introduced in Chapter 4 and referred to in the figure as "exact" solutions are also included for direct comparison. It should be pointed out that in the figure x_1 is the displacement response at the common node joining the two bar elements while x_2 is the displacement response at the free end of the step column. The time step size employed is Δt = 0.83 which is chosen in accordance with Eq. (3.14).

Four main observations should be made. First, the results by the SCD method and "exact" solutions are in excellent agreement.

Second, the ratio of the computational time required by the "exact" solutions to that by the SCD method is approximated 35. The substantially longer time required for the "exact" solutions has to do with the fact that the moment differential equations in Chapter 4 were numerically integrated by the RK4 algorithm. Clearly, even for the above simple system the computational time saved by the SCD method is very significant. Therefore, it is expected that for large-scale FE computation the saving in computational time by appplying the SCD method would even be more impressive.

Third, the largest peak of x_2 is considerably much larger than that of x_1. This is logical in the sense that x_2 is the response at the free end of the step column. The rms value corresponding to the largest peak in Figure 5.1 seems to be large. However, the length of the column is 2ℓ = 9964.8 m and, therefore, the rms value of the largest peak is in fact relatively very small.

Fourth, the SCD method for response statistics presented in this section is considerably much simpler compared with that employing Cumming's approach in which direct numerical integration is required.

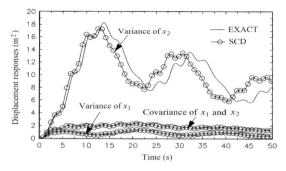

Figure 5.1 Response statistics of step column under external and parametric random excitations.

5.2 Recursive Covariance Matrix of Displacements of Cantilever Pipe Containing Turbulent Fluid

The flow-induced vibration in piping systems that contain a turbulent fluid which is represented as a temporally random process is another case of a quasi-linear structural system. The basic assumption is that the amplitude of oscillation of the pipe is small.

In Sub-section 5.2.1 the recursive covariance matrix of displacements is derived while computed results for a cantilever pipe containing a turbulent fluid and subjected to a nonstationary random excitation applied at the clamped end are presented in Sub-section 5.2.2. The last sub-section, Sub-section 5.2.3 includes concluding remarks.

5.2.1   Recursive covariance matrix of displacements

The matrix equation of motion of a pipe containing a turbulent flow that is represented as a temporally stochastic process is given by

$$M\ddot{x} + \left(C + VC_v\right)\dot{x} + \left(K - V^2 K_v\right)x = F, \qquad (5.11)$$

where C and K are the assembled damping and stiffness matrices of the system. The assembled damping matrix, C of the system is general and it is not limited to the proportional or Rayleigh type. The assembled stiffness matrix, K is constructed from the consistent stiffness matrix in Eq. (2.36). The assembled mass matrix is constructed from the consistent element mass matrix in Eq. (2.35) in which the coefficient $\rho A \ell$ is now replaced with $m_p + m_f$, where m_p and m_f are, respectively the mass per unit length of the pipe without the fluid and mass per unit length of the fluid inside the pipe. The assembled flow-induced damping coefficient matrix C_v is built from the corresponding element flow-induced damping coefficient matrix [2, 3],

$$[c_v] = \frac{m_f}{30} \begin{bmatrix} -30 & 0 & 0 & 30 & 0 & 0 \\ 0 & 0 & 6\ell & 0 & 30 & -6\ell \\ 0 & -6\ell & 0 & 0 & 6\ell & -\ell^2 \\ -30 & 0 & 0 & 30 & 0 & 0 \\ 0 & -30 & -6\ell & 0 & 0 & 6\ell \\ 0 & 6\ell & \ell^2 & 0 & -6\ell & 0 \end{bmatrix}. \qquad (5.12)$$

If the nodal axial dof is disregarded, Eq. (5.12) reduces to the skew symmetric or gyroscopic matrix.

The assembled flow-induced stiffness coefficient matrix K_v is constructed from the corresponding element flow-induced stiffness coefficient matrix [2, 3]

$$[k_v] = \frac{m_f}{30\ell} \begin{bmatrix} 30 & 0 & 0 & -30 & 0 & 0 \\ 0 & 36 & 3\ell & 0 & -36 & 3\ell \\ 0 & 3\ell & 4\ell^2 & 0 & -3\ell & -\ell^2 \\ -30 & 0 & 0 & 30 & 0 & 0 \\ 0 & -36 & -3\ell & 0 & 36 & -3\ell \\ 0 & 3\ell & -\ell^2 & 0 & -3\ell & 4\ell^2 \end{bmatrix}. \tag{5.13}$$

This is a symmetric matrix.

The assembled nodal displacement vector is defined as

$$x = \begin{bmatrix} u_1 & v_1 & \theta_1 & u_2 & v_2 & \theta_2 & u_3 & v_3 & \theta_3 & \cdots \end{bmatrix}^T, \tag{5.14}$$

where u_i and v_i are the nodal displacement along the axial axis and transversal displacement, respectively while θ_i is the angular displacement or rotation about an axis perpendicular to the plane of the 2D beam element.

The turbulent fluid flow velocity is represented as

$$V = V_m + \tilde{V}, \tag{5.15}$$

where V_m is the mean fluid velocity which is constant whereas the second term on the rhs of Eq. (5.15) is the zero-mean Gaussian white noise process.

The temporally discretized Eq. (5.11) becomes

$$M\ddot{x}_s + \left(C + V_s C_v\right)\dot{x}_s + \left(K - V_s^2 K_v\right)x_s = F_s. \tag{5.16}$$

At this stage it should be pointed out that since the second term on the lhs of Eq. (5.16) has a time-dependent coefficient, the Wong-Zakai correction term is required [4, 5]. This will naturally complicate the solution of Eq. (5.16). To circumvent this complication one can move the time-dependent coefficient terms to the rhs of the equation such that they are treated as pseudo-forces applied to the system. Thus, Eq. (5.16) can be re-written as

$$M\ddot{x}_s + C\dot{x}_s + K x_s = F_s - V_s C_v \dot{x}_s + V_s^2 K_v x_s = f_s^r + f_s^{(1)} + f_s^{(2)}, \tag{5.17}$$

where the forcing terms on the rhs of Eq. (5.17) are

$$f_s^r = F_s \; , \quad f_s^{(1)} = -V_s C_v \dot{x}_s \; , \quad f_s^{(2)} = V_s^2 K_v x_s \; . \qquad (5.18a,b,c)$$

Then following the steps for the derivation of recursive covariance matrix as in Sub-section 5.1.1, one has Eq. (5.5) in which the temporally discretized loading vector is now defined by

$$\beta_s = f_s^r + f_s^{(1)} + f_s^{(2)} \; , \qquad (5.19)$$

and the recursive covariance matrix of random excitations, after some lengthy algebraic manipulation, becomes

$$B_s = B_s^{(r)} + B_s^{(v)} \; , \qquad (5.20)$$

where the terms on the rhs of Eq. (5.20) are defined by

$$B_s^{(r)} = \left\langle F_s F_s^{\,T} \right\rangle , \quad B_s^{(v)} = \left\langle \sum_{j=1}^{2} \sum_{i=1}^{2} f_s^{(i)} f_s^{(j)} \right\rangle ,$$

in which

$$\left\langle f_s^{(1)} \left(f_s^{(1)} \right)^T \right\rangle = \frac{\left\langle V_s^2 \right\rangle}{(\Delta t)^2} C_v \left(R_s - D_s^{\,T} - D_s + R_{s-1} \right) C_v^{\,T} \; ,$$

$$\left\langle f_s^{(2)} \left(f_s^{(2)} \right)^T \right\rangle = \left\langle V_s^4 \right\rangle K_v R_s K_v^{\,T} \; ,$$

$$\left\langle f_s^{(1)} \left(f_s^{(2)} \right)^T \right\rangle = \left\langle f_s^{(2)} \left(f_s^{(1)} \right)^T \right\rangle = -\frac{\left\langle V_s^3 \right\rangle}{\Delta t} C_v \left(R_s - D_s^{\,T} \right) K_v^{\,T} \; .$$

In the present investigation the turbulent flow inside the piping system is modeled as a random process and defined in Eq. (5.15). Thus, one can show that

$$\left\langle \tilde{V}_s \right\rangle = 0 \; , \quad \left\langle V_s \right\rangle = V_m \; , \quad \left\langle V_s^2 \right\rangle = V_m^2 + \left\langle \tilde{V}_s^2 \right\rangle ,$$

$$\left\langle V_s^3 \right\rangle = V_m^3 + 3 V_m \left\langle \tilde{V}_s^2 \right\rangle , \quad \left\langle V_s^4 \right\rangle = V_m^4 + \left\langle \tilde{V}_s^4 \right\rangle + 6 V_m^2 \left\langle \tilde{V}_s^2 \right\rangle ,$$

$$\left\langle \tilde{V}_s^4 \right\rangle = 3 \left(\left\langle \tilde{V}_s^2 \right\rangle \right)^2 , \quad \left\langle \tilde{V}_s^2 \right\rangle = 2 \pi S_v \delta_K(0),$$

and S_v is the spectral density of the random component of the turbulent fluid.

Before application of Eq. (5.5) can be made, the remaining correlation terms of random forces and displacements as well as the correlation terms of displacements and delayed displacements on the rhs of Eq. (5.5) have to be derived. In what follows these correlation terms are determined.

First, the correlation between displacements and delayed displacements in Eq. (5.6b) can be written as

$$D_s = \left\langle x_s x_{s-1}^T \right\rangle = N_2 R_{s-1} + N_3 D_{s-1}^T + (\Delta t)^2 N_1 \left\langle \beta_{s-1} x_{s-1}^T \right\rangle . \qquad (5.21)$$

The third term on the rhs of this equation can be expanded to

$$(\Delta t)^2 N_1 \left\langle \beta_{s-1} x_{s-1}^T \right\rangle =$$
$$(\Delta t)^2 N_1 \left\langle V_s^2 \right\rangle K_v R_{s-1} - (\Delta t) N_1 \left\langle V_s \right\rangle C_v \left(R_{s-1} - D_{s-1}^T \right) . \qquad (5.22)$$

Second, the remaining four correlation terms on the rhs of Eq. (5.5) are

$$(\Delta t)^2 N_2 \left\langle x_s \beta_s^T \right\rangle N_1^T =$$
$$(\Delta t)^2 \left\langle V_s^2 \right\rangle N_2 R_s K_v^T N_1^T - (\Delta t) \left\langle V_s \right\rangle N_2 \left(R_s - D_s \right) C_v^T N_1^T , \qquad (5.23)$$

$$(\Delta t)^2 N_1 \left\langle \beta_s x_s^T \right\rangle N_2^T =$$
$$(\Delta t)^2 \left\langle V_s^2 \right\rangle N_1 K_v R_s N_2^T - (\Delta t) \left\langle V_s \right\rangle N_1 C_v \left(R_s - D_s^T \right) N_2^T , \qquad (5.24)$$

$$(\Delta t)^2 N_3 \left\langle x_{s-1} \beta_s^T \right\rangle N_1^T =$$
$$(\Delta t)^2 \left\langle V_s^2 \right\rangle N_3 D_s^T K_v^T N_1^T - (\Delta t) \left\langle V_s \right\rangle N_3 \left(D_s^T - R_{s-1} \right) C_v^T N_1^T , \qquad (5.25)$$

$$(\Delta t)^2 N_1 \left\langle \beta_s x_{s-1}^T \right\rangle N_3^T =$$
$$(\Delta t)^2 \left\langle V_s^2 \right\rangle N_1 K_v D_s N_3^T - (\Delta t) \left\langle V_s \right\rangle N_1 C_v \left(D_s - R_{s-1} \right) N_3^T . \qquad (5.26)$$

With all the terms explicitly obtained now, by making use of Eqs. (5.5) and (5.20) through (5.26), the recursive covariance or mean square matrix of displacements for the piping system containing a turbulent flow and external random excitations can be evaluated.

5.2.2 Cantilever pipe containing turbulent fluid

The cantilever pipe containing a turbulent fluid is idealized by eleven beam finite elements and is shown schematically in Figure 5.2. In this figure the circled integer denotes the element number while the integer without the circle designates the node number. Thus, Node 12 is the free end of the cantilever. Every element has two nodes, each of which has three dof and therefore the assembled nodal displacement vector is defined in Eq. (5.14). The random excitation applied at the clamped end of the pipe is that defined in Eq. (3.15) and illustrated in Figure 5.2. In other words, Eq. (5.11) has to be partitioned as in Eq. (2.29), and due modification to the forcing terms have to be similarly made such that the deterministic modulating function $a_r(t)$ in Eq. (3.15) is defined in Table 5.1 where the subscript $r = 2$ since the applied random excitation is in the v_1 direction at Node 1. Note that the first two elements are of 0.14 m long each whereas the remaining 9 elements are of 0.28 m each. Material and other geometrical properties are included in Table 5.1. In the present investigation the stationary spectral density of the excitation is $S_o = 1.0$ unit.

For simplicity, it is assumed that system damping is proportional such that Eq. (2.19) can be employed. The Rayleigh damping coefficients, λ_m and λ_k based on the damping ratios and first two natural frequencies of the 11-element model are obtained and included in Table 5.1. The TCT described in Section 3.2 is applied in the response statistics computation. Therefore, the highest natural frequency of the system is required. It is provided in the latter table.

The spectral density of the turbulent fluid considered in this sub-section is assumed to be $S_v = 1.0$ unit and the mean flow velocity $V_m = 38$ m/s. Thus, for the inner diameter $D = 2r_i = 0.045$ m, water in the pipe at room temperature having density $\rho = 998.2$ kg/m^3, and viscosity of water $\mu = 1.003 \times 10^{-3}$ Pa/s, the Reynolds number N_R is determined as

$$N_R = \frac{\rho V_m D}{\mu} = 1.702 \times 10^6 , \qquad (5.27)$$

indicating that the flow is highly turbulent since for a uniform pipe with smooth internal surface $N_R = 2300$ is considered critical.

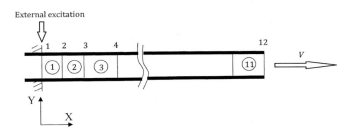

Figure 5.2   Cantilever pipe approximated by eleven beam elements.

Representative computed response statistics by using Eqs. (5.5) and (5.20) through (5.26) as well as TCT in Section 3.2 for the above 11-element model are presented in Figure 5.3 through 5.5 where the case without flowing fluid is denoted as "without", the case with constant mean flow as "constant", and the case with stochastic flow defined by Eq. (5.15) as "stochastic". It should be noted that with the mean flow at $V_m = 38$ m/s both the "constant" and "stochastic" cases are highly turbulent.

As no results have been reported in the literature for direct comparison a more refined 15-element model is considered in the present investigation. The pertinent data provided in Table 5.1 apply in this model except that the first two elements at the clamped end of the cantilever pipe are of 0.10 m long each whereas the remaining 13 elements are of 0.20 m long each. This model has 45 effective dof. The computed first two and the highest natural frequencies for the system without flowing fluid and corresponding exact solutions based on Euler beam theory are presented in Table 5.2 for comparison to those obtained for the 11-element model.

It is observed that the first natural frequency of the 11-element model has 0.0005% discrepancy with respect to the exact solution while the second natural frequency of the same model has 0.0025% discrepancy. The first two natural frequencies are slightly higher than those of the exact solutions. This is consistent with the theory that the FE solutions approach the actual natural frequencies monotonically from above. However, the natural frequencies of the 15-element model are slightly lower than the exact solutions. This is in contradiction to the theory. As the discrepancies are exceptionally small no further investigation as to why this is the case was made.

It should be mentioned that because the natural frequencies for the 11-element representation are so close to those of the 15-element case the Rayleigh damping

Table 5.1 Properties of eleven-element model of cantilever pipe.

Pipe	Pertinent data
Number of elements	11
Effective dof	33

Modulating functions:

Nonstationary random $\quad a_2(t) = 9.4815\,(e^{-45t} - e^{-60t})$

Stationary random $\quad a_2(t) = 9.4815\,(e^{-0.00001\,t} - e^{-100000\,t})$

DGWN of base excitation: $\quad \langle w_s^2(0) \rangle = 2\pi S_o \delta_K(0)$

Material properties: Pipe and water,

$$m_p = 1.49 \text{ kg/m}, \quad E = 68.5 \text{ GPa}$$
$$m_f = 1.59 \text{ kg/m}$$

Geometrical properties of pipe:

Length: $\quad L_p = 2.8 \text{ m}$

Outer radius: $\quad r_o = 26 \text{ mm}$

Thickness: $\quad h = 3.5 \text{ mm}$

Cross-sectional area:
$$A = 5.33 \times 10^{-4} \text{ m}^2$$

Second moment of area:
$$I = 1.576 \times 10^{-7} \text{ m}^4$$

	ω_1	ω_2	Ω
Natural frequencies (rad/s)	38.1738	239.239	118,362.38
Damping ratio ζ_i	0.01	0.05	0.10
Coefficients: λ_m	0.6582	3.2965	6.5818
λ_k	7.204×10^{-5}	3.5766×10^{-5}	7.204×10^{-4}

Table 5.2   Natural frequencies (in rad/s) for response computation.

Pipe with effective dof	Exact	33	45
ω_1	38.17699	38.1738	38.1738
ω_2	239.25305	239.239	239.233
Ω	N/A	118,362.38	231,990.26

coefficients λ_m and λ_k for all the damping ratios included in Table 5.1 are the same. For brevity, only the computed results for the transverse displacements at Node 12 of the 11-element model under nonstationary random and stationary random excitations for damping ratio $\zeta_i = 0.01$ and $\zeta_i = 0.05$ are presented in Figures 5.3 and 5.4. To provide an appreciation of the influence of the stochastic flow on the responses, variances of transverse displacement at Node 12 of the same 11-element model for various values of S_v are included in Figure 5.5. Selected computed transverse displacement responses at the free-end (Node 16) of the 15-element model are presented in Figures 5.6 and 5.7. With reference to the presented results, four main observations should be made.

First, comparing Figures 5.3(b) and 5.4(b), respectively with Figures 5.6 and 5.7, the computed responses for the 11- and 15-element models are in excellent agreement. Therefore, from the computational efficiency point of view, the 11-element model should be employed.

Second, for the plots presented in Figures 5.3 through 5.7, the difference between the constant mean flow and stochastic cases are not very pronounced when the values of the spectral density of the stochastic flow S_v are small. This is because the ratio of the variance of the random component of the turbulent flow to the square of the constant mean flow, $2\pi S_v /(38 \text{ m/s})^2$ is 0.00435 and therefore the influence of S_v on the turbulent flow is relatively very small. As the value of S_v is increased, its influence on the responses is expanded. With reference to Figure 5.5, the plot corresponding to $S_v = 69$ units grows with time. This indicates that the free end of the cantilever pipe experiences instability.

Third, the mean squares of responses presented in Figures 5.3 through 5.7 are large. The reason for such large values is the fact that all the results have been divided by the spectral density of the base random excitation, S_o. Mathematically,

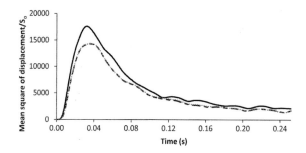

Figure 5.3(a)   Mean squares of displacements of nonstationary random base excitation with $\zeta_i = 0.01$: —, without; - - -, constant; ⋯, stochastic.

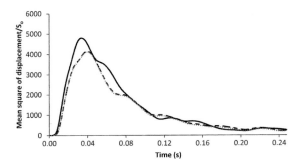

Figure 5.3(b)   Mean squares of displacements of nonstationary random base excitation with $\zeta_i = 0.05$: —, without; - - -, constant; ⋯, stochastic.

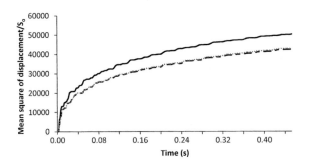

Figure 5.4(a)   Mean squares of displacements of stationary random base excitation with $\zeta_i = 0.01$: —, without; - - -, constant; ⋯, stochastic.

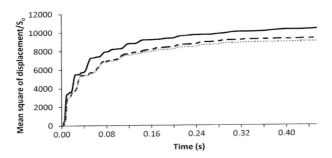

Figure 5.4(b)   Mean squares of displacements of stationary random base excitation with $\zeta_i = 0.05$: — , without; - - -, constant; ⋯, stochastic.

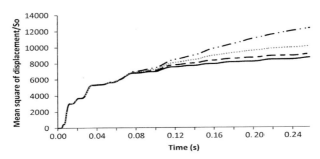

Figure 5.5   Mean squares of displacements of stationary random base excitation and stochastic flow with $\zeta_i = 0.05$: — , $S_v = 1.0$; - - -, $S_v = 23.0$; ⋯, $S_v = 46.0$; − ⋅⋅ −, $S_v = 69.0$.

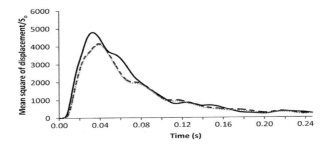

Figure 5.6   Displacements at Node 16 of 15-element model with nonstationary random base excitation and $\zeta_i = 0.05$: — , without; - - -, constant; ⋯, stochastic.

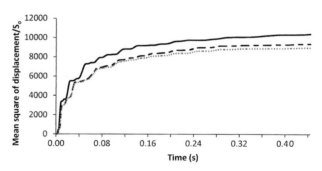

Figure 5.7 Displacements at Node 16 of 15-element model with stationary random base excitation and $\zeta_i = 0.05$: — , without; - - -, constant; ⋯, stochastic.

one observes, for example in Eq. (3.15), that the external random excitation contains elements or entries of the assembled stiffness matrix. These elements or entries are, in general, several orders of magnitudes higher than the variance of the DGWN. Thus, the value along the vertical axis is high.

Fourth, in the above investigation the stochastic model of the fluctuating component of the turbulent flow is relatively simple. More sophisticated models can be considered and investigated but are left to the interested readers.

5.3 Quasi-linear System under Narrow-band Random Excitations

Another class of quasi-linear systems is considered in this section. These systems contain a flowing medium that is approximated as a constant mean flow and under narrow-band random excitations [6]. For simplicity and illustration, a simply-supported uniform piping system containing a constant mean flow and subjected to a transverse random excitation applied at the mid-point of the pipe is considered in this section. Representative computed results along with MCS data are included for comparison. The governing matrix equations are discussed in Sub-section 5.3.1 while representative results, comparison to MCS data, and influence of bandwidth on responses are included in Sub-section 5.3.2.

5.3.1   Recursive covariance matrix of pipe with mean flow and under narrow-band random excitation

When the fluctuating component of the turbulent flow is relatively small, the flowing fluid can be dealt with as constant and hence the coefficient matrices on the lhs of Eq. (5.11) become constant. This renders the Wong-Zakai correction [4, 5] unnecessary. Thus, the flow-related coefficient matrices can be kept on the lhs of Eq. (5.11) and the damping and stiffness matrices can be treated in a similar manner as that in Section 3.4. That is, the governing matrix equations for the filter and system are similar to those defined in Eq. (3.21) where now

$$C = C + V_m C_v , \qquad K = K - V_m^2 K_v . \qquad (5.28a,b)$$

In these equations the C and K on the rhs are just the assembled damping and stiffness matrices of the system when there is no flowing fluid. Consequently, the ESCD method introduced in Section 3.4 and Eqs. (3.22), (3.23), (3.26), and (3.28) for the system under narrow-band random excitations can be applied.

The particular system considered in this sub-section is a simply-supported straight uniform pipe containing flowing water and subjected to a transverse random excitation applied at the mid-span of the pipe, as shown in Figure 5.8. For illustration purposes rather than attempting to present an accurate practical solution to the problem, the pipe is approximated by two 2-node beam finite elements. Each node of the element has 2 dof which are the transverse deflection and angular displacement or rotation. Thus, the consistent element mass and stiffness matrices defined by Eqs. (2.35) and (2.36), and the corresponding flow-induced damping and stiffness coefficient matrices defined by Eqs. (5.12) and (5.13) can be used except that the first and fourth rows, and first and fourth columns of the matrices in Eqs. (2.35), (2.36), (5.12), and (5.13) have to be deleted. After assembling all the system matrices and remembering Eq. (5.28), the recursive covariance matrix of displacement responses derived in Eq. (3.22) and its associated relations, Eqs. (3.23), (3.26), and (3.28) for the system under narrow-band random excitations can be applied.

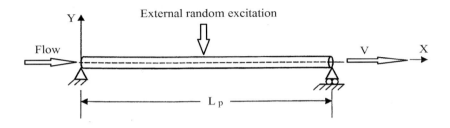

Figure 5.8 Simply-supported pipe containing flowing water.

5.3.2   Responses of pinned pipe with mean flow and under narrow-band random excitation

After assembling and applying the boundary conditions, every system matrix has an order of 4 × 4. Numerical tests are conducted to examine the efficiency and accuracy of the ESCD method in terms of the response statistics of the flow-induced vibration problem. For simplicity, it is assumed that system damping is due to the flowing fluid. In other words, the system damping matrix given by Eq. (5.28a) becomes $C = V_m C_v$. The material and geometrical properties of the simply-supported pipe are those presented in Table 5.1.

5.3.2.1   Pipe under narrow-band nonstationary random excitation

The undamped natural frequencies of the piping system are [6] $\omega_1 = 68.5$ rad/s, $\omega_2 = 325.9$ rad/s, $\omega_3 = 828.3$ rad/s, and $\omega_4 = 1514.7$ rad/s. The center frequency and damping ratio of the single dof filter are $\omega_f = 89.0$ rad/s and $\zeta_f = 0.05$ while the filter modulated Gaussian white noise excitation defined by Eq. (3.21a) is

$$F(t) = e(t)w(t) , \qquad \langle w^2(0) \rangle = 2\pi \times 10^9 \delta(0) ,$$

$$e(t) = 9.4815\left(e^{-45t} - e^{-60t}\right) . \qquad (5.29a,b,c)$$

Computed results by using Eq. (3.22) and the TCT with the dimensionless time step size, $\Delta\tau = 1.0$ and after converting them back in the t-domain they are presented in Figures 5.9 through 5.11 and denoted as SCD in the plots. Results obtained by the MCS are included in these figures for comparison.

It is observed that the SCD results are in excellent agreement with those obtained by the MCS. However, the ratio of computational time for every MCS plot to that by the SCD method is about 480. This means that the ESCD method is much more efficient than the MCS.

5.3.2.2   Effect of bandwidth of excitation process

The purpose of the study here is to examine the effect of bandwidth of the excitation process from the single dof filter on the responses of the simply-supported pipe considered in the foregoing. The filter characteristics adopted above are applied here. The modulating function $e(t)$ of the narrow-band nonstationary random excitation defined by Eq. (3.30) is the same as that given by Eq. (5.29c), while I_f in Eq. (3.30) is such that $I_f = 2\pi \times 10^4$. The above aluminum pipe whose properties are included in Table 5.1 is studied.

Selected obtained results by using Eq. (3.22) and the TCT are presented in

Figures 5.12 through 5.14. It may be appropriate to mention that a steel pipe has been investigated and results can be found in [6] but they are not included here for brevity. With reference to these figures one can observe that the system responses increase with decreasing bandwidth of the random excitation process, which are consistent with the results presented in Figures 5.9 through 5.11.

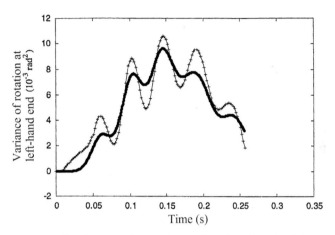

Figure 5.9   Variances of rotation at left-hand end of pipe: $\omega_f = 89.0$ rad/s, and $\zeta_f = 0.05$; •, MCS; +, SCD.

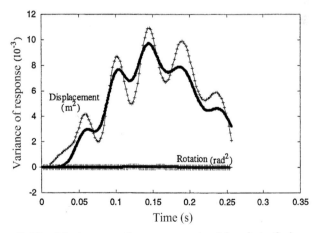

Figure 5.10   Variances of responses at mid-point of pipe: $\omega_f = 89.0$ rad/s, and $\zeta_f = 0.05$; •, MCS; +, SCD.

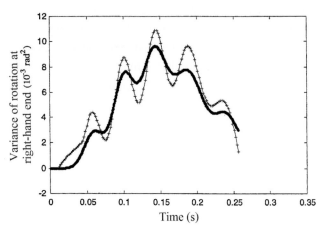

Figure 5.11   Variances of rotation at right-hand end of pipe:
$\omega_f = 89.0$ rad/s, and $\zeta_f = 0.05$; •, MCS; +, SCD.

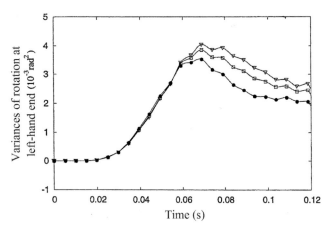

Figure 5.12   Effect of bandwidth on responses at left-hand end of pipe:
$\zeta_f = 0.025, \triangledown$; $\zeta_f = 0.05, \square$; $\zeta_f = 0.10$, •.

5.4 Concluding Remarks

The direct integration methods for response analysis of quasi-linear structural systems have been presented in this chapter. These methods have been motivated by the fact that modal analysis and its complex counterpart are not suitable for dealing with structural systems that have closely spaced modes and large dampings. Furthermore, another motivation is that many structural systems do not possess proportional type damping. The SCD method for quasi-linear systems subjected to broad-band stationary and nonstationary random excitations presented in Section 5.2 and that to narrow-band stationary and nonstationary random excitations in Section 5.3 are very efficient and accurate.

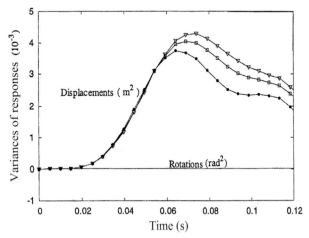

Figure 5.13   Effect of bandwidth on responses at mid-point of pipe:
$\zeta_f = 0.025, \triangledown\,; \zeta_f = 0.05, \square\,; \zeta_f = 0.10, \bullet.$

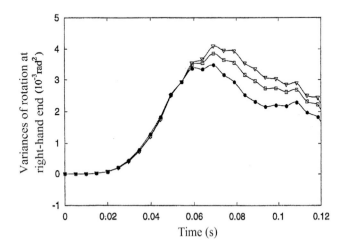

Figure 5.14   Effect of bandwidth on responses at right-hand end of pipe:
$\zeta_f = 0.025, \triangledown\,; \zeta_f = 0.05, \square\,; \zeta_f = 0.10, \bullet.$

References

[1] To, C. W. S. (2001). On computational stochastic structural dynamics applying finite elements, *Archives of Computational Methods in Engineering*, **8(1)**, 3-40.

[2] To, C. W. S. and Healy, J.W. (1986). Further comment on vibration analysis of straight and curved tubes conveying fluid by means of straight beam finite elements, *Journal of Sound and Vibration*, **105(3)**, 513-514.

[3] Dai, X. (2009). Random vibration of piping systems containing a turbulent fluid, M.S. Thesis, University of Nebraska, Lincoln, Nebraska.

[4] To, C. W. S. (1988). On dynamic systems disturbed by random parametric excitations, *Journal of Sound and Vibration*, **123(2)**, 387-390.

[5] Gray, Jr., A. H. and Caughey, T. K. (1965). A controversy in problems involving random parametric excitation, *Journal of Mathematics and Physics*, **44**, 288-296.

[6] Chen, Z. and To, C. W. S. (2005). Responses of discretized systems under narrow band nonstationary random excitations, Part 1: linear problems, *Journal of Sound and Vibration*, **287**, 433-458.

6

Direct Integration Methods for Temporally Stochastic Nonlinear Structural Systems

In this chapter the statistical linearization (SL) technique is briefly outlined in Section 6.1 for completeness. Symplectic algorithms of the Newmark family of integration schemes are identified in Section 6.2. Section 6.3 is concerned with the presentation of the SCD method with TCT and adaptive time schemes (ATS) for response analysis of highly nonlinear structural systems to stationary and nonstationary random excitations. An outline of steps in the computer program is provided in Section 6.4. Applications of plate and shell structures are made in Section 6.5. Concluding remarks are included in Section 6.6.

6.1   Statistical Linearization Techniques

The pioneered application of the SL technique for nonlinear dynamic systems was made by Malhotra and Penzien in 1970 in the response analysis of discretized offshore tower structures in deep water [1]. The particular structure was modeled as a space frame consisting of uniform beam finite elements. Lumped element mass matrix and modal analysis were employed. The nonlinearities due to drag effects were linearized through the SL technique.

Since the 1970's many improved versions and applications of the SL techniques have been proposed and applied because of their simplicity. Many books and chapters in books on random vibration have included these techniques. Recent notable additions are the books by Socha [2] and To [3], for example. While the SL techniques are popular because of their simplicity, they have various limitations. The most important limitation in the context of application of the FEM is the fact that the nonlinearities in the mdof system have to be explicitly defined. The second most important limitation is concerned with accuracy [3]. The third most important limitation is to do with the existence of a solution for mdof nonlinear systems with large deformations.

Stochastic Structural Dynamics: Application of Finite Element Methods, First Edition. C.W.S. To.
© 2014 John Wiley & Sons, Ltd. Published 2014 by John Wiley & Sons, Ltd.

As the focus in this chapter is concerned with the application of the direct integration methods for highly nonlinear systems with large deformations in the context of the FEM, but for its popularity and simplicity the SL technique of Atalik and Utku [4] will be briefly outlined in the following for completeness.

Consider the matrix equation of motion for a mdof nonlinear system under an amplitude modulated zero mean Gaussian white noise excitation vector

$$G(\ddot{x}, \dot{x}, x) = F(t) , \tag{6.1}$$

where the excitation vector $F(t)$ or simply F is defined as $F(t) = a(t) w(t)$, in which now $a(t)$ is a vector of deterministic amplitude modulated functions and $w(t)$ is the zero mean Gaussian white noise; x is the generalized displacement vector; and G_i, for conciseness the arguments are disregarded, is the total internal force acting in the direction of the i'th dof. The assumptions are that G_i is a single valued odd function of its arguments and that a stationary solution to Eq. (6.1) exists. The equivalent linear equation to (6.1) is

$$M^e \ddot{x} + C^e \dot{x} + K^e x = F(t) , \tag{6.2}$$

where the matrices M^e, C^e, and K^e are the equivalent assembled mass, damping, and stiffness matrices of the mdof nonlinear system, respectively. The deficiency vector between Eqs. (6.1) and (6.2) is defined as

$$D = G(\ddot{x}, \dot{x}, x) - M^e \ddot{x} - C^e \dot{x} - K^e x . \tag{6.3}$$

A criterion of minimizing the deficiency D is to require the mean square value of D satisfying the following necessary conditions [3, 4]

$$\frac{\partial \langle D^T D \rangle}{\partial m_{ij}^e} = 0 , \quad \frac{\partial \langle D^T D \rangle}{\partial c_{ij}^e} = 0 ,$$

$$\frac{\partial \langle D^T D \rangle}{\partial k_{ij}^e} = 0 , \quad i,j = 1, 2, \ldots, n ; \tag{6.4a,b,c}$$

where m_{ij}^e, c_{ij}^e, and k_{ij}^e are the elements of the equivalent assembled mass, damping, and stiffness matrices of the mdof nonlinear system, respectively. Substituting Eq. (6.3) into (6.4) and performing the partial differentiations, one obtains

$$\langle U U^T \rangle (M^e, C^e, K^e)^T = \langle U G^T(U) \rangle \tag{6.5}$$

where

$$U^T = [u_1, u_2, \ldots, u_{3n}] = [\ddot{x}^T, \dot{x}^T, x^T].$$

As the excitation vector is Gaussian, the vector U is also Gaussian. The rhs of Eq. (6.5) may be obtained as [4]

$$\langle U G^T(U) \rangle = \langle U U^T \rangle \langle \nabla^T G(U) \rangle \tag{6.6}$$

where

$$\nabla^T = (\partial/\partial u_1, \partial/\partial u_2, \ldots, \partial/\partial u_{3n}).$$

The proof of Eq. (6.6), with different notation, has been provided by Atalik and Utku [4]. It was also included in Ref. [5]. An equation similar to Eq. (6.6) has also been provided by Kazakov [6]. Upon equating (6.6) to (6.5) and assuming that $\langle U U^T \rangle$ is positive definite, one can obtain the linearized mass, damping, and stiffness matrices as

$$m_{ij}^e = \left\langle \frac{\partial G_i(\ddot{x}, \dot{x}, x)}{\partial \ddot{x}_j} \right\rangle, \qquad c_{ij}^e = \left\langle \frac{\partial G_i(\ddot{x}, \dot{x}, x)}{\partial \dot{x}_j} \right\rangle,$$

$$k_{ij}^e = \left\langle \frac{\partial G_i(\ddot{x}, \dot{x}, x)}{\partial x_j} \right\rangle, \qquad i, j = 1, 2, \ldots, n. \tag{6.7a,b,c}$$

It should be noted that even if $\langle UU^T \rangle$ is only positive, Eq. (6.7) will be assumed to be the global minimum solution among all the possible minima.

Substituting Eq. (6.7) into (6.2), and writing the state variables as $z_1 = x$, $z_2 = dx/dt$, and $Z = (z_1, z_2)^T$ one can show that

$$\frac{dZ}{dt} = AZ + P, \tag{6.8}$$

where the coefficient matrix A and the forcing vector P are defined by

$$A = \begin{bmatrix} [0] & \lceil 1 \rfloor \\ -(M^e)^{-1} K^e & -(M^e)^{-1} C^e \end{bmatrix}, \qquad P = \left\{ \begin{matrix} (0) \\ (M^e)^{-1} F(t) \end{matrix} \right\}.$$

Note that the zero matrix in A, $[0]$ is of order n and the unit or identity matrix $\lceil 1 \rfloor$ is also of order n. The zero vector (0) is of $n \times 1$.

Post-multiplying both sides of Eq. (6.8) by the transpose of Z, one has

$$\left(\frac{d}{dt}Z\right)Z^T = AZZ^T + PZ^T .$$

(6.9)

Taking the transpose of Eq. (6.9) yields

$$Z\frac{d}{dt}Z^T = (ZZ^T)^T A^T + ZP^T .$$

(6.10)

Adding Eqs. (6.9) and (6.10), it results in

$$\frac{d}{dt}(ZZ^T) = AZZ^T + (ZZ^T)^T A^T + PZ^T + ZP^T .$$

(6.11)

Taking the expectation of Eq. (6.11) and writing $R = \langle ZZ^T \rangle$, one obtains

$$\frac{dR}{dt} = AR + R^T A^T + B = AR + RA^T + B ,$$

(6.12)

where the matrix B is defined as

$$B = \langle PZ^T \rangle + \langle ZP^T \rangle = \begin{bmatrix} [0] & [0] \\ [0] & 2\pi S (M^e)^{-1} a(t) a^T(t)((M^e)^{-1})^T \end{bmatrix},$$

and $R = R^T$. Note that $a(t)$ is a vector of deterministic amplitude modulating functions and S is the spectral density of the zero mean Gaussian white noise $w(t)$. For quasi-linear systems the last relation for matrix B can easily be proved by applying Eq. (6) of Ref. [7] and the solution of Eq. (6.12) requires a numerical integration technique, such as the RK4 scheme.

6.2 Symplectic Algorithms of Newmark Family of Integration Schemes

In the analysis of a Hamiltonian system whose canonical phase space has a natural geometric structure induced by a skew-symmetric bilinear form, commonly referred to as the *symplectic two-form* [8, 9], a mapping on the phase space is symplectic if it preserves the symplectic two-form. Simply put, symplectic maps define canonical transformations that conserve important invariant properties such as the volume in phase space. Symplectic integrators are algorithms that inherit the symplectic character of the Hamiltonian flow, and therefore they preserve exactly the symplectic two-form in the phase space.

In this section the identification of the symplectic algorithms applied to nonlinear structural dynamic systems under random excitations is presented. The symplectic algorithms for deterministic dynamic systems are considered first in Sub-section 6.2.1 while their stochastic counterparts are included in Sub-section 6.2.2. Sub-section 6.2.3 includes remarks. The identification of symplectic algorithms is of crucial importance in that symplectic algorithms provide accurate solutions of nonlinear dynamic systems. It explains why the SCD method in Chapters 3 and 5 have been employed exclusively.

6.2.1 Deterministic symplectic algorithms

De Vogelaere [10] was the first to introduce integration schemes that preserve exactly the symplectic two-form in phase space. In 1984 Feng [11] has shown that the mid-point rule is a symplectic integrator. Sanchez-Serna [12] and Lasagni [13] have delineated the symplectic members of the Runge-Kutta (RK) family of algorithms. Specific symplectic integrators in the nonlinear regime have been derived by Feng [11], and Chanell and Scovel [14]. Subsequently, many symplectic integrators have been constructed by other researchers and their applications made. For instance, Simo *et al.* [15] have performed a complete analysis of exact energy-momentum conserving algorithms and symplectic schemes. In this sub-section, symplectic members of the Newmark family of algorithms for nonlinear Hamiltonian systems are introduced. The presentation follows closely that provided in Ref. [15].

Consider a mdof nonlinear separable Hamiltonian system

$$H(q,p) = T(p) + U(q) , \qquad (6.13)$$

where $U(q)$ is the potential energy and $T(p)$ the quadratic kinetic energy. Thus,

$$T(p) = \frac{1}{2} p^T M^{-1} p$$

in which M is the constant constrained assembled mass matrix while q and p are the generalized displacement and velocity vectors, respectively. The Newmark algorithms are obtained through the following relations

$$q_{s+1} = q_s + (\Delta t) M^{-1} p_s - (\Delta t)^2 H_1 ,$$

$$p_{s+1} = p_s - \Delta t [(1 - \gamma) \nabla U(q_s) + \beta \nabla U(q_{s+1})] , \qquad (6.14a,b)$$

in which the time step size Δt has been defined in Chapter 3, and

$$H_1 = M^{-1}\left[\left(\frac{1}{2} - \beta\right)\nabla U(q_s) + \beta\nabla U(q_{s+1})\right].$$

To obtain the linearized version of Eq. (6.14), one can write

$$z_s = \begin{pmatrix} q_s \\ \Delta t\, p_s \end{pmatrix}, \qquad K_s = \nabla^2 U(q_s),$$

so that the state vector form of the linearized Eq. (6.14) can be derived as

$$\delta z_{s+1} = A\,\delta z_s, \tag{6.15}$$

where the so-called linearized amplification matrix A, which is not to be confused with that in Eq. (6.8) and z is different from Z in Eq. (6.8), is

$$A = \begin{bmatrix} \left[I + (\Delta t)^2\, \beta\, M^{-1} K_{s+1}\right] & [0] \\ \gamma\,(\Delta t)^2\, K_{s+1} & I \end{bmatrix}^{-1} A^{(s)},$$

with matrix $A^{(s)}$ being given as

$$A^{(s)} = \begin{bmatrix} \left[I - \left(\frac{1}{2} - \beta\right)(\Delta t)^2\, M^{-1} K_s\right] & M^{-1} \\ (\gamma - 1)(\Delta t)^2\, K_s & I \end{bmatrix}.$$

The algorithm is symplectic if A satisfies the condition that $\det(A) = 1$ which means volume conservation in phase space.

Performing the matrix operation, one can show that

$$\det(A) = |A| = |A_1|^{-1} |A_2|,$$

in which

$$A_1 = I + (\Delta t)^2\, \beta\, M^{-1} K_{s+1}, \qquad A_2 = I + (\tfrac{1}{2} + \beta - \gamma)(\Delta t)^2\, M^{-1} K_s.$$

For the present purpose, it suffices to consider a nonlinear single dof system. Thus, applying the above equation, one has

$$\det(A) = 1 - \left[\frac{\left(\gamma - \frac{1}{2}\right)\omega_s^2 + \beta\left(\omega_{s+1}^2 - \omega_s^2\right)}{1 + \beta\,\omega_{s+1}^2(\Delta t)^2} \right](\Delta t)^2 , \qquad (6.16)$$

in which

$$\omega_s = \sqrt{\frac{k_s}{m}} , \qquad \omega_{s+1} = \sqrt{\frac{k_{s+1}}{m}} \qquad (6.17a,b)$$

are the linearized natural frequencies of the nonlinear single dof system at t_s and t_{s+1}, respectively, while m and k_s are the mass and stiffness coefficients at t_s of the nonlinear single dof system. Note that the mass m has been assumed to be constant for the nonlinear system.

For an arbitrary nonlinear Hamiltonian system $\omega_s \neq \omega_{s+1}$, and therefore, with $\beta = 0$ and $\gamma = 1/2$ Eq. (6.16) gives $\det(A) = 1$. In other words, for an arbitrary nonlinear Hamiltonian system the central difference method is the only symplectic scheme in the family of Newmark algorithms. Equation (6.16) was first obtained by Simo et al. [15].

6.2.2 Symplectic members of stochastic version of Newmark family of algorithms

The symplectic members of the stochastic version of Newmark family of algorithms for nonlinear Hamiltonian systems under random excitations can be identified with the steps similar to those in Sub-section 6.2.1 [16]. To proceed, one can apply Eq. (6.15) which is the linearized form of the nonlinear state vector z_{s+1}. Of course, it is understood that now the nonlinear mdof system is under random excitations and Eq. (6.15) is for such a system. One can write

$$z_{s+1} = \delta z_{s+1} + n_1\left(z_{s+1}\right) , \qquad (6.18)$$

where $n_1(z_{s+1})$ is the nonlinear component and δz_{s+1} the linearized part of the state vector z_{s+1}.

Substituting Eq. (6.15) into (6.18), one obtains

$$z_{s+1} = A\delta z_s + n_1\left(z_{s+1}\right) . \qquad (6.19)$$

By making use of Eq. (6.18), one has the state vector at the current time step as

$$z_s = \delta z_s + n_1(z_s), \qquad \text{or} \qquad \delta z_s = z_s - n_1(z_s). \tag{6.20}$$

Substituting Eq. (6.20) into (6.19), it gives

$$z_{s+1} = A\left[z_s - n_1(z_s)\right] + n_1(z_{s+1}) = A z_s + n(z_s), \tag{6.21}$$

where the nonlinear component of z_{s+1} is defined by

$$n(z_s) = n_1(z_{s+1}) - A n_1(z_s).$$

The transpose of Eq. (6.21) is

$$z_{s+1}^T = z_s^T A^T + n^T(z_s), \tag{6.22}$$

Post-multiplying Eq. (6.21) by Eq. (6.22) and taking ensemble average of the resulting equation, one can show that

$$\mathbb{R}_{s+1} = A \, \mathbb{R}_s \, A^T + \mathbb{R}^{(n)}(z_s), \tag{6.23}$$

in which $\mathbb{R}^{(n)}(z_s)$ represents the ensemble average of all the nonlinear product terms of z_s and z_{s+1} while the mean square matrices of the state vectors at time steps t_{s+1} and t_s are respectively, defined as

$$\mathbb{R}_{s+1} = \left\langle z_{s+1} z_{s+1}^T \right\rangle, \qquad \mathbb{R}_s = \left\langle z_s z_s^T \right\rangle.$$

The linearized equation of Eq. (6.23) becomes

$$\delta \mathbb{R}_{s+1} = A \delta \mathbb{R}_s A^T. \tag{6.24}$$

The algorithm is symplectic if the condition that $\det(AA^T) = 1$ is satisfied. Since $\det(AA^T) = \det(A)\det(A^T)$ it is not difficult to show that

$$\det(A)\det(A^T) = |A_1|^{-1} |A_2| |A_1|^{-1} |A_2|.$$

It suffices to consider a nonlinear single dof system so that

$$\det(A)\det(A^T) = 1 - 2\mu(\Delta t)^2 + \mu^2(\Delta t)^4, \tag{6.25}$$

where the coefficient is given by

$$\mu = \left[\left(\gamma - \frac{1}{2}\right)\omega_s^2 + \beta\left(\omega_{s+1}^2 - \omega_s^2\right)\right]\left[1 + \beta\omega_{s+1}^2(\Delta t)^2\right]^{-1}. \tag{6.26}$$

For an arbitrary nonlinear Hamiltonian system $\omega_s \neq \omega_{s+1}$, and similar to the deterministic case in the last sub-section, with $\beta = 0$ and $\gamma = 1/2$ Eq. (6.25) gives

$\det(AA^T) = 1$. That is, the SCD method is the only symplectic scheme in the stochastic version of Newmark family of algorithms.

6.2.3 Remarks

From the foregoing sub-sections it is clear that within the deterministic and stochastic versions of Newmark algorithms only the central difference and SCD method possess the important symplectic property. This is one of the main justifications for exclusive use of the SCD method in Chapters 3 and 5, and for nonlinear mdof systems in this chapter and Chapter 7.

6.3 Stochastic Central Difference Method with Time Co-ordinate Transformation and Adaptive Time Schemes

Aside from the SL techniques introduced in Section 6.1 there are other techniques available in the literature for large-scale FEA of nonlinear systems. Typically, these techniques combine the use of direct numerical integration and MCS. As pointed out in [3] these latter techniques have three major inherent problems. For example, one of the major problems is the determination of plastic deformations where the plastic potential function can swing from positive at one time step to negative at the subsequent step. Still, it can swing back to positive in the third step and so on. The second major problem is concerned with the relatively long computation time required for the averaging process even if the first major problem can be avoided by some artificial means. This is a penalty for large-scale FEA with MCS. The third major problem is that system nonlinearities have to be defined explicitly for the entire time duration of interest. This posts a difficulty in the FEA where the nonlinearities are evaluated incrementally.

Therefore, in this section the SCD method with TCT and adaptive time schemes (ATS) are employed to deal with nonlinear structural systems under stationary and nonstationary random excitations. As the focus in this chapter is on large deformations of large strain and finite rotation in plate and shell structures whose parameter matrices are computed incrementally at every time step, it is important to point out the factors that require careful consideration. As mentioned in [3, 17] even in deterministic environments, general nonlinear shell analysis by the FEM is challenged by many conceptual, theoretical, and computational issues. Even for static and quasi-static analyses these challenging issues may include [17]: (a) *consistent linearization* of the underlying variational form of the governing equations; (b) use of *objective measures* of stress and strain, and their rate, that are suitable for the particular form of the constitutive relation employed;

(c) treatment of *large rotations*, in both stiffness derivation and configuration updating; (d) the proper representation of *nonlinear material behaviours*; and (e) *efficient algorithmic developments*, for both the element level calculation and the global level incremental, iterative solution strategies. For dynamic problems, in addition to the foregoing challenging issues, the *construction of mass matrix* has to be carefully dealt with, especially when encountering large angular velocity and acceleration. Owing to their crucial importance, the above challenging issues have been examined and elaborated by Liu and To [18], and are briefly outlined in the following sub-section. Then in Sub-section 6.3.2 the development of the SCD method for computations of time-dependent covariances and mean squares of responses of nonlinear systems is presented. The average deterministic central difference (ADCD) scheme [3, 19] is included in this sub-section. Sub-section 6.3.3 is concerned with the TCT introduced earlier in Section 3.2 and ATS. The computational strategy for the implementation of the SCD method for nonlinear systems involving large deformations is presented in Sub-section 6.3.4.

6.3.1   Issues in general nonlinear analysis of shells

In deterministic environments, general nonlinear shell dynamic analysis by the FEM is challenged by [3, 18]: (a) *consistent linearization* of the underlying variational form of the governing equations; (b) use of *objective measures* of stress and strain, and their rate, that are suitable for the particular form of the constitutive relation employed; (c) treatment of *large rotations*, in both stiffness derivation and configuration updating; (d) the proper representation of *nonlinear material behaviours*; (e) *efficient algorithmic developments*, for both the element level calculation and the global level incremental, iterative solution strategies; and (f) *construction of mass matrix*, especially when encountering large angular velocity and acceleration.

6.3.1.1   Incremental formulations and variational principles

For structural problems with both geometric and material nonlinearities, the total and updated Lagrangian formulations are two natural choices. The use of either the total Lagrangian or the updated Lagrangian formulation in a FE solution depends, in practice, on their relative numerical effectiveness. A general observation is that the total Lagrangian formulation involves with a more complex strain-displacement relation. This complexity enables ones to include the so-called "initial displacement effects", and to have more control over the accuracy of total displacements. On the other hand, the updated Lagrangian formulation needs to update all the kinematic and static variables at every time step. Although the updated Lagrangian formulation itself does not have a device to consider the "initial displacement effects", by adding a corresponding strain energy term to the variational principle one is able to enforce the compatibility of the total displacements.

Concerning the updated Lagrangian formulation combined with mixed or hybrid finite elements, it can be shown that the modified Hellinger-Reissner's principle in Refs. [20, 21] reduces to the incremental complementary energy functional. It should be mentioned that both Refs. [20] and [21] assumed that the equilibrium at the reference configuration is satisfied. On the other hand, Refs. [22, 23] derived their incremental variational principles without invoking such an assumption.

In the total Lagrangian formulation, the compatibility of the total strains and displacements is accounted for through the part of its linear stiffness matrix that contains the initial displacement effects. As a result, the total Lagrangian formulation usually does not require the compatibility mismatch correction term. For the updated Lagrangian formulation, Saleeb and co-workers [17] found that the correction term was only necessary for the first iteration of every load increment. However, completely disregarding the correction term resulted in convergence difficulties. Another concern of the correction term is that, direct summation of incremental strains has little physical basis, since the incremental strains obtained at every time step are measured with respect to different configurations. Saleeb and co-workers [17] outlined an approximate scheme to sum up such incremental strains.

6.3.1.2 *Linearization of incremental variational principles*

As Gadala and co-workers [24, 25] pointed out, the most general approach for a Lagrangian formulation is to consider fully nonlinear kinematic relations within a linear increment. Starting from the energy balance equation in the rate form, Gadala and co-workers [24] derived an incremental equilibrium equation whose lhs consists of four stiffness matrices: the usual small displacement (or incremental) stiffness matrix, the initial stress (or the geometric) stiffness matrix, the initial displacement (or the initial rotation) stiffness matrix, and the initial load matrix; and whose rhs includes the incremental load vector. In fact, Oden [26] arrived at the same four component stiffness matrices. Note that if the development in Ref. [24] is followed, the updated Lagrangian formulation would lose its attractiveness: being simpler in terms of formulation. Furthermore, the formulation developed in Ref. [24], when employed for dynamic analysis, would be extremely expensive, if not computationally infeasible. Therefore, confining our attention to the updated Lagrangian formulation, linearizing the incremental variational principle would be a more practical approach which is adopted in the present chapter.

Before leaving the present discussion on the issue of linearization of incremental variational principles, a remark should be made that, as observed in Section 3 of [18], owing to the existence of drilling dof (ddof) the stress tensor has its symmetric and skew-symmetric components. Consequently, the initial stress (or the geometric) stiffness matrix becomes non-symmetric. However, on physical reasoning and analytically proved by Bufler [27], a conservative system cannot have non-symmetric stiffness matrix. In the present chapter, only conservative systems with symmetric stiffness matrices are considered. Every of these symmetric stiffness matrices consists of two main components, one is concerned with the nodal displacement dof based on the modified Hellinger-Reissner's variational principle and the other the nodal rotational dof derived through the displacement formulation [18].

6.3.1.3   Stress and strain measures

The most commonly used stress measures are the Cauchy stress tensor, the second Piola-Kirchhoff (PK2) stress tensor and the Jaumann stress rate tensor. The Cauchy stresses represent the true state of stresses of a deformed body and yield simpler equations of motion compared with the PK2 stresses. The Cauchy stresses should be used when determining the occurrence of yielding. The PK2 stresses, on the other hand, describe the state of stresses with respect to the undeformed configuration and therefore have little physical meaning. However, the PK2 stresses are invariant under a rigid body rotation of the material. That is, the PK2 stress tensor is *objective*. The Jaumann stress rate tensor also preserves invariance under rigid body rotations of the material. The three most frequently used strain measures are the Green or the Green-Lagrange strain tensor, the Almansi strain tensor and the velocity strain or the rate of strain tensor. Both the Green strain tensor and the velocity strain tensors are objective and they are energetically conjugate to their corresponding stress tensor, the PK2 stress tensor and the Jaumann stress rate tensor, respectively. The Almansi strain tensor, however, is not invariant under rigid body rotations. Another strain measure is the so-called co-rotational or convected strain. Belytschko [28] compared the accuracy of the foregoing four strain measures (for the velocity strain tensor, its accumulative or integrated result was applied, which is the so-called logarithmic strain) and showed that for finite strain problems, there is a large difference among different strain measures. In general, the Almansi strain tensor performs relatively poorly. The co-rotational strain seems to be the most accurate measure, followed by the Green strain tensor. The co-rotational strain tensor has little physical meaning (except for specific cases such as uniaxial deformation) and should be transformed to the Green, or Almansi strain tensor, as pointed out in [28].

When pairing stress and strain measures, two factors must be taken into account. First, the stress and strain measures should be energetically conjugate. Second, they should be in accordance with the constitutive law used to model material behaviors. Thus, the PK2 stress tensor and the Green strain tensor are suited for elastic and hyper-elastic materials whose behaviors are determined from the total current strains while the Jaumann stress rate tensor and the velocity strain tensor are more effective for the analysis of rate form constitutive relations, or path-dependent materials.

Further, there are incremental stress and strain measures. Such incremental measures are necessary since incremental formulation is used in numerical computation. The incremental stress is usually the incremental PK2 stress tensor, and the incremental strain is the incremental Green strain tensor. Note that the actual form of the incremental Green strain tensor may be different depending on whether the total or the updated Lagrangian formulation is used. The incremental Green strain tensor is generally separated into "linear" and "quadratic" parts. The "quadratic" part remains the same while the "linear" part differs in whether the initial displacement is included. Chiou and co-workers [29] used the term "incremental Green strain" when it is applied with the total Lagrangian formulation, and the term "incremental Washizu strain" with the updated Lagrangian formulation. Other alternative stress and strain measures were introduced by Atluri [30].

Regarding the constitutive law for elasto-plastic deformation, the classical small deformation elasto-plasticity theory decomposes the total infinitesimal strain as the sum of the "elastic" and "plastic" parts. Such a theory establishes a relation between the rate of the PK2 stress tensor and the rate of the Green strain tensor, or in its incremental form, a relation between the incremental PK2 stress tensor and the incremental Green strain tensor or the incremental Washizu strain tensor, depending on whether the total or the updated Lagrangian formulation is used. The additive decomposition which is usually accredited to Green and Naghdi finds physical ground and can be deduced mathematically only when the strain used is infinitesimally small. In finite elasto-plastic deformation problems, two majors approaches appear in the literature. These are the Nemat-Nasser's additive decomposition of total strain rate into its "elastic" and "plastic" parts [31], and Lee's multiplicative decomposition of total deformation gradient into a product of "elastic" and "plastic" deformation gradients [32]. In all the above elasto-plastic theories, the "plastic" part of the strain, or the strain rate, or the deformation gradient, is defined on the physical grounds that unloading is completely elastic. This implies that these theories are applicable to elasto-plastic materials which recovers the "elastic" part of the deformation, as opposed to some other types of material, such as "elastro-plastic" materials which, upon unloading from a deformed configuration, recovers only parts of the strains [33].

6.3.1.4   *Finite rotations*

Nonlinear analysis involving large rotations has attracted much attention since the late 1960's, as noted by Wempner [34]. It seems that the treatment of finite rotations can be classified into the following four approaches. The first one was based on the observation that motion of an individual element, to a large extent, is of the rigid body type [35]. Therefore, if the rigid body motion can be eliminated from the total displacements, the deformation part of the motion is always a small quantity relative to the local element axes. Note that, though the principle of splitting total displacements into its rigid body motion part and deformation part holds true theoretically, it is practically not an easy task with large strain cases. Furthermore, owing to the non-commutative nature of large rotations, the sequence by which the rigid rotations are separated from the total displacements (rotations) will no doubt affect the resulting deformation part of the motion, and consequently other quantities involved in the incremental and updating process. In Ref. [36] Argyris proposed to circumvent the non-commutativity by using the "semi-tangential" angles to represent finite rotations. The second approach [37, 38] retained the nonlinear nodal rotation terms in the displacement interpolation function, which are mainly sines and cosines of the nodal rotations. Thus, it removed the restriction of small nodal rotations between two successive load increments. Yet this approach complicates the derivation of element matrices, and numerical integration has to be employed because of the existence of sine and cosine terms. The third approach was proposed by Bathe and Bolourchi [23]. This approach is in fact equivalent to evaluating the transformation matrix using Euler angles, as presented by Besseling [39] who expressed the transformation matrix in terms of Euler angles. The drawback of this approach is that it may be difficult to generalize to triangular elements, and a relatively large amount of algebra is involved. In addition, for zero nutation angles the corresponding precession and spin angles become undefined. The fourth approach [40-42] starts by noting that the spatial configuration of a shell can be defined by the position vector field for points on the mid-surface of the shell as well as the director vector field that gives the orientations of thickness vectors emanating from the mid-surface points. The description of the position vector field is straightforward. It is not so for the director field due to the fact that finite rotations in space can not be treated as vectors. That is, finite rotations are not commutative. Therefore, the fourth approach focuses on defining a transformation matrix for any magnitude of the director rotation increment from time instant "t" to that at "$t+\Delta t$" in the updating procedure of the director vector field. This transformation matrix or the so-called *exponential mapping* provides a means to map an infinitesimal rotation into a large rotation, and a large rotation into another large rotation. Consequently, the change of the orientations of the directors can be correctly represented and the director field can be updated accurately. The exponential mapping also preserves the length of the director so that the inextensibility of the director will not be violated. Another concern is the representation of the director vector field which is complicated by the non-commutativity of large rotations. In short, the fourth approach is based on the notion that the director vector field exists and is unique. In reality, this may not always be the case because intersecting shell structures can have discontinuous directors, for instance. Furthermore, when the shell structures are complex, the determination of the directors becomes difficult. Formulations that can accommodate such applications have been developed and presented by Liu and To [18]. The formulations from the latter reference are applied in this chapter and Chapter 7.

6.3.1.5 *Dynamic analysis*
When extending a FE formulation to dynamic analysis from a static one, the common approach is to add the consistent or lumped mass matrix. Originally, this approach took into account only the effect of translational inertia [43]. The effects of rotary inertia due to bending and associated with the ddof were included in Ref. [18] by the authors. The approach in the latter reference is accurate if the rotational motions involved are of relatively low angular velocity and acceleration. For those problems exhibiting considerable amount of angular velocity and acceleration, Ref. [44] concluded that the part of mass matrix associated with the

rotational dof is nonsymmetric and configuration dependent. It should be emphasized that the focus in the present book is on plate and shell problems with symmetric stiffness and mass matrices. Cases with non-symmetric stiffness and mass matrices are beyond the scope of this book.

6.3.1.6   *Algorithmic developments*

It is generally agreed that implicit integration techniques, such as the trapezoidal rule, would be more accurate and effective with consistent element mass matrices while explicit integration schemes such as the central difference (CD) method would be more accurate with lumped element mass matrices [45, 46]. However, it should be pointed out that when mass and damping matrices are not diagonal, the CD method is implicit. As pointed out in Section 6.2, the CD method is symplectic and preserves momentum, and its stochastic version, the SCD method is also symplectic, therefore, the SCD method in conjunction with element consistent mass matrices is applied to determine the mean squares and covariances of responses of nonlinear structural systems in the following.

6.3.1.7   *Closing remarks*

The stochastic nonlinear dynamic analysis to be presented in the following sub-section applies the mixed or hybrid strain based lower order triangular shell finite elements. While shell elements based on other approaches can also be applied, it is believed that the ones presented in this chapter have various advantages. First, they can easily deal with shell structures of arbitrary shapes. Second, each of the finite elements is of flat triangular geometry with 3 corner nodes and 18 dof, including 3 translational and 3 rotational dof for every node. This means that with the adopted formulation each of the elements can be used to predict the correct 6 rigid body modes. Third, the derivation of these elements starts with the assumption that every node on the shell mid-surface has a unique director. When simplified, the linear parts of these elements reduce to those identified as AT+k_d and AT+$(k_t^3)'$ in Ref. [18], or those denoted by NFORMU = 14 and 16 in Ref. [47]. Note that in Ref. [18] the director of a node is not uniquely defined. Rather, the three nodes of an element share the same "director". Fourth, the updated Lagrangian approach is adopted in conjunction with an incremental modified Hellinger-Reissner variational principle which takes incremental displacements and incremental strains as the two independently assumed fields. Corresponding to the selected incremental formulation and variational principle, the incremental PK2 stress tensor and the incremental Washizu strain tensor are chosen as the incremental stress and strain measure. Fifth, explicit expressions for the element consistent mass, linear or small displacement stiffness, the initial stress or the geometric stiffness matrices and the pseudo-load vector are obtained and expressed in terms of variables of the reference configuration. Sixth, to account for large rotations, exponential mapping is employed for configuration updating. Seventh, both geometric and material nonlinear problems are studied. In the latter case, the J_2 flow theory of plasticity is employed in conjunction with Ilyushin's yield criterion for thin shells [18]. For deterministic excitations, the correctness, conceptual adequacy, accuracy, and applicability of the theory and derived element matrices were examined. One important conclusion from these deterministic studies is that the mixed formulation-based triangular shell finite elements developed are superior to those based on the displacement formulation. Therefore, one of the developed mixed formulation-based flat triangular shell finite elements, $AT+(k_t^3)'$ whose derivation has been outlined in Sections 1.4 through 1.5, is applied for the computation in this and the next chapters. For more discussion on the cases with deterministic excitations, the interested readers can refer to Refs. [3, 18]. Meanwhile, for direct reference and application, explicit expressions for the consistent linear element stiffness and mass matrices have been presented in Appendices 1B, and 1C.

6.3.2 Time-dependent variances and mean squares of responses

Consider the governing matrix equation of motion for a general nonlinear structural system discretized by the FEM and under combined deterministic and nonstationary random excitations

$$M(x,\dot{x})\ddot{x} + C(x,\dot{x})\dot{x} + K(x)x = F(t) = F^d + F^r \qquad (6.27)$$

where M, K and C, without the arguments for conciseness, are assembled nonlinear mass, stiffness and damping matrices, respectively; whereas the remaining symbols have already been defined in previous chapters. These assembled matrices may be treated as deterministic functions of time t and at every time step they can be considered as constants.

As pointed out in Sub-section 6.3.1, the incremental updated Lagrangian formulation of the FEM for large deformations is adopted and therefore, the local co-ordinates of every element, and the element consistent matrices have to be updated at every time step before computing the assembled M, C and K. On the rhs of Eq. (6.27), the generalized excitation vector $F(t)$ or F is a sum of F^d and F^r, where the superscripts d and r denote, respectively, deterministic and random.

Accordingly, the temporally discretized Eq. (6.27) is

$$M_s \ddot{x}_s + C_s \dot{x}_s + K_s x_s = F_s = F_s^d + F_s^r . \tag{6.28}$$

By following similar steps of derivation for the recursive expressions of the SCD method introduced in Chapter 3, one can show that

$$R_{s+1} = (\Delta t)^4 N_1 \left(B_s^r + B_s^d \right) N_1^T + N_2 R_s N_2^T$$
$$+ N_3 R_{s-1} N_3^T + N_2 D_s N_3^T + N_3 D_s N_2^T + R_s^{(r)} , \tag{6.29}$$

in which

$$R_s^{(r)} = (\Delta t)^2 N_1 F_s^d \left\langle x_s^T \right\rangle N_2^T + (\Delta t)^2 N_2 \left\langle x_s \right\rangle \left(F_s^d \right)^T N_1^T$$
$$+ (\Delta t)^2 N_1 F_s^d \left\langle x_{s-1}^T \right\rangle N_3^T + (\Delta t)^2 N_3 \left\langle x_{s-1} \right\rangle \left(F_s^d \right)^T N_1^T ,$$

with N_1, N_2 and N_3 being identical in form to those defined in Eq. (3.3) for linear systems but the assembled mass, damping, and stiffness matrices are nonlinear and have to be updated at every time step so that

$$N_1 = \left[M_s + \frac{1}{2} (\Delta t) C_s \right]^{-1} , \quad N_2 = N_1 \left[2 M_s - (\Delta t)^2 K_s \right] , \text{ and}$$

$$N_3 = N_1 \left[\frac{1}{2} (\Delta t) C_s - M_s \right] , \text{ while}$$

$$R_s = \left\langle x_s x_s^T \right\rangle , \quad D_s = \left\langle x_s x_{s-1}^T \right\rangle , \tag{6.30a,b}$$

$$D_s = (\Delta t)^2 N_1 F_{s-1}^d \left\langle x_{s-1}^T \right\rangle + N_2 R_{s-1} + N_3 D_{s-1}^T , \tag{6.30c}$$

$$\left\langle x_{s+1} \right\rangle = (\Delta t)^2 N_1 F_s^d + N_2 \left\langle x_s \right\rangle + N_3 \left\langle x_{s-1} \right\rangle , \tag{6.31}$$

$$B_s^r = \left\langle F_s^r \left(F_s^r \right)^T \right\rangle = 2 \pi S_0 \left(e_s e_s^T \right) , \quad B_s^d = F_s^d \left(F_s^d \right)^T . \tag{6.32a,b}$$

In the foregoing derivation the random excitation vector similar to Eq. (3.4) has been adopted. That is,

$$F_s^r = e_s w_s \qquad (6.33)$$

in which w_s is the zero mean DGWN and e_s the deterministic modulating vector, the element or entry of e_s is defined by

$$e_s^i = E_i\left(e^{-\alpha_{i1} t_s} - e^{-\alpha_{i2} t_s}\right) \qquad (6.34)$$

where the subscript or superscript i designates the i'th element of the vector e_s while α_{i1} and α_{i2} are positive constants satisfying $\alpha_{i1} < \alpha_{i2}$. The factor E_i is the constant applied to normalize $e^i(t)$ such that every maximum value is unity. That is, $max\{e^i(t)\} = 1$. Equation (6.29) is the recursive expression for the time-dependent mean square of generalized displacements.

Applying Eq. (6.29), one can show that the recursive expression for the time-dependent covariance matrix of generalized displacements is

$$U_{s+1} = (\Delta t)^4 N_1 B_s^r N_1^T + N_2 U_s N_2^T + N_3 U_{s-1} N_3^T$$
$$\qquad (6.35)$$
$$+ N_2 A_s^m N_3^T + N_3 \left(A_s^m\right)^T N_2^T ,$$

where

$$A_s^m = N_2 U_{s-1} + N_3 \left(A_{s-1}^m\right)^T , \qquad (6.36)$$

$$U_s = R_s - \left\langle x_s \right\rangle\left\langle x_s^T \right\rangle , \qquad A_s^m = D_s - \left\langle x_s \right\rangle\left\langle x_{s-1}^T \right\rangle . \qquad (6.37\text{a,b})$$

Remark 6.3.2.1

Equations (6.29), (6.31), and (6.35) constitute the SCD method. In general, the system assembled mass, damping, stiffness matrices, and forcing vector have to be updated at every time step. For large deformations the updating of stresses and configuration at every time step in the present updated Lagrangian formulation approach is by applying Eq. (6.31) which is called the averaged deterministic central difference (ADCD) scheme. The latter is one of the efficient features of the SCD method.

Superficially, when the deterministic forcing vector on the rhs of Eq. (6.31) is zero, it leads to zero ensemble average of the generalized displacement vector. This seems to post two major problems in the evaluation of the responses of the nonlinear system. First, the co-ordinate updating and stress updating that are required in the large nonlinear response problems seem unable to be performed.

Second, it seems that systems with asymmetrical nonlinearities cannot be dealt with by applying Eq. (6.31) since the mean square responses of such systems are not zero. However, a closer inspection will reveal that the case of asymmetrical nonlinearities can be dealt with by removing the nonlinearities at every time step to the rhs of the equation and treating these nonlinearities as a deterministic excitation at every time step. This strategy has been applied successfully to a two dof asymmetrical nonlinear system in Refs. [48, 3].

Remark 6.3.2.2
One relatively efficient route of implementing the SCD method is to employ Eq. (6.35) to compute the covariance matrix U_{s+1} by way of using Eqs. (6.31), (6.37b), (6.36), and (6.37a). The time-dependent mean square of generalized displacements R_s is subsequently evaluated by applying Eq. (6.37a).

6.3.3   Time co-ordinate transformation and adaptive time schemes
When the discretized system under consideration is stiff, which is usually the case in the FEA, application of the TCT technique is necessary. The TCT has been introduced in Section 3.2 and therefore it is not repeated here. However, it should be pointed out that the TCT applied in the FEM context enables one to select much coarser meshes for accurate response computations because the dimensionless highest natural frequency Ω is always equal to unity.

For nonlinear systems, the natural frequencies are time dependent and therefore the time step sizes for the computation of mean squares and covariances of responses are changing in time. Naturally, application of the SCD method to nonlinear systems requires time step size updating. Three representative computational strategies for discretized nonlinear structural systems under random excitations are included in the following.

SATS   is the SCD method with ATS. In this strategy the lowest natural frequency of the linearized system at every time step is evaluated so that the time step size for the application of the SCD method can be determined in accordance with Eq. (3.14).

ATST   is the SCD method with the ATS and TCT. In this strategy the recursive algorithms at every time step are first transformed into their corresponding dimensionless time co-ordinate and the ATS is performed subsequently so that the responses are evaluated in the dimensionless time co-ordinate and afterward to be converted back to the original *t*-domain.

TATS   is the SCD method with TCT and ATS. It involves with the TCT being performed once at the start of the computation before the application of ATS.

Liu and To [49, 19] found that in terms of computational effectiveness and accuracy, the TATS is a better approach for stiff systems. It was also found that in a large class of nonlinear structural systems discretized by the FEM the dimensionless fundamental natural frequencies of the nonlinear systems are each well below 0.1, and accordingly the dimensionless time step size for computation applying the TATS is $\Delta \tau = 1.0$, which is unique to the SCD method. The rationale of selecting this time step size is based on the corresponding deterministic case considered in Ref. [3]. It should be noted that the pitfall with this approach in the deterministic response computation is that the time step size may well exceed the critical time step size of the CD method, if the fundamental and the highest natural frequencies are far apart from each other. Indeed, the wide separation of the fundamental and highest natural frequencies occurs in many structural systems approximated by the FEM. On the other hand, the SCD method with the TCT scheme has eliminated this pitfall because the dimensionless time step size associated with the dimensionless fundamental natural frequency reaches 1.0 as the dimensionless fundamental natural frequency approaches zero. The dimensionless time step size, $\Delta \tau = 1.0$ as given by Eq. (3.43) is well below the dimensionless critical time step, $(\Delta \tau)_c = (2)^{1/2}$. The detailed derivation of the latter has been presented in Appendix 4A of Ref. [3]. Meanwhile, for completeness and direct reference, the deterministic critical time step size is given in Appendix 3A.

6.4 Outline of steps in computer program

The digital computer program has been written in Fortran and developed for the implementation of the SCD method with TCT and ATS presented in Subsections 6.3.2 and 6.3.3. It may be noted that when the system is linear, the ATS is no longer required. This class of problems has been included in Chapter 3. In the following an outline of the important main steps of the program is included.

Step 1: Create an input data file. This includes element geometric and material properties of the discretized system, system damping parameters and external random excitation data.

Step 2: The consistent element mass and stiffness matrices and force vectors are formed by the FEM. Assembled mass and stiffness matrices, and force vectors of the entire system are constructed.

Step 3: The highest natural frequency of the linear system is computed. The general assembled damping matrix is constructed. If proportional damping is assumed, the lowest natural frequencies are also obtained at this stage. With these frequencies and the damping parameters input in Step 1, the associated damping coefficients are determined.

Step 4: Based on the highest natural frequency in Step 3 the TCT is applied so that the resulting matrix equation of motion is expressed in the dimensionless time domain. The dimensionless time step size for the computation of mean square and covariance matrices of responses is given by Eq. (3.43), $\Delta\tau = 1$.

Step 5: The SATS or ATST or TATS is (are) performed.

Step 6: For large deformation problems, the system parameter matrices, such as the assembled stiffness matrix is updated at every dimensionless time step. This requires, in turn, updating of configuration and stresses, as outlined in Sub-section 1.5.3 in the incremental updated Lagrangian formulation. In updating configuration and stresses, Eq. (6.31) which is the ADCD scheme is applied.

Step 7: Step 6 is repeated until the time-dependent quantities, such as the mean squares of responses for all dimensionless time steps are obtained.

Step 8: Applying Eq. (3.13) and the relation, $U_s = \Omega U_{\tau s}$ with $U_{\tau s}$ being the covariance matrix of displacements at dimensionless time step τ_s in the τ-domain, to convert the computed quantities in Steps 5 and 7 into their corresponding results in the original time domain. Repeat this step for all the time steps of the responses. These converted solutions are stored in an output file for further analysis and presentation.

6.5 Large Deformations of Plate and Shell Structures

Applications of the developed digital computer program have been made and representative results for a plate and shell structures are presented in Sub-sections 6.5.1 and 6.5.2. Note that both the plate and shell structures have been idealized by the lower order triangular shell finite elements mentioned in Sub-section 6.3.1. The random excitations are treated as combinations of deterministic and random components. The latter are products of deterministic modulating functions and zero mean DGWN.

6.5.1 Responses of cantilever plate structure

This cantilever plate structure, identified as CF3, is selected as a basis for comparison and as an illustration. Its linear nonstationary random responses have been reported by To and Liu [50], and presented in Sub-section 3.3.2 where the exact solutions were included for comparison.

Three cases are considered here. In the first case, the entire plate is represented by 8 high precision triangular plate bending elements, TBH6 as in Sub-section 3.3.2, while in the second case the entire plate is idealized by 8 shell elements, which is identified as AT+$(k^3_t)'$ [47, 18]. For linear analysis, the stiffness matrix of this element is $k = (k_m)' + k_b + k_s + (k^3_t)'$ where the subscripts m, b, and s denote, respectively, the membrane, bending and shear components of the stiffness matrix, whereas $(k^3_t)'$ is the component associated with the ddof which vary linearly over an element. The stiffness matrix, $(k_m)' + k_b + k_s$ is based on the hybrid strain formulation whereas $(k^3_t)'$ has been derived by employing the displacement formulation. In the third case, a half of the cantilever plate is represented by 4 shell elements. It is used to briefly examine the effect of geometric symmetry on computational time and validity of the application of geometric symmetry in the nonstationary random response analysis. The deterministic modulating function of the nonstationary random excitation is similar to that defined by Eq. (3.20) while the damping in the system is proportional. Thus, Eq. (2.19) applies here. The 8-element idealization is given in Figure 3.7 and the 4-element representation is presented in Figure 6.1 in which the deterministic component of the nonstationary random excitation for the nonlinear case is included.

The geometrical and material properties are those given in Sub-section 3.3.2. In the TBH6 case there are 27 unknown generalized displacements to be determined while, in the 8 shell triangular element case there are 24. The boundary conditions applied are: $U = V = W = \Theta_x = \Theta_y = \Theta_z = 0.0$, which is the clamped side, where U, V, and W are deformations along the global axes X, Y, and Z, respectively. The global angular deformations, Θ_x, Θ_y, and Θ_z are about the global axes X, Y and Z, respectively. In addition, $U = \Theta_z = 0.0$ have been imposed on all the nodes so that no twisting would be allowed to occur. For the 8 triangular shell element model, the first two and the highest natural frequencies as well as Rayleigh damping coefficients corresponding to 5% damping for the first and second modes are listed in Table 6.1.

In the third case, the 4 shell element model in Figure 6.1 (for half of the cantilever plate), in addition to the boundary conditions applied along the clamped edge and at every node as for the second case, $U = \Theta_y = \Theta_z = 0.0$ are imposed on the symmetry line in order to circumvent twisting of the plate. The resulting FE model has 14 unknown generalized displacements.

The nonstationary random excitation is assumed to be a concentrated load and is applied transversally at the middle of the free end. With reference to Figure 3.7 the point of load application is Node 8 or Node 6 for the 4 shell element case in Figure 6.1. The amplitude modulating function is

$$e^{i}(t) = 9.4815(e^{-45t} - e^{-60t}) , \qquad (6.38)$$

where the superscript i is the integer corresponding to the dof. For the linear response cases spectral density of the DGWN process is $S_o = 1.0 \ N^2$.

Representative computed variances of transversal generalized displacements (henceforth, in this sub-section displacements are meant to be the transversal displacements, unless it is stated otherwise) at Node 8 are presented in Figure 6.2. Excellent agreement can be observed between the results by applying the TBH6 elements and shell elements. Two observations should be mentioned at this stage. First, recall that in Sub-section 3.3.2 the first 6 modes have damping ratios of 5% each. Second, the computed results applying the SCD method with TCT strategy in Sub-section 3.3.2 are in excellent agreement with the corresponding exact solutions. Consequently, in this problem all the modes that have a natural frequency less than or in the neighborhood of 136.5 rad/s are assumed to have 5% critical damping. This is why the results from Sub-section 3.3.2 can be directly adopted for comparison to those using the shell elements with the same amount of damping in the linear case.

The computed maximum variance and covariance, with $S_o = 1.0 \ N^2$, of displacements at Nodes 4 and 6 in Figure 6.1 are presented in Table 6.2. Those for the 8 shell element model are also included in the latter table for comparison. Note that the variances and covariances of generalized displacements from the 4 shell element case are very close to those obtained for the 8-element model. However, this finding is not general in that during the course of computational experiments it was observed that the rule for geometric symmetry of structures under deterministic excitations did not apply to systems excited by nonstationary random excitations. In other words, for accurate response statistics computation it is necessary to consider the entire structure.

Another important observation from Table 6.2 for the 4-element model is that the computational time is substantially reduced. The highest natural frequency is 1.211×10^5 rad/s which is very much reduced compared with that for the 8-element case (see Table 6.1). In fact, the 4-element model requires only 15500 instead 32000 time steps for the 8-element case to cover the same interval in the original time co-ordinate t. The reductions in both the system matrix size and computational time are very impressive if one can apply the rule of geometric symmetry. Further, the triangular shell element applied can give very accurate solutions even with a relatively coarse mesh.

Table 6.1   Natural frequencies and damping coefficients.

Model	8 shell elements	4 shell elements
Unknown displacements	24	14
ω_1 (rad/s)	28.1	28.21
ω_2 (rad/s)	136.5	276.9
Ω (rad/s)	2.492×10^5	1.211×10^5
λ_m	2.33	2.56
λ_k	6.075×10^{-4}	3.278×10^{-4}

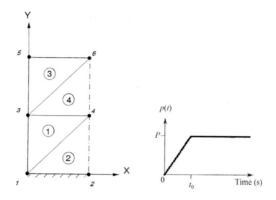

Figure 6.1   Four shell element model of half cantilever plate.

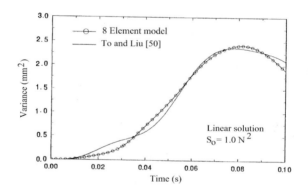

Figure 6.2   Time-dependent variances of displacements.

Applying the 4 shell element model, variances and covariances of displacements of the CF3 plate with geometrical nonlinearities are computed. The single point nonstationary random excitation is applied at Node 6 and is perpendicular to the plane of the plate. The deterministic component of the nonstationary random excitation had a time function $p(t)$ described in Figure 6.1 where t_0 is 6 ms and P either 500 N or 1250 N. The latter two values for P are selected so as to achieve different degrees of geometrical nonlinearity. The spectral density, S_0 of the DGWN is varied with nonlinearity. For example, in the linear analysis the

Table 6.2 Maximum variance and covariance of displacements.

Model	8 shell elements	4 shell elements
Unknown displacements	24	14
$S_o(N^2)$	1	1
Variance at Node 6 (mm^2)	4.782	4.836
Variance at Node 4 (mm^2)	0.4696	0.4828
Covariance at Nodes 4 and 6 (mm^2)	1.451	1.515
Computing time (minutes)	100	10

spectral density is $S_o = 10^5$ N^2. In the nonlinear analysis $S_o = 2.5 \times 10^5$ N^2 with $P = 500$ N or 5×10^5 N^2 with $P = 1250$ N. Note that P and S_o are referred to the half plate case and they lead to very large nonstationary random responses. These excitation parameters are chosen to demonstrate the capability of the developed digital computer program for responses of discretized plate and shell structures under very intensive nonstationary random excitations. Some representative computed results are shown in Figures 6.3 through 6.7. It can be seen that the peak values of variances and covariances are reduced with increasing degree of nonlinearity indicating stiffness hardening of the structure. It is interesting to note that the covariances of rotations about the X-axis at Nodes 4 and 6 become negative when nonlinearity is present, as shown in Figure 6.7. This negative value of covariances of rotations simply means that the rotation at Node 4 is out of phase with that at Node 6. For clear comparison, the time-dependent mean square and variance of linear transversal displacement at Node 6 are presented in Figure 6.8. The time-dependent mean square and variance are close to one another because the ensemble average of displacement is relatively small. Computed results of the corresponding nonlinear case are included in Figure 6.9. With the presence of nonlinearity the difference between the time-dependent mean square and variance remained small prior to the occurrence of peak values. After that, the time-dependent mean square became much larger than the variance due to the relatively large value of the ensemble average of displacement which is a result of the deterministic component of the nonstationary random excitation.

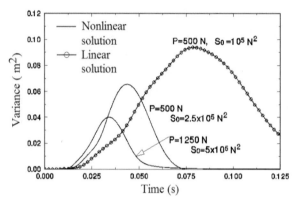

Figure 6.3   Time-dependent variance of displacement at Node 4.

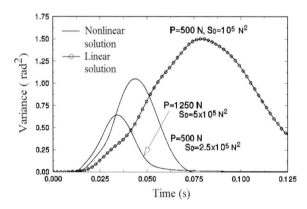

Figure 6.4   Time-dependent variance of rotation about X-axis at Node 4.

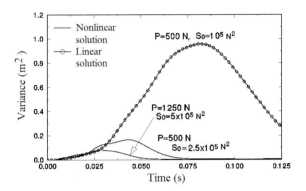

Figure 6.5 Time-dependent variance of displacement at Node 6.

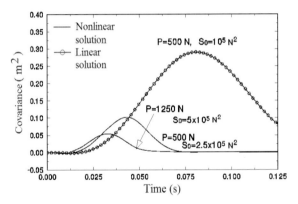

Figure 6.6 Covariance of displacements at Nodes 4 and 6.

It should be pointed out that in all the above calculations, TATS was used exclusively. This is because the change of the highest natural frequency is insignificant. For example, with a load level of $P = 500$ N, it was observed that the fundamental and the highest natural frequencies were both increased: ω_1 from 28.21 to 32.46 rad/s and Ω from 1.211×10^5 to 1.217×10^5 rad/s. The percentage increases for ω_1 and Ω were 15.1 and 0.5, respectively. Therefore, the exclusive use of TATS is justified from the viewpoint of accuracy, computational effectiveness and physics.

The cpu time required to perform every of the foregoing nonlinear problem computations applying TATS was 95 minutes and the time to compute the natural frequencies approximately 5 minutes. These times were accumulated over the required 15500 time steps. The computing machine employed was a Silicon Graphics engineering workstation. It was observed during the computational experiments that for the present plate structure problem the computing time using ATST was not substantially more than that of TATS. The reason was that the total unknown displacements were only 14. However, the portion of computing time used to calculate natural frequencies at every time step is expected to be drastically increased for systems having very large number of dof.

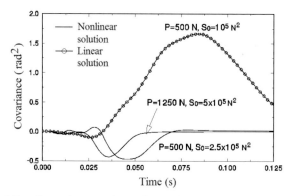

Figure 6.7   Covariance of rotations about X-axis at Nodes 4 and 6.

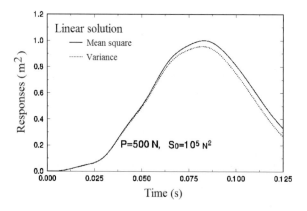

Figure 6.8   Mean square and variance of displacement at Node 6.

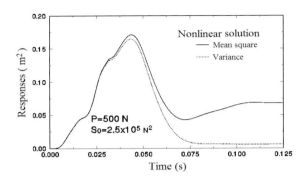

Figure 6.9 Nonlinear mean square and variance of displacement at Node 6.

6.5.2 Responses of clamped spherical cap

The shell structure considered in this sub-section is the spherical cap [3, 19]. It has been studied in Ref. [51] for deterministic static and dynamic analysis with geometric and material as well as geometric nonlinearities. The geometry of the cap is shown in Figure 6.10, where the radius of the cap is $R = 120.90$ mm (4.76 in), thickness $h = 0.4$ mm (0.01575 in) and sagitta $H = 2.18$ mm (0.08589 in). The material is either elastic with Young's modulus $E = 69.85$ GPa (10^7 psi), Poisson's ratio $v = 0.3$, tangent modulus $E_T = 0$, and density $\rho = 2621.54$ kg/m^3 (0.000245 lb-sec^2/in^4), or elastic-perfectly plastic with yield stress $\sigma_y = 344.75$ MPa (50 ksi) which was incorrectly reported as 139.7 MPa (20 ksi) in [51].

In order to reduce the computation time and because of geometrical symmetry, boundary conditions, and loading, only one quadrant of the cap is approximated by the triangular shell finite element [18]. The 28-node, 36-element mesh in Figure 6.11 is employed. The boundary conditions for this case are: on the clamped circumference, $U = V = W = \Theta_x = \Theta_y = \Theta_z = 0.0$; on the symmetrical side parallel to the X-axis, $V = \Theta_x = \Theta_z = 0.0$; on the symmetrical side parallel to the Y-axis, $U = \Theta_y = \Theta_z = 0.0$; and at the apex, $U = V = \Theta_x = \Theta_y = \Theta_z = 0.0$. The total number of unknowns is 85.

For this shell structure, the first two and the highest natural frequencies have been found as 5.858×10^4, 7.348×10^4 and 1.225×10^7 rad/s, respectively. With the damping ratio for the first two modes being chosen to be 0.05 or 5% critical, the two coefficients in the Rayleigh damping relation, Eq. (2.19), were found as $\lambda_m = 3.259 \times 10^3$ s^{-1} and $\lambda_k = 7.572 \times 10^{-7}$ s. For the external point

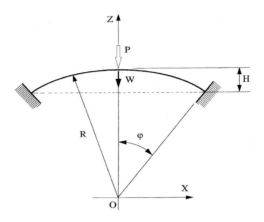

Figure 6.10   Clamped spherical cap ($\varphi = 10.9°$).

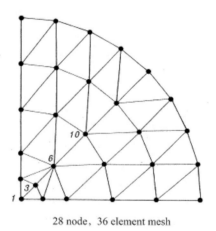

28 node, 36 element mesh

Figure 6.11   Finite element idealization of quadrant spherical shell
(Only Nodes 1, 3, 6 and 10 are shown)

loading on the quadrant cap, the deterministic component had the same time
function as that depicted in Figure 6.1. The time $t_0 = 0.00003$ s or 30 μs while the
load level P is 38.9725 N (8.75 lb). The modulating function $e^i(t)$ applied is

$$e^i(t) = 9.4815(e^{-18000t} - e^{-24000t}),\qquad(6.39)$$

and the spectral density S_o of the DGWN is kept at $S_o = 19.84\,\text{N}^2\,(1.0\,\text{lb}^2)$. It should be noted that in this problem $S_o = 1.00\,\text{N}^2$ was incorrectly provided in Ref. [19]. The computed results presented there, however, are correct.

Three cases are presented in this sub-section. These include the linear, geometrically nonlinear, and materially as well as geometrically nonlinear cases. Representative computed variances and covariances of responses are presented in Figures 6.12 through 6.14.

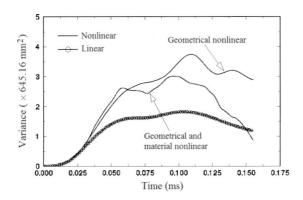

Figure 6.12   Variance of displacement at Node 1.

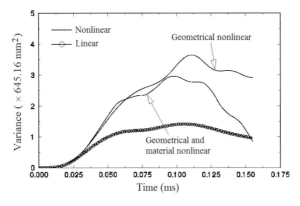

Figure 6.13   Variance of displacement at Node 3.

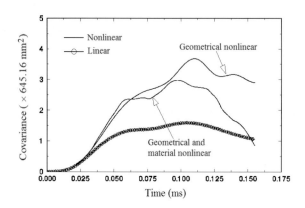

Figure 6.14   Covariance of displacements at Nodes 1 and 3.

Two main observations should be mentioned. First, variances for the two nonlinear cases are considerably increased compared with the linear solutions. Second, with reference to Figure 6.12, when both material and geometrical nonlinearities are present in the shell structure, the largest peak of the variance of transversal displacement at Node 1 is about 2.9×645.16 mm^2 or 1870.964 mm^2. When the shell structure has geometrical nonlinearity only, the largest peak of the variance at the same dof is about 3.75×645.16 mm^2 or 2419.35 mm^2. Note that the responses are very large. In every of the above computations 1900 time steps were performed. The cpu times were approximately 6 hours for the linear solution, 9.5 hours for the geometrically nonlinear results, and 10 hours for results with both material and geometrical nonlinearities.

Similar to the plate structure in Sub-section 6.5.1, the TATS was employed. During the computational experiments, natural frequencies were monitored and it was found that for the clamped spherical cap the fundamental frequency was reduced from 5.858×10^4 rad/s to its minimum 8.194×10^3 rad/s. This is a 86.0% reduction. Meanwhile, the highest natural frequency Ω was increased from 1.225×10^7 rad/s to 1.238×10^7 rad/s. This increase is about 1.1%. Since the change of Ω was small, application of the TATS was well justified.

6.6 Concluding Remarks

In this chapter the SL techniques for mdof nonlinear systems have been introduced. The symplectic integrator in the SN family of algorithms has been identified. For discretized nonlinear systems whose parameter matrices are not defined explicitly in the entire time duration, typically encountered in the FEA, the SCD method with TCT and ATS has been presented in Sections 6.3 and 6.4. It has been applied to evaluate the mean squares and covariances of displacement responses of plate and shell structures under point nonstationary random excitations in Section 6.5. These systems were represented by the mixed formulation-based flat triangular shell finite elements, or simply called shell elements, developed by the author and his associate [18]. Each element has three nodes. Every node has three translational and three rotational dof so that the six rigid body modes can correctly be represented. The emphasis was on large deformations involving with finite strains and finite rotations.

While it is understood that there are other approaches employing direct integration schemes for the computation of random responses of nonlinear systems idealized by the FEM in the literature, they cannot be applied to deal with large deformation problems without introducing further approximations or computational complications. For example, they require either the normal mode or complex modal analysis with some reduction techniques, such as the Karhunen-Loéve expansion [52] or analytical procedure in addition to some forms of linearization to the original system of equations. Furthermore, some orthogonalization techniques, such as the Gram-Schmidt scheme, are also necessary before solving the associated eigenvalue problems at every time step. Owing to the required applications of the linearization technique, the orthogonalization scheme, and eigenvalue solution at every time step, the presumably efficient feature of employing the Karhunen-Loéve expansion is lost. In terms of digital computer implementation, this category of techniques is believed to be far more expensive to employ and computationally it involves with more levels of approximations. On the other hand, the approaches presented in Section 6.3 are truly direct integration schemes and they do not require some other reduction techniques. These approaches consist of the TCT for stiff systems, and ATS for highly nonlinear systems with the ADCD for co-ordinate and stress updatings at every time step. The application of the TCT enables one to deal with very stiff systems or systems with a very large number of dof yet a relatively large dimensionless time step size can be retained. In particular, when the TATS is applied, the linear eigenvalue problem has essentially to be solved only once at the initial stage of the computation. Thus, the ADCD and TCT are the two very efficient features in the foregoing introduced SCD method.

In principle, many other direct integration schemes of implicit and explicit types can be applied. However, in this chapter only the SN family of algorithms has been considered and in particular, the SCD method has been exclusively employed. As pointed out in Section 6.2, this is due to the fact that the SCD method is a symplectic integrator.

References

[1]   Malhotra, A.K. and Penzien, J. (1970). Nondeterministic analysis of offshore structures, *Proc. A.S.C.E. Journal of Eng. Mech. Div.*, **96**, 985-1003.

[2]   Socha, L. (2008). *Linearization Methods for Stochastic Dynamic Systems*, Springer, New York.

[3]   To, C.W.S. (2010). *Nonlinear Random Vibration: Computational Methods*, Zip Publishing, Columbus, Ohio.

[4]   Atalik, T.S. and Utku, S. (1976). Stochastic linearization of multi-degree-of-freedom nonlinear systems, *Earthquake Engineering and Structural Dynamics*, **4**, 411-420.

[5]   To, C.W.S. (2012). *Nonlinear Random Vibration: Analytical Techniques and Applications*, Second Edition, CRC Press, Boca Raton, Florida.

[6]   Kazakov, I.E. (1965). Generalization of the method of statistical linearization to multi-dimensional systems, *Automatic Remote Control*, **26**, 1201-1206.

[7]   Wan, F.Y.M. (1973). Nonstationary response of linear time-varying dynamical systems to random excitation, *Trans. A.S.M.E. Journal of Applied Mechanics*, **40**, 422-428.

[8]   Flanders, H. (1963). *Differential Forms with Applications to Physical Sciences*, Academic Press, New York.

[9]   Arnold, V.I. (1988). *Mathematical Methods of Classical Mechanics*, Springer, New York.

[10] De Vogelaere, R. (1956). Methods of integration which preserve the contact transformation property of Hamiltonian equations, Report 4, Department of Mathematics, University of Notre Dame, Notre Dame, Indiana.

[11] Feng, K. (1986). Difference schemes for Hamiltonian formalism and symplectic geometry, *J. Comput. Math.*, **4**, 279-289.

[12] Sanchez-Serna, J.M. (1973). Runge-Kutta schemes for Hamiltonian systems, *BIT*, **28**, 877-883.

[13] Lasagni, F.M. (1988). Canonical Runge-Kutta methods, *Z. Angew. Math. Phys.*, **39**, 952-953.

[14] Chanell, P.J. and Scovel, J.C. (1989). Symplectic integration of Hamiltonian systems, Preprint.

[15] Simo, J.C., Tarnow, N. and Wong, K.K. (1992). Exact energy-momentum conserving algorithms and symplectic schemes for nonlinear dynamics, *Computer Methods in Applied Mechanics and Engineering*, **100**, 63-116.

[16] To, C.W.S. (2010). Symplectic algorithms in computational stochastic nonlinear structural dynamics. In *Proc. of 2nd Int. Symp. on Computational Mechanics*, Nov. 30-Dec. 3, 2009, Hong Kong and Macau, Part One, pp. 334-338.

[17] Saleeb, A.F., Chang, T.Y., Graf, W. and Yingyeunyong, S. (1990). A hybrid/mixed model for non-linear shell analysis and its applications to large rotation problems, *International Journal for Numerical Methods in Engineering*, **29**, 407-446.

[18] Liu, M.L., and To, C.W.S. (1995). Hybrid strain based three-node flat triangular shell elements, Part I: Nonlinear theory and incremental formulation, *Computers and Structures*, **54(6)**, 1031-1056.

[19] To, C.W.S. and Liu, M.L. (2000). Large nonstationary random responses of shell structures with geometrical and material nonlinearities, *Finite Elements in Analysis and Design*, **35**, 59-77.

[20] Horrigmoe, G. (1978). Hybrid stress finite element model for non-linear shell problems, *Int. J. for Numerical Methods in Engineering,* **12**, 1819-1839.

[21] Horrigmoe, H. and Bergan, P.G. (1976). Incremental variational principles and finite element models for nonlinear problems, *Computer Methods in Applied Mechanics and Engineering,* **7**, 201-217.

[22] Boland, P.L. and Pian, T.H.H. (1977). Large deflection analysis of thin elastic structures by the assumed stress hybrid finite element method, *Computers and Structures*, **7**, 1-12.

[23] Bathe, K.J. and Bolourchi, S. (1979). Large displacement analysis of three-dimensional beam structures, *Int. J. for Num. Meth. in Eng.*, **14**, 861-886.

[24] Gadala, M.S., Dokainish, M.A. and Oravas, G.A.E. (1984). Formulation methods of geometric and material nonlinearity problems, *International Journal for Numerical Methods in Engineering*, **20**, 887-914.

[25] Gadala, M.S. and Oravas, G.A.E. (1984). Numerical solutions of nonlinear problems of continua, I: Survey of formulation methods and solution techniques, *Computers and Structures*, **19**, 865-877.

[26] Oden, J.T. (1972). *Finite Elements of Nonlinear Continua*. McGraw-Hill Book Co., New York.

[27] Bufler, H. (1993). Nonlinear conservative systems and tangent operators, In *Proceedings of The Second International Conference on Nonlinear Mechanics*, pp. 110-113, August 23-26, Beijing, China.

[28] Belytschko, T. (1975). Nonlinear analysis descriptions and numerical stability, *Shock and Vibration Computer Programs, Reviews and Summaries*, pp. 537-562, Pilkey, W. and Pilkey, B., Eds., U.S. Department of Defense.

[29] Chiou, J.H., Lee, J.D. and Erdman, A.G. (1990). Development of a three-dimensional finite element program for large strain elastic-plastic solids, *Computers and Structures,* **36**, 631-645.

[30] Atluri, S.N. (1984). Alternate stress and conjugate strain measures, and mixed variational formulations involving rigid rotations, for computational analysis of finitely deformed solids with applications to plates and shells, I: Theory, *Computers and Structures*, **18**, 93-116.

[31] Nemat-Nasser, S. (1982). On finite deformation elasto-plasticity, *International Journal of Solids and Structures*, **18**, 857-872.

[32] Lee, E.H. (1969). Elastic plastic deformation at finite strains, *Journal of Applied Mechanics,* **36**, 1-6.

[33] Hutchinson, J.W. (1970). Elastic-plastic behaviour of polycrystalline metals and composite, *Proceedings of the Royal Society of London,* **A319**, 247-272.

[34] Wempner, G. (1969). Finite elements, finite rotations and small strains of flexible shells, *International Journal of Solids and Structures*, **5**, 117-153.

[35] Hsiao, K.M. (1987). Nonlinear analysis of general shell structures by flat triangular shell element, *Computers and Structures*, **25**, 665-675.

[36] Argyris, J.H. (1982). An excursion into large rotations, *Computer Methods in Applied Mechanics and Engineering*, **32**, 85-155.

[37] Surana, K.S. (1983). Geometrically nonlinear formulation for curved shell elements, *Int. J. for Numerical Methods in Engineering*, **19**, 581-615.

[38] Ramm, E. (1976). A plate/shell element for large deflections and rotations, *Formulations and Computational Algorithms in Finite Element Analysis*, pp. 255-291, Bathe, K.J., Oden, J.T. and Wunderlich, W., Eds., M.I.T. Press, Cambridge, Massachusetts.

[39] Besseling, J.F. (1977). Derivatives of deformation parameter for bar elements and their use in buckling and postbuckling analysis, *International Journal for Numerical Methods in Engineering,* **12**, 97-124.

[40] Simo, J.C. (1985). A finite strain beam formulation, Part I: The three-dimensional dynamic problem, *Computer Methods in Applied Mechanics and Engineering,* **49**, 55-70.

[41] Basar, Y. (1987). A consistent theory of geometrically nonlinear shells with an independent rotation vector, *Int. J. of Solids and Structures,* **23**, 1401-1415.

[42] Sansour, C. and Bufler, H. (1992). An exact finite rotation shell theory, its mixed variational formulation and its finite element implementation, *Int. Journal for Numerical Methods in Engineering,* **34**, 73-115.

[43] MacNeal, R.H. (1978). A simple quadrilateral shell element, *Computers and Structures*, **8**, 175-183.

[44] Simo, J.C. and Vu-Quoc, L. (1988). On the dynamics in space of rods undergoing large motions: A geometrically exact approach, *Computer Methods in Applied Mechanics and Engineering*, **66**, 125-161.

[45] Dokainish, M.A. and Subbaraj, K. (1989). A survey of direct time-integration methods in computational structural dynamics, I: Explicit methods, *Computers and Structures*, **32**, 1371-1386.

[46] Subbaraj, K. and Dokainish, M.A. (1989). A survey of direct time-integration methods in computational structural dynamics, II: Implicit methods, *Computers and Structures*, **32,** 1387-1401.

[47] Liu, M.L., and To, C.W.S. (1995). Vibration analysis of structures by hybrid strain based three-node flat triangular shell elements, *Journal of Sound and Vibration,***184(5)**, 801-821.

[48] To, C.W.S., and Liu, M.L. (1993). Recursive expressions for time dependent means and mean square responses of a multi-degree-of-freedom nonlinear system, *Computers and Structures*, **48(6)**, 993-1000.

[49] Liu, M.L. and To, C.W.S. (1994). Adaptive time schemes for responses of nonlinear multi-degree-of-freedom systems under random excitations, *Computers and Structures,* **52(3),** 563-571.

[50] To, C.W.S. and Liu, M.L. (1994). Random responses of discretized beams and plates by the stochastic central difference method with time co-ordinate transformation, *Computers and Structures,* **53(3),** 727-738.

[51] To, C.W.S., and Liu, M.L. (1995). Hybrid strain based three-node flat triangular shell elements, Part II: Numerical investigation of nonlinear problems, *Computers and Structures,* **54(6)**, 1057-1076.

[52] Balakrishnan, A.V. (1995). *Introduction to Random Processes in Engineering*, John Wiley and Sons, New York.

7

Direct Integration Methods for Temporally and Spatially Stochastic Nonlinear Structural Systems

As remarked in Section 6.6 in the last chapter, it is understood that there are other approaches employing direct integration schemes for the computation of random responses of nonlinear systems idealized by the FEM in the literature. They cannot be applied to deal with large deformation problems without introducing further approximations or computational complications. For techniques applicable to temporally and spatially stochastic nonlinear structural systems represented by the FEM, the so-called probabilistic finite element method (PFEM) [1, 2] or the stochastic finite element method (SFEM) [3-8] are better known examples. Applications of these approximation techniques to computation of response statistics and sensitivity analysis of systems with random parameters have been made. However, the PFEM or SFEM can only be applied to analyze dynamic systems with spatially stochastic properties whose variances are assumed to be small. In other words, the variances of the properties within the random field are small. As a rule of thumb it was suggested by Kleiber and Hien [7] that the variance of the spatially stochastic property should be within 10% of the mean value of that property.

The main objectives of this chapter are: (a) to provide a procedure suitable for the analysis of large nonlinear responses of temporally and spatially stochastic structural systems to stationary and nonstationary random exciations, (b) to apply the developed procedure to structures with homogeneous and non-homogeneous spatially stochastic properties. For completeness, PFEM or SFEM is first introduced in the next section. The SCD method with time co-ordinate transformation (TCT) and adaptive time schemes (ATS) is presented in Section 7.2 for temporally and spatially homogeneous as well as spatially non-homogeneous stochastic structural systems. Applications of the approaches presented in Section 7.2 are made in Section 7.3 for the spherical cap clamped at its circumference. The emphases in the latter section are on large nonlinear deformations and large uncertainties or spatially stochastic properties. Closing remarks are included in Section 7.4.

7.1   Perturbation Approximation Techniques and Stochastic Finite Element Methods

There are various approximation techniques employing direct integration schemes for the computation of random responses of nonlinear systems idealized by the FEM in the literature. However, they cannot be applied to deal with large deformation problems without introducing further approximations or computational complications. For example, these approximation techniques require either the normal mode or complex modal analysis with some reduction techniques, such as the Karhunen-Loéve expansion [9] or analytical procedure in addition to some forms of linearization to the original system of equations. Furthermore, some orthogonalization tehcniques, such as the Gram-Schmidt scheme [10], are also necessary before solving the associated eigenvalue problems at every time step. On the other hand, the PFEM or SFEM can be applied as an alternative technique in dealing with nonlinear dynamic systems with spatially stochastic properties whose variances are small. For completeness, the PFEM or the SFEM is introduced in the following sub-section. Sub-section 7.1.2 deals with statistical moments of responses. Solution procedure and computational steps are outlined in Sub-section 7.1.3.

7.1.1   Stochastic finite element method

Consider the matrix nonlinear equation of motion

$$M\ddot{x} + C\dot{x} + Kx = F , \tag{7.1}$$

where the assembled mass matrix M is a function of displacements, velocities, accelerations, and the u-dimensional spatially stochastic vector φ (not to be confused with the angle in Figure 6.10); the assembled damping matrix C is a function of displacements, velocities and the spatially stochastic vector φ; the assembled stiffness matrix K or $K(x,\varphi)$ is a function x and φ; the assembled external force vector F is a function of time t and φ; and x is the temporally and spatially stochastic displacement vector. The spatially random uncertainty vector is defined as

$$\varphi = \overline{r} + r, \tag{7.2}$$

in which

$$\langle \varphi \rangle^x = \overline{r}, \quad \langle r \rangle^x = (0),$$

where (0) denotes the zero vector and the angular brackets with the superscript x represent the ensemble average or the mathematical expectation in the spatial domain. This is not to be confused with the angular brackets without the superscript in the last chapter and elsewhere in this book, which designate ensemble average in the time domain.

By Taylor series expansion the random functions C, K, x and F are expanded about $\overline{r}$ and retaining up to the second order terms so that

$$M = \overline{M} + \sum_{i=1}^{u} \frac{\partial \overline{M}}{\partial \varphi_i} r_i + \frac{1}{2} \sum_{i=1}^{u} \sum_{j=1}^{u} \frac{\partial^2 \overline{M}}{\partial \varphi_i \partial \varphi_j} r_i r_j, \tag{7.3}$$

$$C = \overline{C} + \sum_{i=1}^{u} \frac{\partial \overline{C}}{\partial \varphi_i} r_i + \frac{1}{2} \sum_{i=1}^{u} \sum_{j=1}^{u} \frac{\partial^2 \overline{C}}{\partial \varphi_i \partial \varphi_j} r_i r_j, \tag{7.4}$$

$$K = \overline{K} + \sum_{i=1}^{u} \frac{\partial \overline{K}}{\partial \varphi_i} r_i + \frac{1}{2} \sum_{i=1}^{u} \sum_{j=1}^{u} \frac{\partial^2 \overline{K}}{\partial \varphi_i \partial \varphi_j} r_i r_j, \tag{7.5}$$

$$x = \overline{x} + \sum_{i=1}^{u} \frac{\partial \overline{x}}{\partial \varphi_i} r_i + \frac{1}{2} \sum_{i=1}^{u} \sum_{j=1}^{u} \frac{\partial^2 \overline{x}}{\partial \varphi_i \partial \varphi_j} r_i r_j, \tag{7.6}$$

$$F = \overline{F} + \sum_{i=1}^{u} \frac{\partial \overline{F}}{\partial \varphi_i} r_i + \frac{1}{2} \sum_{i=1}^{u} \sum_{j=1}^{u} \frac{\partial^2 \overline{F}}{\partial \varphi_i \partial \varphi_j} r_i r_j, \tag{7.7}$$

where φ_i is the i'th element of the vector φ, and the over-bar denotes the random function being evaluated at $\varphi = \overline{r}$. That is, for example,

$$\frac{\partial \overline{C}}{\partial \varphi_i} = \frac{\partial C}{\partial \varphi_i}\bigg|_{\varphi = \overline{r}}.$$

Of course, if the force vector F is a function of time t only, then the second and third terms on the rhs of Eq. (7.7) become zero.

Substituting Eqs. (7.3) through (7.7) into (7.1) and equating equal order terms of r, the governing matrix equations for the zeroth, first and second order perturbation approximations may be obtained.

Thus, the governing matrix equation of motion for the zeroth order approximation becomes

$$\overline{M}\,\overline{a} + \overline{C}\,\overline{v} + \overline{K}\,\overline{d} = \overline{F}\,, \tag{7.8}$$

where the definitions

$$a = \ddot{x}\,, \quad v = \dot{x}\,, \quad d = x\,,$$

have been used.

The governing matrix equation of motion for the first order approximation can be shown to be

$$\overline{M}\,\overline{a}_{\varphi} + \overline{C}\,\overline{v}_{\varphi} + \overline{K}\,\overline{d}_{\varphi} = \overline{F}_{\varphi} + \overline{Q}_{\varphi}\,, \tag{7.9}$$

in which

$$\overline{a}_{\varphi} = \sum_{i=1}^{u} \frac{\partial \overline{a}}{\partial \varphi_{i}}\,, \quad \overline{v}_{\varphi} = \sum_{i=1}^{u} \frac{\partial \overline{v}}{\partial \varphi_{i}}\,, \quad \overline{d}_{\varphi} = \sum_{i=1}^{u} \frac{\partial \overline{d}}{\partial \varphi_{i}}\,,$$

$$\overline{F}_{\varphi} = \sum_{i=1}^{u} \frac{\partial \overline{F}}{\partial \varphi_{i}}\,, \quad \overline{Q}_{\varphi} = -\left(\overline{M}_{\varphi}\,\overline{a} + \overline{C}_{\varphi}\,\overline{v} + \overline{K}_{\varphi}\,\overline{d}\right)\,,$$

$$\overline{M}_{\varphi} = \sum_{i=1}^{u} \frac{\partial \overline{M}}{\partial \varphi_{i}}\,, \quad \overline{C}_{\varphi} = \sum_{i=1}^{u} \frac{\partial \overline{C}}{\partial \varphi_{i}}\,, \quad \overline{K}_{\varphi} = \sum_{i=1}^{u} \frac{\partial \overline{K}}{\partial \varphi_{i}}\,.$$

Equation (7.9) is also known as the sensitivity equation for the system.

Likewise, the governing matrix equation for the second order approximation can also be shown to be

$$\overline{M}\,\overline{a}_{\varphi\varphi} + \overline{C}\,\overline{v}_{\varphi\varphi} + \overline{K}\,\overline{d}_{\varphi\varphi} = \overline{F}_{\varphi\varphi} + \overline{G}_{\varphi} + \overline{Q}_{\varphi\varphi}\,, \tag{7.10}$$

where the symbols are defined as

$$\overline{a}_{\varphi\varphi} = \sum_{i=1}^{u}\sum_{j=1}^{u} \frac{\partial\overline{a}}{\partial\varphi_i}\frac{\partial\overline{a}}{\partial\varphi_j}, \qquad \overline{v}_{\varphi\varphi} = \sum_{i=1}^{u}\sum_{j=1}^{u} \frac{\partial\overline{v}}{\partial\varphi_i}\frac{\partial\overline{v}}{\partial\varphi_j},$$

$$\overline{d}_{\varphi\varphi} = \sum_{i=1}^{u}\sum_{j=1}^{u} \frac{\partial\overline{d}}{\partial\varphi_i}\frac{\partial\overline{d}}{\partial\varphi_j}, \qquad \overline{F}_{\varphi\varphi} = \sum_{i=1}^{u}\sum_{j=1}^{u} \frac{\partial\overline{F}}{\partial\varphi_i}\frac{\partial\overline{F}}{\partial\varphi_j},$$

$$\overline{G}_{\varphi} = -\left(\overline{M}_{\varphi}\,\overline{a}_{\varphi} + \overline{C}_{\varphi}\,\overline{v}_{\varphi} + \overline{K}_{\varphi}\,\overline{d}_{\varphi}\right),$$

$$\overline{Q}_{\varphi\varphi} = -\left(\overline{M}_{\varphi\varphi}\,\overline{a} + \overline{C}_{\varphi\varphi}\,\overline{v} + \overline{K}_{\varphi\varphi}\,\overline{d}\right)$$
$$-\left(\overline{M}_{\varphi}\,\overline{a}_{\varphi} + \overline{C}_{\varphi}\,\overline{v}_{\varphi} + \overline{K}_{\varphi}\,\overline{d}_{\varphi}\right),$$

$$\overline{M}_{\varphi\varphi} = \sum_{i=1}^{u}\sum_{j=1}^{u} \frac{\partial\overline{M}}{\partial\varphi_i}\frac{\partial\overline{M}}{\partial\varphi_j}, \qquad \overline{C}_{\varphi\varphi} = \sum_{i=1}^{u}\sum_{j=1}^{u} \frac{\partial\overline{C}}{\partial\varphi_i}\frac{\partial\overline{C}}{\partial\varphi_j},$$

$$\overline{K}_{\varphi\varphi} = \sum_{i=1}^{u}\sum_{j=1}^{u} \frac{\partial\overline{K}}{\partial\varphi_i}\frac{\partial\overline{K}}{\partial\varphi_j}.$$

In the foregoing, $\overline{M}$, $\overline{C}$, $\overline{K}$ and $\overline{F}$ are, respectively M, C, K and F evaluated at $\varphi = \overline{r}$, and hence they are independent of the spatially stochastic vector φ. In fact they are, respectively, the deterministic assembled mass, damping and stiffness matrices, and the assembled external force vector. Once these matrices are obtained, the zeroth order approximation Eq. (7.8) can be solved. Then the rhs of Eq. (7.9) can be determined. In turn, the first order approximation, Eq. (7.9) can be solved. Finally, the rhs of Eq. (7.10) can also be computed so that the second order approximation, Eq. (7.10), in principle, may be solved. The computational aspects of the solution will be considered in Sub-section 7.1.3.

In the next sub-section statistical moments of responses are determined. At this stage, it may be appropriate to point out that in the literature of SFEM and PFEM whenever the first and second statistical moments are considered, no statement is made as in which domain the mathematical expectation or ensemble average is taken. For clarity, in the next sub-section such a statement is explicitly added.

7.1.2   Statistical moments of responses

The first and second order statistical moments of responses can be obtained conveniently by making use of Eq. (7.6). Thus, the second order approximation to the ensemble average in the spatial domain, of displacement vector becomes

$$\langle x \rangle^x = \bar{x} + \sum_{i=1}^{u} \frac{\partial \bar{x}}{\partial \varphi_i} \langle r_i \rangle^x + \frac{1}{2} \sum_{i=1}^{u} \sum_{j=1}^{u} \frac{\partial^2 \bar{x}}{\partial \varphi_i \partial \varphi_j} \langle r_i r_j \rangle^x ,$$

which can be reduced to

$$\langle x \rangle^x = \bar{x} + \frac{1}{2} \sum_{i=1}^{u} \sum_{j=1}^{u} \frac{\partial^2 \bar{x}}{\partial \varphi_i \partial \varphi_j} \langle r_i r_j \rangle^x , \qquad (7.11)$$

since $\langle r_i \rangle^x = 0$ and $\langle r_i r_j \rangle^x$ is the covariance function of r_i and r_j in the spatial domain. The latter covariance function represents the uncertainty of the system parameters and in general it is given for a particular problem.

Similarly, the first order approximation to the covariance matrix of displacements in the spatial domains may be obtained as

$$\langle (x - \bar{x})(x - \bar{x})^T \rangle^x = \sum_{i=1}^{u} \sum_{j=1}^{u} \left(\frac{\partial \bar{x}}{\partial \varphi_i} \right) \left(\frac{\partial \bar{x}^T}{\partial \varphi_j} \right) \langle r_i r_j \rangle^x . \qquad (7.12)$$

On the other hand, by making use of Eq. (7.6) the second order approximation to the covariance matrix of displacements in the spatial domain becomes

$$\langle (x - \bar{x})(x - \bar{x})^T \rangle^x = \Gamma_1 + \Gamma_2 , \qquad (7.13)$$

where

$$\Gamma_1 = \sum_{i=1}^{u} \sum_{j=1}^{u} \left(\frac{\partial \bar{x}}{\partial \varphi_i} \right) \left(\frac{\partial \bar{x}^T}{\partial \varphi_j} \right) \langle r_i r_j \rangle^x ,$$

$$\Gamma_2 = \frac{1}{4} \left(\sum_{i=1}^{u} \sum_{j=1}^{u} \frac{\partial^2 \bar{x}}{\partial \varphi_i \partial \varphi_j} \right) \left(\sum_{k=1}^{u} \sum_{\ell=1}^{u} \frac{\partial^2 \bar{x}}{\partial \varphi_k \partial \varphi_\ell} \right)^T \langle r_i r_j r_k r_\ell \rangle^x .$$

Since r is small and in Eq. (7.6) up to second order terms are retained, therefore the contribution of the second term on the rhs of Eq. (7.13) is small. If the second term on the rhs of Eq. (7.13) is disregarded, the first order and second order approximations, in the spatial domain, to the covariance matrices of displacements are identical. This indicates that if one is interested in the effect of the second order approximation to the variances or covariances of displacements of the nonlinear system, no significant distinction can basically be observed. This is a major shortcoming of the SFEM or PFEM.

It should be noted that in the derivation of Eq. (7.13) r_i and r_j are assumed to be normally distributed random processes. When elements of φ are uncorrelated, the covariance matrices in Eqs. (7.12) and (7.13) become diagonal.

7.1.3   Solution procedure and computational steps

With reference to Eqs. (7.11), (7.12) and (7.13), the second order ensemble average of displacements, the first order approximation to the covariances of displacements and the second order approximation to the covariances of displacements in the spatial domain are functions of $\bar{x}$ and its derivatives with respect to the spatial stochastic variables φ_i. This requires the solution of Eqs. (7.8), (7.9) and (7.10). As the system parameter matrices are nonlinear, therefore, an incremental approach to the solution is necessary. Thus, within every time step the lhs of Eqs. (7.8), (7.9) and (7.10) can be treated as those for the linear systems. In the following the issue of secularity will be addressed before outlining the computational steps involved in the SFEM for nonlinear systems.

7.1.3.1 Secularity issue in linear or nonlinear systems

At resonance in a linear or nonlinear system, in which at least one of the applied frequency is identical or close to one of the lower natural frequencies of the system, the solution of the governing equation of motion grows with time. This solution is correct and should not be confused with the unbound solution due to a particular numerical technique employed in the analysis for a relatively long duration. The unbound solution as a direct consequence of the particular numerical technique applied is known as *secularity*. It is well known in the approximate solution by perturbation techniques of second order nonlinear differential equations for deterministic nonlinear systems [9].

In the present context, the Taylor series expansion has been applied to the derivation of Eqs. (7.8), (7.9) and (7.10). Therefore the solution of Eq. (7.9) or (7.10) leads to the secularity problem if a relatively long duration is required. For instance, in the solution of the first order and second order equations, (7.9) and (7.10) respectively, the secularity problem appears whether these equations are expressed in coupled or uncoupled form. This is because, on the rhs of the first order equation there are terms associated with the solution of the zero order Eq. (7.8). Likewise, on the rhs of the second order equation there are terms associated with the solutions of the zero order and first order equations. To provide an accurate solution of response statistics for a long duration, numerical schemes have been proposed for the elimination of secularity problem. Liu, Besterfield and Belytschko [10] presented a scheme based on the application of the sine and cosine Fourier transforms. The steps in the scheme are outlined in the following.

To begin with, the forcing terms on the rhs of the first order or second order equation are separated into resonant and non-resonant components. This is achieved by performing the fast Fourier transformation (FFT) on the forcing terms on the rhs of Eq. (7.9) or (7.10). For example, on the rhs of the first order equation, Eq. (7.9), the forcing terms that contribute to the secularity are expressed as

$$\overline{Q}_{\varphi} = \Phi p(t) , \tag{7.14}$$

where the coefficient matrix Φ is independent of time t while $p(t)$ is the time dependent scalar function which can be recovered from the discrete Fourier series,

$$p(t) = a_o + \sum_{j=1}^{N} \left(a_j \cos\overline{\omega}_j t + b_j \sin\overline{\omega}_j t \right) , \tag{7.15}$$

in which a_j and b_j are, respectively, the cosine and sine transform coefficients defined as

$$a_j = \frac{1}{T} \int_0^T p(t) \cos\overline{\omega}_j t \, dt , \qquad b_j = \frac{1}{T} \int_0^T p(t) \sin\overline{\omega}_j t \, dt ,$$

and $\overline{\omega}_j$ the discrete Fourier frequency is given by

$$\overline{\omega}_j = \frac{2\pi j}{T} , \quad j = 0 , 1 , 2 , \cdots , N , \tag{7.16}$$

where T is the period or record length and N the number of Fourier coefficients or number of sampling points in the time domain. That is,

$$T = N\Delta t , \quad f_N = \frac{1}{2\Delta t} , \tag{7.17a,b}$$

where Δt is the sampling interval and f_N is the *Nyquist* or *folding frequency*. For a given period, the number of sampling points have to be chosen in such a way that the *aliasing* effect [11, 12] is avoided. This can be achieved by ensuring that $f_N >$ f_{max}, where f_{max} is the maximum frequency component existed in $p(t)$.

To eliminate secular terms, the coefficients of the Fourier series, a_j and b_j, which lie within the specified range of any of the natural frequencies of the system, are removed or weighted. That is, the resonant components in the Fourier series are removed. When the following condition is satisfied secularities occur

$$\left(\omega_i - \omega_R\right) \le \overline{\omega}_j \le \left(\omega_i + \omega_R\right), \tag{7.18}$$

where ω_i is the lower natural frequency of the system and ω_R the specified frequency range. Whenever the condition in Eq. (7.18) is satisfied, the resonant component of Eq. (7.14) is eliminated or weighted. This is repeated for all natural frequencies of the system. In this way, only the non-resonant components are retained for the computation of the responses.

In principle, many data windows are available for filtering frequencies that cause secularities. However, in Ref. [10] the triangular, cosine and cosine-squared windows were employed and it was found that th best results were obtained by using the cosine-squared weighting function. Thus, only this weighting function defined in the following is introduced

$$a_j = a_j\left(1 - \cos\theta\right)/2, \qquad \theta = \pi\left(\omega_i - \overline{\omega}_j\right)/\omega_R. \tag{7.19}$$

After the elimination of secularities for the first order equation, the foregoing procedure is repeated for the second order equation.

7.1.3.2 *Computational steps in solution of response statistics*
Once all secularities are removed, the following steps are required for the computation of second order ensemble average of displacements, the first order approximation and second order approximation to the covariances of displacements. Equations (7.8) and (7.9) can be expressed as

$$\overline{M}\overline{a}_{s+1} + \overline{C}\overline{v}_{s+1} + \overline{K}\overline{d}_{s+1} = \overline{F}_{s+1}, \tag{7.20}$$

and

$$\overline{M}\left(\overline{a}_\varphi\right)_{s+1} + \overline{C}\left(\overline{v}_\varphi\right)_{s+1} + \overline{K}\left(\overline{d}_\varphi\right)_{s+1} = \overline{G}_{s+1}, \tag{7.21}$$

$$\overline{G} = \overline{F}_\varphi + \overline{Q}_\varphi \ ,$$

in which the subscript $s+1$ denotes the quantity at the next time step such that the next time step is $t_{s+1} = t_s + \Delta t$. For instance, $\overline{G}_{s+1}$ and $\overline{K}$ are respectively the external force vector and *tangent stiffness matrix* calculated at $\overline{r}$, $\overline{d}_{s+1}$ and time step t_{s+1}. In principle, Eqs. (7.20) and (7.21) can be solved by an implicit or explicit or a mixed implicit and explicit algorithms. It was suggested in Ref. [1] that for low frequency response the implicit Newmark β scheme and for impulsive short duration transient problems the explicit numerical integration algorithm can be used to solve for the above equations. The implicit Newmark β scheme was employed [1] because of its unconditionally stable property. In Ref. [1] the *mean value* Eq. (7.20) was solved by Newton-Raphson iteration

$$\overline{K}^\varepsilon \, \Delta a_{s+1}^{\alpha+1} = \Gamma_{s+1}^\alpha \ , \tag{7.22}$$

where the effective stiffness matrix is

$$\overline{K}^\varepsilon = \left[\overline{M} + \beta \, (\Delta t)^2 \, \overline{K} \right]^\alpha \ ,$$

and the residual vector is defined by

$$\Gamma_{s+1}^\alpha = \left(\overline{F}_{s+1} - \overline{M} \overline{a}_{s+1} - \overline{C} \overline{v}_{s+1} - \overline{K} \overline{d}_{s+1} \right)^\alpha \ ,$$

in which the superscript α designates the equilibrium iteration counter at time step $s+1$, and iterations are repeated until the incremental acceleration vector $\Delta a_{s+1}^{\alpha+1}$ approaches a zero vector.

Similarly, applying the Newmark β scheme, the first order Eq. (7.21) can be expressed as

$$\overline{K}^\varepsilon \left(\overline{d}_\varphi \right)_{s+1} = - \, \beta \, (\Delta t)^2 \left(h_\varphi \right)_{s+1} + \overline{M} u_{s+1} \ , \tag{7.23}$$

where

$$\left(h_\varphi \right)_{s+1} = \overline{C} \left(\overline{v}_\varphi \right)_{s+1} + \overline{K} \left(\overline{d}_\varphi \right)_{s+1} \ ,$$

$$u_{s+1} = \left(\overline{d}_\varphi \right)_s + \Delta t \left(\overline{v}_\varphi \right)_s + \left(\frac{1}{2} - \beta \right) (\Delta t)^2 \left(\overline{a}_\varphi \right)_s \ .$$

The effective stiffness matrix $\bar{K}^\varepsilon$ is identical in both Eqs. (7.22) and (7.23).

7.1.3.3 *Closing remarks*

The secularity condition given by Eq. (7.18) is that for the *primary resonance* of linear or nearly linear systems. For nonlinear systems there are additional secularities associated with *secondary resonances* [9]. These secondary resonances include superharmonic and subharmonic resonances, for example. In the context of SFEM for the computation and analysis of response statistics in nonlinear systems, it seems that the issue of secularities of secondary resonances has not been addressed.

7.1.4 Concluding remarks

The idea central to the SFEM and PFEM seems to be first stated in Problem 3.14 of the book by Hart [13] in the sense that the Taylor series expansion was employed to approximate the equation of motion. The PFEM or SFEM can only be applied to linear and nonlinear structural systems with stochastic properties or parameter uncertainties of small variations. In addition, the PFEM or SFEM has other important drawbacks. First, as pointed out in Ref. [7], its accuracy may deteriorate even with a slight structural damping. Second, to ensure validity of the response to a long duration of excitation the secularity problem has to be eliminated in the computation. The secularity problem dealt with and reported in the literature is concerned with primary resonance of the system. The secularity problems associated with secondary resonances have yet to be investigated. Third, the solution is essentially hinged on the sensitivity equation of the system, for example, Eq. (7.9), and therefore the stress and co-ordinate updatings in systems with large deformation or large response with plastic deformation can not be handled even if stochastic properties with small variations are assumed. Finally, even if the application of the FFT in dealing with the secularity problems is disregarded, the computational time can be very formidable for engineering structures approximated by a large number of dof because of the process of double summation operations as indicated in Eqs. (7.11), (7.12), and (7.13) in which quadruple summations also have to be performed.

7.2 Stochastic Central Difference Methods for Temporally and Spatially Stochastic Nonlinear Systems

In Chapter 6, the development and application of the SCD with TCT and ATS to temporally stochastic nonlinear structures discretized by the FEM have been presented. In particular, the deterministic nonlinear finite element formulation and issues for dynamic shell analysis have been outlined in Sub-section 6.3.1. In this section, the formulation and developments of the SCD-MATS with TCT for application to temporally and spatially stochastic nonlinear systems with large stochastic spatial variations or parameter uncertainties are presented. The approach considered in this section is in contrast to those techniques based on the SFEM or SFEM that was introduced in the last section. The techniques in the latter section can only be applied to spatially homogeneous or spatially stationary stochastic systems with small spatial stochastic variations. In the present section the emphasis of the development of the SCD-MATS with TCT is on finite element-based nonlinear stochastic structures with finite deformations including finite strain and finite rotation. The nonlinear parameter matrices, such as the tangent stiffness matrix, are evaluated at every time step. The formulation and development of spatially homogeneous stochastic nonlinear systems are included in Sub-section 7.2.1 whereas Sub-section 7.2.2 is concerned with the formulation and development of the approach for spatially non-homogeneous stochastic nonlinear systems. In each of these two sub-sections, a summary of steps in computer program implementation is presented.

7.2.1   Temporally and spatially homogeneous stochastic nonlinear systems

From the FE standpoint, the following direct integration approach simplifies the computation and analysis of the dynamic response of temporally and spatially stochastic nonlinear structures with large stochastic spatial variations. In addition, no normal mode or complex normal mode analysis or other reduction scheme is required in the computation of variances and covariances of displacements, no restriction is placed on the type of mass, stiffness and damping matrices, and in general the dimensionless time step size is always unity for structural systems discretized by the FEM. In other words, the procedure is general, relatively simple, accurate and efficient compared with other methods based partially or entirely on the MCS.

7.2.1.1 Generalized stochastic central difference method

Consider the matrix equation of motion for a general nonlinear spatially stochastic system under combined deterministic and random excitations [14, 15]

$$M(x,\dot{x})\ddot{x} + C(x,\dot{x})\dot{x} + \left[K(x) + r\Phi(x)\right]x = F^G = F^d + F^r \qquad (7.24)$$

in which the spatially random variable r can be large whereas the superscripts G, d, and r, denote general excitation, deterministic excitation, and random excitation, respectively. The other symbols have similar meaning to those already defined in Eq. (6.27) which is identical to Eq. (7.24) except for the second term inside the square brackets on the lhs of the latter equation. Further, every term on the lhs of Eq. (7.24) is temporally and spatially stochastic in general.

The meaning and origin of the second term inside the square brackets on the lhs of Eq. (7.24) and its derivation have been provided in Chapter 1 for its corresponding linear counterpart.

By disregarding the arguments for conciseness, the discrete time version of Eq. (7.24) can be written as

$$M_s \ddot{x}_s + C_s \dot{x}_s + \left(K_s + r\Phi_s \right) x_s = F_s^G = F_s^d + F_s^r \tag{7.25}$$

where the temporally discretized system matrices M_s, C_s, K_s, and Φ_s are considered linear within time step t_s. These system matrices have to be updated at every time step for finite deformation cases. It is noted that the present finite element procedure adopted the exponential mapping in dealing with rotations in the generalized displacement vector at every time step. Therefore, the present procedure is applicable to cases with small and finite rotations.

In order to understand the meaning of the spatially stochastic term $r\Phi_s$ inside the brackets on the lhs of the last equation the following remarks are helpful.

Remark 7.2.1.1
If $E_h = E + r$, is a spatially homogeneous random variable such that E is the deterministic Young's modulus of elasticity and r is the spatially random component with zero ensemble average in the spatial domain, then Φ_s in Eq. (7.25) is equal to K_s/E, where K_s is the assembled tangent stiffness matrix at time step t_s.

Remark 7.2.1.2
Aside from the spatially homogeneous stochastic properties of the modulus of elasticity, other geometrical properties, such as the second moment of cross-section, can be considered. In this way, Φ_s is equal to K_s divided by the geometrical property. More sophisticated spatially homogeneous stochastic models [16, 17] can be considered but are not included in the present book.

Remark 7.2.1.3
In general, if the spatially random variable r is associated with the individual element stiffness matrix, Eq. (7.25) can still be applied except that in Eq. (7.28) which will follow, the covariance matrix of forcing functions has to be considered one element at a time and then the covariance matrix of the forcing functions for the entire system is assembled accordingly.

By applying a procedure similar to that for the derivation of Eq. (6.31), one can show that the ensemble average of the displacement vector in the spatial domain becomes

$$\left\langle\left\langle x_{s+1}\right\rangle\right\rangle^x = (\Delta t)^2 N_1\left(F_s^d\right) + N_2\left\langle\left\langle x_s\right\rangle\right\rangle^x + N_3\left\langle\left\langle x_{s-1}\right\rangle\right\rangle^x ,\qquad(7.26)$$

where the inner angular brackets designate ensemble average in the time domain, while the outer angular brackets with the superscript x denote ensemble average in the spatial domain. Equation (7.26) shall be referred to as the extended average of the deterministic central difference (EADCD) scheme.

The derivation of the temporally and spatially stochastic recursive mean square matrix of generalized displacements or, for conciseness, simply referred to as mean square matrix, requires similar steps to those for the derivation of Eq. (6.29). The main difference here is that an additional step is required by taking the ensemble average in the spatially domain. In other words, the first two steps in the present procedure are similar to those for the linear system in Chapter 3. Thus, the steps for the present procedure are:

(*a*) post-multiplying Eq. (3.3) by (3.5),
(*b*) taking ensemble average in the time domain,
(*c*) then taking ensemble average in the spatial domain, and
(*d*) simplifying.

After simplification, one obtains the mean square matrix as

$$\left\langle R_{s+1}\right\rangle^x = \Pi_1 + \Pi_2 ,\qquad(7.27)$$

where

$$\Pi_1 = N_2\left\langle R_s\right\rangle^x N_2^T + N_3\left\langle R_{s-1}\right\rangle^x N_3^T$$
$$+ N_2\left\langle D_s\right\rangle^x N_3^T + N_3\left\langle D_s^T\right\rangle^x N_2^T ,\qquad(7.28a)$$

$$\left\langle D_s\right\rangle^x = \left\langle\left\langle x_s x_{s-1}^T\right\rangle\right\rangle^x ,\qquad(7.28b)$$

$$\left\langle D_s\right\rangle^x = (\Delta t)^2 N_1 F_{s-1}^d\left\langle\left\langle x_{s-1}^T\right\rangle\right\rangle^x + N_2\left\langle R_{s-1}\right\rangle^x + N_3\left\langle D_{s-1}^T\right\rangle^x ,\qquad(7.28c)$$

$$\Pi_2 = (\Delta t)^4 N_1 \left\langle B_s^x \right\rangle^x N_1^T + (\Delta t)^2 N_2 \left\langle\!\left\langle x_s \left(F_s^G \right)^T \right\rangle\!\right\rangle^x N_1^T$$

$$+ (\Delta t)^2 N_1 \left\langle\!\left\langle \left(F_s^G \right) x_s^T \right\rangle\!\right\rangle^x N_2^T$$

$$+ (\Delta t)^2 N_3 \left\langle\!\left\langle x_{s-1} \left(F_s^G \right)^T \right\rangle\!\right\rangle^x N_1^T \tag{7.28d}$$

$$+ (\Delta t)^2 N_1 \left\langle\!\left\langle \left(F_s^G \right) x_{s-1}^T \right\rangle\!\right\rangle^x N_3^T \,,$$

$$\left\langle B_s^x \right\rangle^x = B_s^r + F_s^d \left(F_s^d \right)^T + B_2^x \Phi_s \left\langle R_s \right\rangle^x \Phi_s^T \,, \tag{7.28e}$$

$$B_s^r = \left\langle\!\left\langle F_s^r \left(F_s^r \right)^T \right\rangle\!\right\rangle^x = \left\langle F_s^r \left(F_s^r \right)^T \right\rangle, \qquad B_2^x = \left\langle r^2 \right\rangle^x, \tag{7.28f,g}$$

in which the coefficient matrices N_1, N_2, and N_3 have the forms similar to those defined in Eq. (6.29), while Eq. (7.28g) gives the variance of the spatially stochastic variable r. For nonstationary stochastic excitations having uniformly modulating functions, the first term on the rhs of Eq. (7.28e) is similar in form to Eq. (6.32a) and is defined in Eq. (7.28f). Further simplification of Eq. (7.28d) is possible and will be considered in the following.

Remark 7.2.1.4
Equation (7.26) is applied in the updated Lagrangian finite element procedure. More precisely, it is employed in the configuration and stress updatings as indicated in Chapter 1.

Remark 7.2.1.5
Superficially, Eq. (7.26) seems to be independent of the temporally stochastic components of the excitation vector. However, on close inspection it does reveal that Eq. (7.26) is dependent of the temporally stochastic components of the excitation vector because of Eq. (7.28e). Specifically, Eqs. (7.26) and (7.28e) are related through the common discretized deterministic force vector, F_s^d.

Upon further simplification, Eq. (7.27) becomes

$$\left\langle R_{s+1} \right\rangle^x = \Pi_1 + \Pi_{2m} \,, \tag{7.29}$$

where the second term on the rhs of Eq. (7.29) is defined as

$$
\begin{aligned}
\Pi_{2m} = & (\Delta t)^4 N_1 \left\langle B_s^x \right\rangle^x N_1^T + (\Delta t)^2 N_2 \left\langle\!\left\langle x_s \right\rangle\!\right\rangle^x \left(F_s^d\right)^T N_1^T \\
& + (\Delta t)^2 N_1 \left(F_s^d\right) \left\langle\!\left\langle x_s^T \right\rangle\!\right\rangle^x N_2^T \\
& + (\Delta t)^2 N_3 \left\langle\!\left\langle x_{s-1} \right\rangle\!\right\rangle^x \left(F_s^d\right)^T N_1^T \\
& + (\Delta t)^2 N_1 \left(F_s^d\right) \left\langle\!\left\langle x_{s-1}^T \right\rangle\!\right\rangle^x N_3^T .
\end{aligned}
\tag{7.30}
$$

Similar to the derivation of Eq. (6.35), the expression for the recursive covariance matrix of the temporally and spatially stochastic nonlinear system becomes

$$
\begin{aligned}
\left\langle U_{s+1} \right\rangle^x = & (\Delta t)^4 N_1 \left[\left\langle B_s^r \right\rangle^x + B_2^x \Phi_s \left\langle R_s \right\rangle^x \Phi_s^T \right] N_1^T \\
& + N_2 \left\langle U_s \right\rangle^x N_2^T + N_3 \left\langle U_{s-1} \right\rangle^x N_3^T \\
& + N_2 \left\langle A_s^m \right\rangle^x N_3^T + N_3 \left\langle \left(A_s^m\right)^T \right\rangle^x N_2^T ,
\end{aligned}
\tag{7.31}
$$

where

$$
\left\langle A_s^m \right\rangle^x = N_2 \left\langle U_{s-1} \right\rangle^x + N_3 \left\langle \left(A_{s-1}^m\right)^T \right\rangle^x ,
\tag{7.32a}
$$

$$
\left\langle U_s \right\rangle^x = \left\langle R_s \right\rangle^x - \left\langle\!\left\langle x_s \right\rangle\!\right\rangle^x \left\langle\!\left\langle x_s^T \right\rangle\!\right\rangle^x ,
\tag{7.32b}
$$

$$
\left\langle A_s^m \right\rangle^x = \left\langle D_s \right\rangle^x - \left\langle\!\left\langle x_s \right\rangle\!\right\rangle^x \left\langle\!\left\langle x_{s-1}^T \right\rangle\!\right\rangle^x .
\tag{7.32c}
$$

Remark 7.2.1.6
An efficient route for the computation of recursive mean square and covariance matrices of generalized displacements of the temporally and spatially stochastic nonlinear system is to apply Eq. (7.31) for $\langle U_s \rangle^x$ and then determine the mean square $\langle R_s \rangle^x$ by Eq. (7.32b). This route is repeated for all time steps.

Remark 7.2.1.7
In the FE context and similar to the TATS strategy in Chapter 6, the nonlinear system

matrices have to be updated at every time step. The updated Lagrangian formulation is adopted in the present procedure so that it is called the modified TATS. It is similar to the TATS except that during the co-ordinate and stress updatings, Eq. (6.31) is replaced by Eq. (7.26). These co-ordinate and stress updatings are performed in conjunction with Eqs. (7.29) and (7.31) for the recursive mean square and covariance matrices of generalized displacements, respectively.

Remark 7.2.1.8
Unlike other techniques in the literature, the present procedure has no restriction on the finite magnitude of the variation of r defined in Eq. (7.24). Furthermore, the presently introduced procedure is efficient for dealing with nonlinear structures undergoing finite deformations, including finite strain and finite rotation. It is relatively easy to implement in a commercially available FE package.

7.2.1.2 Summary of steps in computer program implementation
The main steps in the digital computer program written for the foregoing procedure are very much similar to those outlined in Section 6.4. They are summarized in the following.

Step 1: Create an input data file. This includes element geometric and material properties of the discretized system, system damping parameters, input force data, and parameters of the spatial variations.

Step 2: The element matrices and force vectors are formed. The matrix associated with the spatially stochastic variable is evaluated in accordance with *Remarks 7.2.1.1, 7.2.1.2* and *7.2.1.3.* In turn, M_s, K_s, and Φ_s matrices, and force vectors are assembled.

Step 3: The highest natural frequency of the discretized linear system is evaluated. For checking purposes, and if proportional damping is assumed, the lowest natural frequencies are computed at this stage. With these lowest natural frequencies and the damping parameters input in Step 1, the associated coefficients of the damping matrix equation are determined.

Step 4: Based on the highest natural frequency in Step 3, the TCT is applied so that the resulting matrix equation of motion is expressed in the dimensionless time domain. It should be noted that the dimensionless time step size for the computations of recursive response statistics is $\Delta\tau = 1.0$ which is given by Eq. (3.43) for the SCD method.

Step 5: The modified TATS or other member(s) of the modified SATS and ATST is (are) initialized and performed.

Step 6: For large deformation problems, the system parameter matrices, such as the assembled tangent stiffness matrix and Φ_s are updated at every dimensionless time step. This requires, in turn, updating co-ordinates and stresses in the incremental updated Lagrangian formulation. During this step, the EADCD scheme is applied.

Step 7: Step 6 is repeated until the temporally and spatially stochastic variances, covariances and mean squares of responses of the discretized nonlinear system for all dimensionless time steps are obtained.

Step 8: Applying Eq. (3.13) and the relation $\langle U_s \rangle^x = \Omega \langle U_{\tau s} \rangle^x$ with $\langle U_{\tau s} \rangle^x$ being the covariance matrix of displacements at dimensionless time step τ_s in the τ-domain, to transform the results in Steps 5 and 7 back into the original time domain of the problem. Repeat this step for all the time steps of the responses. These transformed solutions are stored into an output file for analysis and presentation.

7.2.2 Temporally and spatially non-homogeneous stochastic nonlinear systems

In Sub-sections 7.2.1 a novel approach has been introduced to the computation of temporally and spatially dependent mean squares of generalized displacement responses of relatively sophisticated highly nonlinear structures with large spatially stochastic variations. The structures are represented by the FEM and the time domain solution is achieved by employing the SCD method with TCT and modified SATS or ATST or TATS. In this sub-section, the foregoing approach is extended to deal with temporally and spatially stochastic nonlinear systems with large spatially non-homogeneous variations of system parameters. The variational principle and element matrices for this class of structures are similar to those with large spatially homogeneous variations presented in Section 1.2 for linear systems. The approach in Section 1.2 can be modified and incorporated into Section 1.4 for nonlinear systems. In the following, the new approach is presented.

7.2.2.1 Generalized stochastic central difference method for systems with spatially non-homogeneous variations

The matrix equation of motion for a geometrically and materially nonlinear temporally and spatially stochastic system under combined deterministic and random excitations is similar to Eq. (7.24) and can be written as

$$M(x,\dot{x})\ddot{x} + C(x,\dot{x})\dot{x} + \left[K(x) + r_n\Phi(x)\right]x = F^G = F^d + F^r \qquad (7.33)$$

and its discrete time version is similar to Eq. (7.25). Thus, the discrete time version of Eq. (7.33) is

$$M_s\ddot{x}_s + C_s\dot{x}_s + \left(K_s + r_n\Phi_s\right)x_s = F_s^G = F_s^d + F_s^r, \qquad (7.34)$$

where the superscripts G, d, r, and other symbols have been defined previously, except that the coefficient, r_n is a general non-homogeneous spatially stochastic variable that represents the parameter uncertainty.

Remark 7.2.2.1
If the spatially non-homogeneous random variable is the modulus of elasticity such that $E_n = E + r_n$, in which E is the deterministic Young's modulus of elasticity and r_n is the spatially non-homongeneous random component with zero ensemble average in the spatial domain, then similar to that in **Remark 7.2.1.1**, Φ_s in Eq. (7.34) is equal to K_s/E, where K_s is the assembled tangent stiffness matrix for the system at time step t_s.

Remark 7.2.2.2
While the modulus of elasticity can frequently be measured accurately and easily in practice and therefore is considered in this chapter, other geometrical properties, such as second moment of the cross-section, can be similarly invesstigated. The remaining remarks in Sub-section 7.2.1 can be similarly applied here.

Remark 7.2.2.3
It is understood that other forms of modulating or envelope functions in the spatial domain for the temporally and spatially stochastic parameters can be applied but only the Young's modulus with envelope functions in the spatial domain will be illustrated in this chapter. Assuming the spatially non-homogeneous random component of the Young's modulus of elasticity is defined by

$$r_n = e_x r \qquad (7.35)$$

in which the spatial envelope function, e_x is defined by

$$e_x = H_x \left(e^{-\alpha_1 x} - e^{-\alpha_2 x} \right) \tag{7.36}$$

while r is the spatially homogeneous stochastic component of the Young's modulus of elasticity, and the subscript and power x designate the spatial domain and spatial coordinate, respectively. The parameters α_1 and α_2 are positive constants satisfying $\alpha_1 < \alpha_2$. The factor H_x is constant and applied to normalize the envelope function e_x such that the maximum value of the rhs of Eq. (7.36) is unity. That is, $max\{e_x\} = 1.0$. Equations (7.35) and (7.36) are similar in form to Eqs. (6.33) and (6.34). In other words, the non-homogeneous spatial random variable in Eq. (7.35) is defined as a product of an envelope function in the spatial domain and a stationary or homogeneous spatial random variable with zero ensemble average in the spatial domain. That is, $\langle r \rangle^x = 0$.

Following similar steps in Sub-section 7.2.1, the equations for the mean square matrix and covariance matrix of displacement vectors can be easily shown to be in the forms of Eqs. (7.29) and (7.31), respectively, except that Eq. (7.28g) now is defined by

$$B_2^x = e_x^2 \langle r^2 \rangle^x . \tag{7.37}$$

Remark 7.2.2.4

Other forms of the spatially non-homogeneous random component of the Young's modulus can be easily considered. For example, Eq. (7.35) may be replaced by the following relation

$$r_n = e_{xy} r , \tag{7.38}$$

in which now the spatially deterministic envelope function e_{xy} is dependent of two perpendicular global directions such that

$$e_{xy} = H_{xy} \left(e^{-\alpha_{1x} x} - e^{-\alpha_{2x} x} \right) \left(e^{-\alpha_{1y} y} - e^{-\alpha_{2y} y} \right), \tag{7.39}$$

where the subscripts and powers x and y denote the spatial co-ordinates along the global co-ordinates X and Y of the system, the factor H_{xy} is constant and applied to normalize e_{xy} so that $max\{e_{xy}\} = 1.0$, and Eq. (7.37) is replaced by

$$B_2^x = e_{xy}^2 \langle r^2 \rangle^x . \tag{7.40}$$

Remark 7.2.2.5
Similar to the reasoning in Sub-section 7.2.1, an efficient route of computation for recursive mean square and covariance of generalized displacements of the temporally and non-homogeneous spatially stochastic nonlinear system is to first apply Eq. (7.31) for the covariance matrix $\langle U_{s+1} \rangle^x$ and then determine mean square matrix $\langle R_s \rangle^x$ by Eq. (7.32b). That is, the steps are identical to those of the temporally and spatially homogeneous stochastic nonlinear systems considered in Sub-section 7.2.1.

7.2.2.2 Computer program implementation
The computations of nonlinear recursive covariance and mean square matrices of displacements in the temporal and spatial domains can be performed by following the steps outlined in Sub-section 7.2.1. In the present case, Eqs. (7.31) and (7.29) can be employed with Eq. (7.28g) being replaced by either Eq. (7.37) or (7.40), depending on whether the stochastic non-homogeneous spatial variation is in one or two global directions.

The application of TCT, modified SATS or ATST or TATS, and updating the configuration and stresses at every time step for the finite deformations is similar to that for the spatially stochastic homogeneous case in Sub-section 7.2.1.

As the computer program developed for the temporally and spatially stochastic case with large homogeneous random variation can be easily modified for large non-homogeneous random variation problems, and their steps in the computer program implementation are similar, they are not repeated here.

7.3 Finite Deformations of Spherical Shells with Large Spatially Stochastic Parameters
For application of the approaches introduced in Section 7.2 the spherical cap investigated in Sub-section 6.5.2 is adopted in this section. The sketch of the clamped cap and its FE representation for one quarter of the cap are provided in Figures 6.10 and 6.11. Thus, the geometrical and material properties, and the boundary as well as symmetry conditions are those included in Sub-section 6.5.2. The point load applied at the apex of the cap is shown in Figure 6.10 and is defined by Eq. (6.69). Recall that this cap is represented by 36 elements and 28 nodes as indicated in Figure 6.11. The low order mixed or hybrid strain formulation based triangular shell finite element identified as $AT+(k^3_t)'$ is employed for the studies.

While the damping of the system can be general, for simplicity and readily available data, proportional damping identical to that considered in Sub-section 6.5.2 is employed in this section. This spherical cap with spatially stochastic homogeneous properties is studied in the following Sub-section 7.3.1 whereas its spatially stochastic non-homogeneous case is included in Sub-section 7.3.2.

7.3.1   Spherical cap with spatially homogeneous properties

Before the spatially homogeneous stochastic spherical cap is considered, the case without spatially stochastic properties is studied first in this sub-section. This case is similar to that in Sub-section 6.5.2 except that the magnitudes of the deterministic component of the point loading is now $P = 3.8977$ N (0.875 lb) and the spectral density S_0 of the nonstationary random component of the point load with the deterministic modulating function being defined by Eq. (6.69) equals to 0.99 N^2 (0.05 lb^2). Of course, for this case the first two and the highest natural frequencies are those presented in Sub-section 6.5.2 in which the Rayleigh damping coefficients for the first two modes with 5% damping each are given. Representative computed results by the approach presented in Sub-section 7.2.1 for the spherical cap without spatial uncertainty are plotted in Figure 7.1. Note that the node numbers in Figure 7.1 are those indicated in Figure 6.11.

With reference to Figure 7.1, the largest mean square of displacement is 62.39 mm^2 at Node 1 and occurs at 0.104 ms whereas the peak value of the displacement at Node 10 has not been reached within the 0.20 ms range of the plots. The peak mean square of displacement at Node 6 is 20.99 mm^2 and occurs at 0.112 ms. This indicates that the peak mean square of displacement takes 8 μs to reach Node 10 from the point of load application. It may be noted that the root-mean-square (rms) of displacement at Node 1 is 7.8987 mm, which is 3.62 times the sagitta H. This means that the deflection is large.

Now, the spherical cap with spatially homogeneous stochastic properties or uncertainties is considered. The latter are the modulus of elasticity and thickness of the shell. The other pertinent data are identical to those of the temporally stochastic and spatially deterministic case whose computed results have been presented in Figure 7.1. Selected computed results by applying the approach in Sub-section 7.2.1 are included in Figure 7.2 in which the legend NSU means no spatial uncertainty, SH Modulus denotes spatial uncertainty of the modulus of elasticity, and SH Thickness indicates spatial uncertainty of thickness for all the elements representing the cap. Specifically, the spatial uncertainty of the modulus of elasticity is $\langle r^2 \rangle^x = 0.040 E^2$ in the present investigation. The latter magnitude corresponds to a spatial standard deviation of $0.20E$ or 20% of the deterministic Young's modulus of elasticity. In other words, the uncertainty is large. Other large values of $\langle r^2 \rangle^x$ were studied in the computation but not included in this sub-section for brevity.

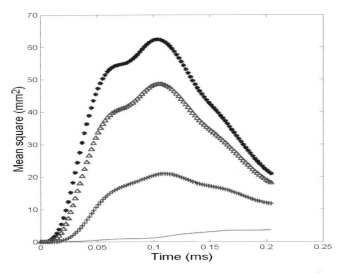

Figure 7.1   Mean squares of displacements of cap without uncertainty: **, at Node 1; △△, at Node 3; + +, at Node 6; —, at Node 10.

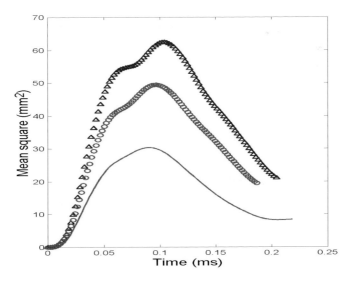

Figure 7.2   Mean squares of displacements at Node 1 of cap: △△, NSU; ○ ○, SH Modulus; —, SH Thickness.

The specific uncertainty of thickness presented in Figure 7.2 is $\langle r^2 \rangle^x = 0.049h^2$ which corresponds to a thickness standard deviation of $0.22h$ or 22% of the deterministic thickness of 0.4 mm (0.01575 in). The first two natural frequencies and corresponding Rayleigh damping coefficients required for the application of the approach in Sub-section 7.2.1 are included in Table 7.1. The highest natural frequency for the uncertainty in the modulus of elasticity case has been found to be $\Omega = 1.3414 \times 10^7$ rad/s whereas that for the uncertainty in thickness system $\Omega = 1.1463 \times 10^7$ rad/s.

With reference to the computed results presented in Figure 7.2, one can conclude that the presence of either uncertainty in the modulus of elasticity or uncertainty in the thickness increases the stiffness of the system and therefore, the peak value of the mean square of displacement at Node 1 is reduced.

Before leaving this sub-section, it may be appropriate to note that the plots presented in Figures 7.1 and 7.2 are very different from those included in [18] and [19]. This is due to the fact that the density of the spherical cap was applied as 0.094511 lb/in^3 in the latter references instead of 2.45×10^{-4} lb-sec^2/in^4 which is 2621.54 kg/m^3 in the present studies.

Table 7.1   Natural frequencies and Rayleigh coefficients of damping.

System	Natural frequencies (rad/s)		Rayleigh coefficients	
	ω_1	ω_2	λ_m (10^3/s)	λ_k (10^{-7} s)
SH Modulus	64180	80510	3.571	6.910
SH Thickness	60250	80250	3.441	7.110

7.3.2   Spherical cap with spatially non-homogeneous properties

The geometrical and material properties, boundary conditions, symmetry conditions, and the nonstationary random excitation in the last sub-section are applied in the present studies. The approach in Sub-section 7.2.2 for spatially non-homogeneous stochastic systems is employed. To limit the scope only the spatially non-homogeneous stochastic modulus of elasticity is included although other non-homogeneous uncertainties can be similarly investigated. For illustration, simplicity and without loss of generality, non-homogeneous uncertainty in one global direction is examined. Thus, the uncertain modulus of elasticity is given as $E_n = E + r_n$ in which r_n is defined by Eq. (7.35) with the deterministic spatial envelope function, e_x now defined as

$$e_x = e^{-\alpha_1 x}. \tag{7.41}$$

In words, the spatially non-homogeneous stochastic modulus of elasticity is treated as a product of the spatially deterministic envelope function and a zero mean spatially homogeneous stochastic modulus of elasticity. The latter is kept identical to that in the last sub-section for direct comparison. That is, $\langle r^2 \rangle^x = 0.040E^2$ is applied in the present studies. The spatially stochastic variable r_n is assumed to be non-homogeneous along the global X-axis only, in which $X = 11.86$ mm (0.45 in). This is at the middle of the radius of the projected circle of the spherical cap. That is, it is at half of the value of 120.9 mm (4.76 in) multiplied by $\sin 10.9°$. In Eq. (7.41), $\alpha_1 = 7.61/\text{m}$ (0.19/in) has been applied. The first two natural frequencies for the present system are $\omega_1 = 61450$ rad/s and $\omega_2 = 77080$ rad/s, while the highest natural frequency is $\Omega = 1.2843 \times 10^7$ rad/s. The corresponding Rayleigh damping coefficients for 5% damping of the first two modes are $\lambda_m = 3.419 \times 10^3 \text{ s}^{-1}$ and $\lambda_k = 7.210 \times 10^{-7}$ s.

Selected computed mean squares of transverse displacements by using the approach in Sub-section 7.2.2 are included in Figure 7.3 in which the legend SH refers to spatially homogeneous uncertain modulus of elasticity, and SNH denotes spatially non-homogeneous uncertain modulus of elasticity along the X axis. For the SNH case, the transverse displacement at Node 6 is also included for comparison. It is observed that the displacement at Node 6 is considerably smaller than that at Node 1. This indicates that the amplitude of wave propagation from Node 1, the point of load application, to Node 6 is significantly attenuated.

7.4 Closing Remarks

Several remarks are in order. First, in this chapter relatively simple and straightforward approaches for dealing with large random responses of highly nonlinear systems with uncertain properties approximated by the FEM have been

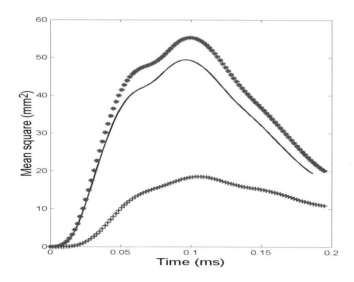

Figure 7.3   Mean squares of displacements of cap with uncertainties: —, SH at Node 1; **, SNH at Node 1; + +, SNH at Node 6.

presented. It includes the important features of being able to deal efficiently with large deformations of finite strain and finite rotation, and a relatively wide class of nonstationary random excitations.

Second, they can be applied to systems with homogeneous and non-homogeneous spatially stochastic properties of large stochastic variations. They are computationally very efficient compared with those based partially or entirely on the MCS.

Third, the proposed approaches are free from the limitations of (*a*) secularities due to primary and secondary resonances that are inherent to the second order approximation of the PFEM or SFEM, and (*b*) the size effect of random field elements [20].

Fourth, future work can be pursued for the above cases with narrow band nonstationary stochastic and non-Gaussian random excitations.

Fifth, during the derivations of the recursive expressions for the mean squares and covariances of displacements, the time and space have been assumed to be uncorrelated. However, partially and fully correlated time and space problems can be similarly dealt with. This is left to the interested readers.

Sixth, if the excitation vector on the rhs of Eq. (7.24), for example, is spatially and temporally stochastic, the foregoing procedures can be similarly applied with due modifications to Eqs. (7.28d) through (7.28g).

Seventh, it may be logical, in term of accuracy, to refine the above finite element representation of the spherical cap. However, early results [21] of the shell structure having the same mesh of 28 nodes and 36 elements, and subject to a deterministic apex point load were compared with those of 46 nodes and 72 elements, and 51 nodes and 80 elements indicating that they were very close to those presented by Oliver and Onate [22]. Thus, for computational efficiency, the 28 nodes and 36 elements mesh was used in the present chapter.

Eighth, for large deformation problems in which the co-ordinates and stresses have to be updated at every time step, Eq. (7.26), for example, is applied. Superficially, this latter equation leads to zero EADCD when the deterministic component of the forcing vector is zero. This seems to post two major problems in the evaluation of the responses of the nonlinear system. However, as pointed out in Ref. [18] this issue can be easily dealt with.

Finally, it should be mentioned that there are other techniques in the literature for dealing with large stochastic variations of the stiffness matrix. The non-parametric approach presented in Ref. [23] is a good example. However, it requires the modal analysis and can only deals with systems undergoing small deformations. Even if the approach in the latter reference is modified to be applied to the cases dealt with in this chapter, it is likely to be of low efficiency in comparison to the proposed approaches in that MCS is employed in [23].

References

[1] Liu, W.K., Belytschko, T. and Mani, A.(1986). Probabilistic finite elements for non-linear structural dynamics, *Computer Methods in Applied Mechanics and Engineering*, **56**, 61-81.

[2] Liu, W.K., Belytschko, T. and Mani, A. (1987). Applications of probabilistic finite element methods in elastic/plastic dynamics, *Trans. A.S.M.E. Journal of Engineering for Industry*, **109(1)**, 2-8.

[3] Sun, T.C. (1979). A finite element method for random differential equations with random coefficients, *S.I.A.M. J. of Num. Analysis* **16(6)**, 1019-1035.

[4] Hisada, T. and Nakagiri, S. (1980). A note on stochastic finite element method, Part 3: An extension of the methodology to non-linear problems, *Seisan-Kenkyu*, **32(12)**, 572-575.

[5]   Nakagiri, S. and Hisada, T. (1985). *Stochastic Finite Element Method* (in Japanese), Baifukan, Tokyo.

[6]   Ghanem, R. and Spanos, P.D. (1991). *Stochastic Finite Elements: A Spectral Approach,* Springer-Verlag, New York.

[7]   Kleiber, M. and Hien, T.D. (1992). *Stochastic Finite Element Method,* John Wiley and Sons, New York.

[8]   Köylüoğlu, H.U., Nielsen, S.R.K. and Cakmak, A.S. (1996). Stochastic dynamics of geometrically nonlinear structures with random properties subject to stationary random excitation, *J. of Sound and Vibr.,* **190(5)**, 821-841.

[9]   Nayfeh, A.H. and Mook, D.T. (1979). *Nonlinear Oscillations,* John Wiley and Sons., Chichester, U.K.

[10] Liu, W.K., Besterfield, G. and Belytschko, T. (1988). Transient probabilistic systems, *Comput. Methods in Appl. Mech. and Engineering,* **67**, 27-54.

[11] Bloomfield, P. (1976). *Fourier Analysis of Time Series: An Introduction,* John Wiley and Sons., Chichester, U.K.

[12] Bendat, J.S. and Piersol A.G. (1986). *Random Data: Analysis and Measurement Procedures,* 2nd edition, Wiley and Sons., Chichester, U.K.

[13] Hart, G. C. (1982). *Uncertainty Analysis, Loads, and Safety in Structural Engineering,* Prentice-Hall, Englewood Cliffs, New Jersey.

[14] To, C.W.S. (2001). On computational stochastic structural dynamics applying finite elements, *Archives of Comput. Meth. in Eng.,* **8(1)**, 3-40.

[15] To, C.W.S. (2002). Nonlinear random responses of shell structures with spatial uncertainties, *Proc. Des. Eng. Tech. Conf. and Comput. Inf. Eng. Conf.,* Montreal, Canada, Sept. 29-Oct. 2, DETC2002/CIE-34472, pp. 1-11.

[16] Vanmarcke, E. (1983). *Random Fields: Analysis and Synthesis,* The M.I.T. Press, Cambridge, Massachusetts.

[17] Balakrishnan, A.V. (1995). *Introduction to Random Processes in Engineering,* John Wiley and Sons, New York.

[18] To, C.W.S. (2009). Large nonlinear responses of spatially nonhomogeneous stochastic shell structures under nonstationary random excitations, *Journal of Computing and Information Science in Engineering,* **9(4)**, 041002-1-8.

[19] To, C.W.S. (2010). *Nonlinear Random Vibration: Computational Methods,* Zip Publishing, Columbus, Ohio.

[20] Liu, P.L. (1991). Size effect of random field elements on finite-element reliability methods, *Proceedings of the 7th International Conference on Structural Safety and Reliability*, pp. 223-239.

[21] To, C.W.S., and Liu, M.L. (1995). Hybrid strain based three-node flat triangular shell elements, Part II: Numerical investigation of nonlinear problems, *Computers and Structures,* **54(6)**, 1057-1076.

[22] Oliver, J., and Onate, E. (1984). A total Lagrangian formulation for the geometrically nonlinear analysis of structures using finite elements, Part I: Two-dimensional problems, shell and plate structures, *International Journal for Numerical Methods in Engineering,* **20**, 2253-2281.

[23] Soize, C. (2001). Nonlinear dynamical systems with nonparametric model of random uncertainties, *Uncertainties in Engineering Mechanics,* **1**, 1-38.

Appendix 1A   Mass and Stiffness Matrices of Higher Order Tapered Beam Element

As the consistent element mass and stiffness matrices are symmetric, only half of the explicit expressions for these matrices are given in this appendix. Two additional symbols are used in this appendix. They are:

$$\eta = \frac{\rho A_{i-1} \ell}{441080640}, \qquad Q = \frac{EI_{i-1}}{1081080 \ell^3}, \qquad \text{(1A.1a, b)}$$

in which E is the Young's modulus of elasticity and ρ the density of the material.

Explicit expressions for the consistent mass matrix of the higher order tapered beam element are:

$$m_{11} = \eta \left(178\,557\,120 + 40\,269\,600 \gamma_1 + 13\,144\,320 \gamma_2 \right),$$

$$m_{12} = \eta \ell \left(33\,268\,320 + 10\,208\,160 \gamma_1 + 3\,805\,920 \gamma_2 \right),$$

$$m_{13} = \eta \ell^2 \left(3\,353\,760 + 1\,168\,920 \gamma_1 + 468\,648 \gamma_2 \right),$$

$$m_{14} = \eta \ell^3 \left(156\,264 + 58\,344 \gamma_1 + 24\,468 \gamma_2 \right),$$

$$m_{15} = \eta \left(41\,983\,200 + 20\,991\,600 \gamma_1 + 11\,360\,160 \gamma_2 \right),$$

$$m_{16} = - \eta \ell \left(13\,990\,320 + 6\,695\,280 \gamma_1 + 3\,470\,760 \gamma_2 \right),$$

$$m_{17} = \eta \ell^2 \left(1\,897\,200 + 878\,220 \gamma_1 + 440\,784 \gamma_2 \right),$$

$$m_{18} = - \eta \ell^3 \left(106\,284 + 47\,940 \gamma_1 + 23\,472 \gamma_2 \right),$$

$$m_{22} = \eta \ell^2 \left(8\,078\,400 + 2\,864\,160 \gamma_1 + 1\,161\,216 \gamma_2 \right),$$

$$m_{23} = \eta \ell^3 \left(899\,640 + 345\,168 \gamma_1 + 147\,420 \gamma_2 \right),$$

$$m_{24} = \eta \ell^4 \left(44\,064 + 17\,748 \gamma_1 + 7\,848 \gamma_2 \right),$$

$$m_{25} = \eta \ell \left(13\,990\,320 + 7\,295\,040 \gamma_1 + 4\,070\,520 \gamma_2 \right),$$

$$m_{26} = -\eta\ell^2\left(4\,565\,520 + 2\,282\,760\,\gamma_1 + 1\,221\,912\,\gamma_2\right),$$

$$m_{27} = \eta\ell^3\left(608\,940 + 294\,984\,\gamma_1 + 153\,072\,\gamma_2\right),$$

$$m_{28} = -\eta\ell^4\left(33\,660 + 15\,912\,\gamma_1 + 8\,064\,\gamma_2\right),$$

$$m_{33} = \eta\ell^4\left(105\,264 + 42\,840\,\gamma_1 + 19\,080\,\gamma_2\right),$$

$$m_{34} = \eta\ell^5\left(5\,304 + 2\,244\,\gamma_1 + 1\,029\,\gamma_2\right),$$

$$m_{35} = \eta\ell^2\left(1\,897\,200 + 1\,018\,980\,\gamma_1 + 581\,544\,\gamma_2\right),$$

$$m_{36} = -\eta\ell^3\left(608\,940 + 313\,956\,\gamma_1 + 172\,044\,\gamma_2\right),$$

$$m_{37} = \eta\ell^4\left(80\,172 + 40\,086\,\gamma_1 + 21\,312\,\gamma_2\right),$$

$$m_{38} = -\eta\ell^5\left(4\,386 + 2\,142\,\gamma_1 + 1\,113\,\gamma_2\right),$$

$$m_{44} = \eta\ell^6\left(272 + 119\,\gamma_1 + 56\,\gamma_2\right),$$

$$m_{45} = \eta\ell^3\left(106\,284 + 58\,344\,\gamma_1 + 33\,876\,\gamma_2\right),$$

$$m_{46} = -\eta\ell^4\left(33\,660 + 17\,748\,\gamma_1 + 9\,900\,\gamma_2\right),$$

$$m_{47} = \eta\ell^5\left(4\,386 + 2\,244\,\gamma_1 + 1\,215\,\gamma_2\right),$$

$$m_{48} = -\eta\ell^6\left(238 + 119\,\gamma_1 + 63\,\gamma_2\right),$$

$$m_{55} = \eta\left(178\,557\,120 + 138\,287\,520\,\gamma_1 + 111\,162\,240\,\gamma_2\right),$$

$$m_{56} = -\eta\ell\left(33\,268\,320 + 23\,060\,160\,\gamma_1 + 16\,657\,920\,\gamma_2\right),$$

$$m_{57} = \eta\ell^2\left(3\,353\,760 + 2\,184\,840\,\gamma_1 + 1\,484\,568\,\gamma_2\right),$$

$$m_{58} = -\eta\ell^3\left(156264 + 97920\gamma_1 + 64044\gamma_2\right),$$

$$m_{66} = \eta\ell^2\left(8078400 + 5214240\gamma_1 + 3511296\gamma_2\right),$$

$$m_{67} = -\eta\ell^3\left(899640 + 554472\gamma_1 + 356724\gamma_2\right),$$

$$m_{68} = \eta\ell^4\left(44064 + 26316\gamma_1 + 16416\gamma_2\right),$$

$$m_{77} = \eta\ell^4\left(105264 + 62424\gamma_1 + 38664\gamma_2\right),$$

$$m_{78} = -\eta\ell^5\left(5304 + 3060\gamma_1 + 1845\gamma_2\right),$$

$$m_{88} = \eta\ell^6\left(272 + 153\gamma_1 + 90\gamma_2\right). \tag{1A.2}$$

The diagonal and sub-diagonal elements of the consistent element stiffness matrix of the tapered beam are:

$$k_{11} = Q\left(27518400 + 13759200\delta_1 + 8467200\delta_2 + 5821200\delta_3 + 4233600\delta_4\right),$$

$$k_{21} = Q\ell\left(13759200 + 5896800\delta_1 + 3250800\delta_2 + 2116800\delta_3 + 1512000\delta_4\right),$$

$$k_{22} = Q\ell^2\left(8424000 + 3229200\delta_1 + 1555200\delta_2 + 907200\delta_3 + 604800\delta_4\right),$$

$$k_{31} = Q\ell^2\left(1310400 + 704340\delta_1 + 420840\delta_2 + 283500\delta_3 + 206640\delta_4\right),$$

$$k_{32} = Q\ell^3\left(886860 + 421200\delta_1 + 219780\delta_2 + 131400\delta_3 + 88020\delta_4\right),$$

$$k_{33} = Q\ell^4\left(234000 + 70200\delta_1 + 34440\delta_2 + 20250\delta_3 + 13392\delta_4\right),$$

$$k_{41} = Q\ell^3\left(49\,140 + 32\,760\,\delta_1 + 21\,420\,\delta_2 + 15\,120\,\delta_3 + 11\,340\,\delta_4\right),$$

$$k_{42} = Q\ell^4\left(37\,440 + 21\,060\,\delta_1 + 11\,880\,\delta_2 + 7\,380\,\delta_3 + 5\,040\,\delta_4\right),$$

$$k_{43} = Q\ell^5\left(10\,920 + 3\,900\,\delta_1 + 1\,986\,\delta_2 + 1\,188\,\delta_3 + 792\,\delta_4\right),$$

$$k_{44} = Q\ell^6\left(624 + 234\,\delta_1 + 120\,\delta_2 + 72\,\delta_3 + 48\,\delta_4\right),$$

$$k_{51} = -\,Q\left(27\,518\,400 + 13\,759\,200\,\delta_1 + 8\,467\,200\,\delta_2 + 5\,821\,200\,\delta_3 + 4\,233\,600\,\delta_4\right),$$

$$k_{52} = -\,Q\ell\left(13\,759\,200 + 5\,896\,800\,\delta_1 + 3\,250\,800\,\delta_2 + 2\,116\,800\,\delta_3 + 1\,512\,000\,\delta_4\right),$$

$$k_{53} = -\,Q\ell^2\left(1\,310\,400 + 704\,340\,\delta_1 + 420\,840\,\delta_2 + 283\,500\,\delta_3 + 206\,640\,\delta_4\right),$$

$$k_{54} = -\,Q\ell^3\left(49\,140 + 32\,760\,\delta_1 + 21\,420\,\delta_2 + 15\,120\,\delta_3 + 11\,340\,\delta_4\right),$$

$$k_{55} = Q\left(27\,518\,400 + 13\,759\,200\,\delta_1 + 8\,467\,200\,\delta_2 + 5\,821\,200\,\delta_3 + 4\,233\,600\,\delta_4\right),$$

$$k_{61} = Q\ell\left(13\,759\,200 + 7\,862\,400\,\delta_1 + 5\,216\,400\,\delta_2 + 3\,704\,400\,\delta_3 + 2\,721\,600\,\delta_4\right),$$

$$k_{62} = Q\ell^2\left(5\,335\,200 + 2\,667\,600\,\delta_1 + 1\,695\,600\,\delta_2 + 1\,209\,600\,\delta_3 + 907\,200\,\delta_4\right),$$

$$k_{63} = Q\ell^3\left(423\,540 + 283\,140\,\delta_1 + 201\,060\,\delta_2 + 152\,100\,\delta_3 + 118\,620\,\delta_4\right),$$

$$k_{64} = Q\ell^4\left(11\,700 + 11\,700\,\delta_1 + 9\,540\,\delta_2 + 7\,740\,\delta_3 + 6\,300\,\delta_4\right),$$

$$k_{65} = -\,Q\ell\left(13\,759\,200 + 7\,862\,400\,\delta_1 + 5\,216\,400\,\delta_2 + 3\,704\,400\,\delta_3 + 2\,721\,600\,\delta_4\right),$$

$$k_{66} = Q\ell^2 \big(8\,424\,000 + 5\,194\,800\,\delta_1 + 3\,520\,800\,\delta_2 + 2\,494\,800\,\delta_3 + 1\,814\,400\,\delta_4\big),$$

$$k_{71} = -Q\ell^2 \big(1\,310\,400 + 606\,060\,\delta_1 + 322\,560\,\delta_2 + 176\,400\,\delta_3 + 90\,720\,\delta_4\big),$$

$$k_{72} = -Q\ell^3 \big(423\,540 + 140\,400\,\delta_1 + 58\,320\,\delta_2 + 25\,200\,\delta_3 + 7\,560\,\delta_4\big),$$

$$k_{73} = Q\ell^4 \big(-2\,340 - 1\,170\,\delta_1 + 660\,\delta_2 + 1\,575\,\delta_3 + 2\,196\,\delta_4\big),$$

$$k_{74} = Q\ell^5 \big(1\,950 + 780\,\delta_1 + 471\,\delta_2 + 348\,\delta_3 + 294\,\delta_4\big),$$

$$k_{75} = Q\ell^2 \big(1\,310\,400 + 606\,060\,\delta_1 + 322\,560\,\delta_2 + 176\,400\,\delta_3 + 90\,720\,\delta_4\big),$$

$$k_{76} = -Q\ell^3 \big(886\,860 + 465\,660\,\delta_1 + 264\,240\,\delta_2 + 151\,200\,\delta_3 + 83\,160\,\delta_4\big),$$

$$k_{77} = Q\ell^4 \big(234\,000 + 163\,800\,\delta_1 + 128\,040\,\delta_2 + 106\,470\,\delta_3 + 92\,232\,\delta_4\big),$$

$$k_{81} = Q\ell^3 \big(49\,140 + 16\,380\,\delta_1 + 5\,040\,\delta_2 - 2\,520\,\delta_4\big),$$

$$k_{82} = -Q\ell^4 \big(-11\,700 + 2\,160\,\delta_2 + 2\,520\,\delta_3 + 2\,520\,\delta_4\big),$$

$$k_{83} = -Q\ell^5 \big(1\,950 + 1\,170\,\delta_1 + 861\,\delta_2 + 675\,\delta_3 + 558\,\delta_4\big),$$

$$k_{84} = -Q\ell^6 \big(234 + 117\,\delta_1 + 75\,\delta_2 + 54\,\delta_3 + 42\,\delta_4\big),$$

$$k_{85} = -Q\ell^3 \big(49\,140 + 16\,380\,\delta_1 + 5\,040\,\delta_2 - 2\,520\,\delta_4\big),$$

$$k_{86} = Q\ell^4 \big(37\,440 + 16\,380\,\delta_1 + 7\,200\,\delta_2 + 2\,520\,\delta_3\big),$$

$$k_{87} = -Q\ell^5 \big(10\,920 + 7\,020\,\delta_1 + 5\,106\,\delta_2 + 3\,990\,\delta_3 + 3\,276\,\delta_4\big),$$

$$k_{88} = Q\ell^6 \big(624 + 390\,\delta_1 + 276\,\delta_2 + 210\,\delta_3 + 168\,\delta_4\big). \qquad (1A.3)$$

Appendix 1B Consistent Stiffness Matrix of Lower Order Triangular Shell Element

The consistent element stiffness matrix of the lower order triangular shell element k_L is defined by Eq. (1.89) whose component matrices are given by Eq. (1.90). This particular triangular shell element is identified as $AT+(k^3_t)'$ in Refs. [1, 2], for example, and in Chapters 6 and 7 of this book. It provides the correct six rigid-body modes and has superior features as outlined in [1, 2] and in Chapter 6. The component stiffness matrix $(k^3_t)'$ is represented by k_t in Eq. (1.90d). As indicated in Eq. (1.90), k_L consists of the element leverage matrix G_L and inverse of the generalized element stiffness matrix H, therefore in this appendix explicit expressions for H^{-1}, matrices G_L, and k_t are presented. These expressions are translated from those in [3]. For conciseness, let $Z = H^{-1}$. Section 1B.1 presents the non-zero entries or elements of Z while Section 1B.2 provides the non-zero entries of G_L or simply G and Section 1B.3 deals with the non-zero entries of k_t. It is understood that all quantities are with respect to C^t.

1B.1 Inverse of Element Generalized Stiffness Matrix

Since the element generalized stiffness matrix H is symmetric, its inverse is also symmetric. That is, the matrix $Z = [z_{i,j}]_{10\times10} = H^{-1}$ is symmetric. Only explicit expressions for the non-zero entries are presented in this section. They are:

$$z_{1,1} = \frac{1}{AEh} = z_{2,2}, \quad z_{1,2} = \frac{-v}{AEh} = z_{2,1}, \quad z_{3,3} = \frac{2(1+v)}{AEh} = z_{10,10},$$

$$z_{4,4} = \frac{12}{AEh^3} = z_{5,5}, \quad z_{4,5} = \frac{-12v}{AEh^3} = z_{5,4}, \quad z_{6,6} = \frac{24(1+v)}{AEh^3},$$

$$z_{7,7} = \frac{6(1+v)}{AEh\kappa_s\left(s_3^2 + r_3^2\right)}, \quad z_{8,8} = \frac{6(1+v)}{AEh\kappa_s\left(s_3^2 + r_3^2 - 2r_3r_2 + r_2^2\right)},$$

$$z_{9,9} = \frac{6(1+v)}{AEh\kappa_s r_2^2},$$

where A, E, h, v, and κ_s are the area of the shell element, Young's modulus of elasticity, the thickness of the shell element, Poisson's ratio, and the form factor of shear.

1B.2 Element Leverage Matrices

This section is concerned with the expressions for the leverage matrix G_L. For identification purpose and conciseness the matrix is divided into the component matrices associated with membrane, bending, and transverse shear.

For conciseness, the first subscript in Eqs. (1.89) and (1.90) is disregarded in this and the next sections. Thus, for example,

$$G_m = G_{L_m}, \qquad G_b = G_{L_b}, \qquad G_s = G_{L_s},$$

and so on, are applied. Let $(G_m)_{i,j}$ be the i'th row and j'th column element or entry of matrix G_m, and so. The non-zero entries of the latter matrix are:

$$\left(G_m\right)_{1,1} = \frac{AEh}{r_2\left(v^2 - 1\right)}, \qquad \left(G_m\right)_{1,2} = \frac{AEhv\left(r_2 - r_3\right)}{r_2 s_3\left(v^2 - 1\right)},$$

$$\left(G_m\right)_{1,6} = \frac{AEh\left(s_3 a_{31} - vr_2 b_{12} - vr_3 b_{31}\right)}{3 r_2 s_3\left(v^2 - 1\right)}, \qquad \left(G_m\right)_{1,7} = \frac{AEh}{r_2\left(1 - v^2\right)},$$

$$\left(G_m\right)_{1,8} = \frac{AEhvr_3}{r_2 s_3\left(v^2 - 1\right)}, \qquad \left(G_m\right)_{1,12} = \frac{AEh\left(vr_2 b_{12} + vr_2 b_{23} + s_3 a_{23} - vr_3 b_{23}\right)}{3 r_2 s_3\left(v^2 - 1\right)},$$

$$\left(G_m\right)_{1,14} = \frac{AEhv}{s_3\left(1 - v^2\right)},$$

$$\left(G_m\right)_{1,18} = \frac{AEh\left(vr_2 b_{23} - vr_3 b_{31} - vr_3 b_{23} + s_3 a_{31} + s_3 a_{23}\right)}{3 r_2 s_3\left(1 - v^2\right)},$$

$$\left(G_m\right)_{2,1} = \frac{AEhv}{r_2\left(v^2 - 1\right)}, \qquad \left(G_m\right)_{2,2} = \frac{AEh\left(r_2 - r_3\right)}{r_2 s_3\left(v^2 - 1\right)},$$

$$\left(G_m\right)_{2,6} = \frac{AEh\left(- r_2 b_{12} + vs_3 a_{31} - r_3 b_{31}\right)}{3 r_2 s_3\left(v^2 - 1\right)}, \qquad \left(G_m\right)_{2,7} = \frac{AEhv}{r_2\left(1 - v^2\right)},$$

$$\left(G_m\right)_{2,8} = \frac{AEhr_3}{r_2 s_3\left(v^2 - 1\right)}, \quad \left(G_m\right)_{2,12} = \frac{AEh\left(r_2 b_{12} + r_2 b_{23} + v s_3 a_{23} - r_3 b_{23}\right)}{3 r_2 s_3\left(v^2 - 1\right)},$$

$$\left(G_m\right)_{2,14} = \frac{AEh}{s_3\left(1 - v^2\right)},$$

$$\left(G_m\right)_{2,18} = \frac{AEh\left(r_2 b_{23} + v s_3 a_{31} + v s_3 a_{23} - r_3 b_{31} - r_3 b_{23}\right)}{3 r_2 s_3\left(1 - v^2\right)},$$

$$\left(G_m\right)_{3,1} = \frac{AEh\left(r_3 - r_2\right)}{2 r_2 s_3\left(1 + v\right)}, \quad \left(G_m\right)_{3,2} = \frac{-AEh}{2 r_2\left(1 + v\right)},$$

$$\left(G_m\right)_{3,6} = \frac{AEh\left(- s_3 b_{31} + r_3 a_{31}\right)}{6 r_2 s_3\left(1 + v\right)}, \quad \left(G_m\right)_{3,7} = \frac{-AEhr_3}{2 r_2 s_3\left(1 + v\right)},$$

$$\left(G_m\right)_{3,8} = \frac{AEh}{2 r_2\left(1 + v\right)}, \quad \left(G_m\right)_{3,12} = \frac{-AEh\left(r_2 a_{23} + s_3 b_{23} - r_3 a_{23}\right)}{6 r_2 s_3\left(1 + v\right)},$$

$$\left(G_m\right)_{3,13} = \frac{AEh}{2 s_3\left(1 + v\right)},$$

$$\left(G_m\right)_{3,18} = \frac{AEh\left(r_2 a_{23} + s_3 b_{31} + s_3 b_{23} - r_3 a_{31} - r_3 a_{23}\right)}{6 r_2 s_3\left(1 + v\right)}.$$

The non-zero elements $\left(G_b\right)_{i,j}$ of the component leverage matrix associated with bending have been found to be

$$\left(G_b\right)_{4,4} = \frac{AEh^3 v\left(r_2 - r_3\right)}{12 r_2 s_3\left(1 - v^2\right)}, \quad \left(G_b\right)_{4,5} = \frac{AEh^3}{12 r_2\left(v^2 - 1\right)},$$

$$\left(G_b\right)_{4,10} = \frac{AEh^3 v r_3}{12 r_2 s_3\left(1 - v^2\right)}, \quad \left(G_b\right)_{4,11} = \frac{AEh^3}{12 r_2\left(1 - v^2\right)},$$

$$\left(G_b\right)_{4,16} = \frac{AEh^3v}{12s_3\left(v^2 - 1\right)} \quad, \qquad \left(G_b\right)_{5,4} = \frac{AEh^3\left(r_2 - r_3\right)}{12r_2s_3\left(1 - v^2\right)} \quad,$$

$$\left(G_b\right)_{5,5} = \frac{AEh^3v}{12r_2\left(v^2 - 1\right)} \quad, \qquad \left(G_b\right)_{5,10} = \frac{AEh^3r_3}{12r_2s_3\left(1 - v^2\right)} \quad,$$

$$\left(G_b\right)_{5,11} = \frac{AEh^3v}{12r_2\left(1 - v^2\right)} \quad, \qquad \left(G_b\right)_{5,16} = \frac{AEh^3}{12s_3\left(v^2 - 1\right)} \quad,$$

$$\left(G_b\right)_{6,4} = \frac{AEh^3}{24r_2\left(1 + v\right)} \quad, \qquad \left(G_b\right)_{6,5} = \frac{AEh^3\left(r_3 - r_2\right)}{24r_2s_3\left(1 + v\right)} \quad,$$

$$\left(G_b\right)_{6,10} = \frac{-AEh^3}{24r_2\left(1 + v\right)} \quad, \quad \left(G_b\right)_{6,11} = \frac{-AEh^3r_3}{24r_2s_3\left(1 + v\right)} \quad, \quad \left(G_b\right)_{6,17} = \frac{AEh^3}{24s_3\left(1 + v\right)} \quad.$$

The non-zero elements $(G_s)_{i,j}$ of the component leverage matrix associated with shear have been derived as

$$\left(G_s\right)_{7,3} = \frac{AEh\kappa_s}{6\left(1 + v\right)} \quad, \quad \left(G_s\right)_{7,4} = \frac{AEh\kappa_s s_3}{12\left(1 + v\right)} \quad, \quad \left(G_s\right)_{7,5} = \frac{-AEh\kappa_s r_3}{12\left(1 + v\right)} \quad,$$

$$\left(G_s\right)_{7,15} = -\left(G_s\right)_{7,3} \quad, \qquad \left(G_s\right)_{7,16} = \left(G_s\right)_{7,4} \quad, \qquad \left(G_s\right)_{7,17} = \left(G_s\right)_{7,5} \quad,$$

$$\left(G_s\right)_{8,9} = -\left(G_s\right)_{7,3} \quad, \quad \left(G_s\right)_{8,10} = -\left(G_s\right)_{7,4} \quad, \quad \left(G_s\right)_{8,11} = \frac{AEh\kappa_s\left(r_3 - r_2\right)}{12\left(1 + v\right)} \quad,$$

$$\left(G_s\right)_{8,15} = \left(G_s\right)_{7,3} \quad, \quad \left(G_s\right)_{8,16} = -\left(G_s\right)_{7,4} \quad, \quad \left(G_s\right)_{8,17} = \left(G_s\right)_{8,11} \quad,$$

$$\left(G_s\right)_{9,3} = -\left(G_s\right)_{7,3} = -\left(G_s\right)_{9,9} \quad, \quad \left(G_s\right)_{9,5} = \left(G_s\right)_{9,11} = \frac{AEh\kappa_s r_2}{12\left(1 + v\right)} \quad.$$

The above symbols have already been defined in Section 1B.1.

Finally, the non-zero entries of the component element leverage matrix associated with the ddof and based on the hybrid strain formulation are included here for completeness although the particular triangular shell element considered in Chapter 1 is based on the mixed formulation and the hybrid strain-based component. Note that the element stiffness matrix k_L consists of two parts, one of which is based on the hybrid strain formulation and the other, k_t defined by Eq. (1.90d) is based on the the displacement formulation. The ddof part of the element leverage matrix by the hybrid strain formulation is G_d. The non-zero entries $(G_d)_{i,j}$ of G_d are presented in the following:

$$(G_d)_{10,1} = \frac{AEh(r_3 - r_2)}{4r_2 s_3(1 + v)}, \qquad (G_d)_{10,2} = \frac{-AEh}{4r_2(1 + v)} = -(G_d)_{10,8},$$

$$(G_d)_{10,6} = \frac{AEh}{6(1 + v)} = (G_d)_{10,12} = (G_d)_{10,18}, \qquad (G_d)_{10,7} = \frac{-AEhr_3}{4r_2 s_3(1 + v)}$$

$$(G_d)_{10,13} = \frac{AEh}{4s_3(1 + v)}.$$

The above entries complete the definition of G_d and therefore, in turn, it completes the definition of the element leverage matrix G_L or G in Eq. (1.69) such that the linear element stiffness matrix can be formed as

$$k_L = G_L^T H^{-1} G_L.$$

1B.3   Element Component Stiffness Matrix Associated with Torsion

Since k_t is symmetric the non-zero entries $(k_t)_{i,j}$ of the upper triangular matrix of k_t defined by Eq. (1.90d) are given below:

$$(k_t)_{1,1} = \frac{Ah(r_2 - r_3)^2}{r_2^2 s_3^2}, \qquad (k_t)_{1,2} = \frac{Ah(r_3 - r_2)}{r_2^2 s_3},$$

$$(k_t)_{1,6} = \frac{Ah(r_3 - r_2)(2r_2 s_3 + r_3 a_{31} + s_3 b_{31})}{3r_2^2 s_3^2},$$

$$\left(k_t\right)_{1,7} = \frac{Ah\left(r_2 - r_3\right)r_3}{r_2^2 s_3^2} \quad , \qquad \left(k_t\right)_{1,8} = \frac{Ah\left(r_2 - r_3\right)}{r_2^2 s_3} \quad ,$$

$$\left(k_t\right)_{1,12} = \frac{Ah\left(r_3 - r_2\right)\left(2r_2 s_3 - r_2 a_{23} + s_3 b_{23} + r_3 a_{23}\right)}{3 r_2^2 s_3^2} \quad , \quad \left(k_t\right)_{1,13} = \frac{Ah\left(r_3 - r_2\right)}{r_2 s_3^2} \quad ,$$

$$\left(k_t\right)_{1,18} = \frac{Ah\left(r_3 - r_2\right)\left(r_2 a_{23} - r_3 a_{31} - r_3 a_{23} + 2r_2 s_3 - s_3 b_{31} - s_3 b_{23}\right)}{3 r_2^2 s_3^2} \quad ,$$

$$\left(k_t\right)_{2,2} = \frac{Ah}{r_2^2} \quad , \qquad \left(k_t\right)_{2,6} = \frac{Ah\left(2r_2 s_3 + r_3 a_{31} + s_3 b_{31}\right)}{3 r_2^2 s_3} \quad ,$$

$$\left(k_t\right)_{2,7} = \frac{-Ah r_3}{r_2^2 s_3} \quad , \qquad \left(k_t\right)_{2,8} = \frac{-Ah}{r_2^2} \quad ,$$

$$\left(k_t\right)_{2,12} = \frac{Ah\left(2r_2 s_3 - r_2 a_{23} + s_3 b_{23} + r_3 a_{23}\right)}{3 r_2^2 s_3} \quad , \qquad \left(k_t\right)_{2,13} = \frac{Ah}{r_2 s_3} \quad ,$$

$$\left(k_t\right)_{2,18} = \frac{Ah\left(r_2 a_{23} - r_3 a_{31} - r_3 a_{23} + 2r_2 s_3 - s_3 b_{31} - s_3 b_{23}\right)}{3 r_2^2 s_3} \quad ,$$

$$\left(k_t\right)_{6,6} = \frac{Ah\left(n_{61} + n_{62} + n_{63}\right)}{6 r_2^2 s_3^2} \quad ,$$

$$n_{61} = 2r_3 a_{31} r_2 s_3 - r_2 a_{31} b_{31} s_3 + 2 b_{31} s_3^2 r_2 - 2 r_2 r_3 a_{31}^2 \quad ,$$

$$n_{62} = 2r_3 a_{31} b_{31} s_3 + r_3^2 a_{31}^2 + r_2^2 a_{31}^2 + a_{31} b_{12} s_3 r_2 + 2 r_2^2 s_3 a_{31} \quad ,$$

$$n_{63} = 2r_2 b_{12} s_3^2 + b_{31}^2 s_3^2 + s_3^2 b_{12}^2 + 4 s_3^2 r_2^2 \quad ,$$

$$(k_t)_{6,7} = \frac{-Ahr_3(2r_2s_3 + r_3a_{31} + s_3b_{31})}{3r_2^2s_3^2},$$

$$(k_t)_{6,8} = \frac{-Ah(2r_2s_3 + r_3a_{31} + s_3b_{31})}{3r_2^2s_3},$$

$$(k_t)_{6,12} = \frac{-Ah(n_{71} + n_{72} + n_{73} + n_{74})}{12r_2^2s_3^2},$$

$$n_{71} = -2r_3^2a_{31}a_{23} + r_2a_{23}b_{31}s_3 - 2a_{23}s_3r_2r_3 - 2r_2s_3^2b_{23},$$

$$n_{72} = -2r_3a_{31}b_{23}s_3 + r_2s_3a_{31}b_{23} + 2r_2r_3a_{31}a_{23} - 2a_{23}b_{31}s_3r_3,$$

$$n_{73} = -r_2s_3a_{23}b_{12} - 2b_{31}b_{23}s_3^2 + 2r_2^2s_3a_{23} - 2r_2r_3s_3a_{31},$$

$$n_{74} = -2r_2b_{31}s_3^2 + a_{31}b_{12}r_2s_3 + 4r_2s_3^2b_{12} + 2s_3^2b_{12}^2 - 4r_2^2s_3^2,$$

$$(k_t)_{6,13} = \frac{Ah(2r_2s_3 + r_3a_{31} + s_3b_{31})}{3r_2s_3^2},$$

$$(k_t)_{6,18} = \frac{-Ah(n_{81} + n_{82} + n_{83} + n_{84} + n_{85} + n_{86})}{12r_2^2s_3^2},$$

$$n_{81} = 2r_3^2a_{31}a_{23} - r_2a_{23}b_{31}s_3 + 2a_{23}s_3r_2r_3 + 2r_2s_3^2b_{23},$$

$$n_{82} = 2r_3a_{31}b_{23}s_3 - r_2s_3a_{31}b_{23} - 2r_2r_3a_{31}a_{23} + 2a_{23}b_{31}s_3r_3,$$

$$n_{83} = r_2s_3a_{23}b_{12} + 2b_{31}b_{23}s_3^2 - 2r_2^2s_3a_{23} - 2r_2r_3s_3a_{31},$$

$$n_{84} = -2r_2s_3a_{31}b_{31} - 2b_{31}r_2s_3^2 - 2r_2r_3a_{31}^2,$$

$$n_{85} = 4r_3s_3a_{31}b_{31} + 2a_{31}^2r_3^2 + 2r_2^2a_{31}^2 + r_2s_3a_{31}b_{12},$$

$$n_{86} = 4r_2^2 s_3 a_{31} + 2b_{31}^2 s_3^2 - 4r_2^2 s_3^2,$$

$$\left(k_t\right)_{7,7} = \frac{Ahr_3^2}{r_2^2 s_3^2}, \qquad \left(k_t\right)_{7,8} = \frac{Ahr_3}{r_2^2 s_3},$$

$$\left(k_t\right)_{7,12} = \frac{-Ahr_3\left(2r_2 s_3 - r_2 a_{23} + s_3 b_{23} + r_3 a_{23}\right)}{3r_2^2 s_3^2}, \qquad \left(k_t\right)_{7,13} = \frac{-Ahr_3}{r_2 s_3^2},$$

$$\left(k_t\right)_{7,18} = \frac{Ahr_3\left(-2r_2 s_3 - r_2 a_{23} + r_3 a_{31} + r_3 a_{23} + s_3 b_{31} + s_3 b_{23}\right)}{3r_2^2 s_3^2},$$

$$\left(k_t\right)_{8,8} = \frac{Ah}{r_2^2}, \qquad \left(k_t\right)_{8,12} = \frac{Ah\left(-2r_2 s_3 + r_2 a_{23} - s_3 b_{23} - r_3 a_{23}\right)}{3r_2^2 s_3},$$

$$\left(k_t\right)_{8,13} = \frac{-Ah}{r_2 s_3},$$

$$\left(k_t\right)_{8,18} = \frac{Ah\left(-2r_2 s_3 - r_2 a_{23} + r_3 a_{31} + r_3 a_{23} + s_3 b_{31} + s_3 b_{23}\right)}{3r_2^2 s_3},$$

$$\left(k_t\right)_{12,12} = \frac{Ah\left(n_{91} + n_{92} + n_{93} + n_{94}\right)}{6r_2^2 s_3^2},$$

$$n_{91} = 2r_2 r_3 s_3 a_{23} + 2r_2 s_3^2 b_{23} - a_{23} b_{12} s_3 r_2,$$

$$n_{92} = -4r_2^2 a_{23} s_3 + 2r_2 s_3^2 b_{12} + s_3^2 b_{12}^2 + 4r_2^2 s_3^2,$$

$$n_{93} = -r_2 s_3 a_{23} b_{23} - a_{23}^2 r_2 r_3 + 2r_3 s_3 a_{23} b_{23},$$

$$n_{94} = r_2^2 a_{23}^2 + b_{23}^2 s_3^2 + r_3^2 a_{23}^2,$$

$$\left(k_t\right)_{12,13} = \frac{Ah\left(2r_2s_3 - r_2a_{23} + s_3b_{23} + r_3a_{23}\right)}{3r_2s_3^2},$$

$$\left(k_t\right)_{12,18} = \frac{Ah\left(n_{11} + n_{12} + n_{13} + n_{14} + n_{15}\right)}{12r_2^2s_3^2},$$

$$n_{11} = -2r_3^2a_{31}a_{23} + r_2a_{23}b_{31}s_3 + 2a_{23}s_3r_2r_3 + 2r_2s_3^2b_{23},$$

$$n_{12} = -2r_3a_{31}b_{23}s_3 + r_2s_3a_{31}b_{23} + 2r_2r_3a_{31}a_{23} - 2a_{23}b_{31}s_3r_3,$$

$$n_{13} = r_2s_3a_{23}b_{12} - 2b_{31}b_{23}s_3^2 + 2r_2^2s_3a_{23} - 2r_2r_3s_3a_{31},$$

$$n_{14} = -2r_2s_3^2b_{31} + a_{31}b_{12}r_2s_3 + 4r_2^2s_3^2 + 2r_2s_3a_{23}b_{23},$$

$$n_{15} = 2r_2r_3a_{23}^2 - 4a_{23}b_{23}r_3s_3 - 2r_2^2a_{23}^2 - 2s_3^2b_{23}^2 - 2r_3^2a_{23}^2,$$

$$\left(k_t\right)_{13,13} = \frac{Ah}{s_3^2}, \qquad\qquad \left(k_t\right)_{13,18} = \left(\frac{-r_2}{s_3}\right)\left(k_t\right)_{8,18},$$

$$\left(k_t\right)_{18,18} = \frac{Ah\left(n_{21} + n_{22} + n_{23} + n_{24} + n_{25} + n_{26}\right)}{6r_2^2s_3^2},$$

$$n_{21} = 2r_3^2a_{31}a_{23} - r_2a_{23}b_{31}s_3 - 4a_{23}s_3r_2r_3 - 4r_2s_3^2b_{23},$$

$$n_{22} = 2r_3a_{31}b_{23}s_3 - r_2s_3a_{31}b_{23} - 2r_2r_3a_{31}a_{23} + 2a_{23}b_{31}s_3r_3,$$

$$n_{23} = 2s_3^2b_{23}b_{31} + 2a_{23}r_2^2s_3 - 4r_2r_3s_3a_{31} - r_2s_3a_{31}b_{31},$$

$$n_{24} = -4r_2s_3^2b_{31} - a_{31}^2r_2r_3 + 2r_3s_3a_{31}b_{31} + r_3^2a_{31}^2 + r_2^2a_{31}^2,$$

$$n_{25} = 2r_2^2s_3a_{31} + b_{31}^2s_3^2 + 4r_2^2s_3^2 - r_2s_3a_{23}b_{23} - r_2r_3a_{23}^2,$$

$$n_{26} = 2r_3s_3a_{23}b_{23} + a_{23}^2r_2^2 + s_3^2b_{23}^2 + r_3^2a_{23}^2.$$

References

[1] Liu, M.L., and To, C.W.S. (1995). Vibration analysis of structures by hybrid strain based three-node flat triangular shell elements, *Journal of Sound and Vibration,* **184(5)**, 801-821.

[2] Liu, M.L., and To, C.W.S. (1995). Hybrid strain based three-node flat triangular shell elements, Part I: Nonlinear theory and incremental formulation, *Computers and Structures*, **54(6)**, 1031-1056.

[3] Liu, M.L. (1993). Response statistics of shell structures with geometrical and material nonlinearities, Ph.D. Thesis, University of Western Ontario, Canada.

Appendix 1C Consistent Mass Matrix of Lower Order Triangular Shell Element

The consistent element mass matrix of the lower order triangular shell element defined by Eq. (1.96) has been derived and explicit expressions for all entries or elements $m_{i,j}$ of this consistent element mass matrix m are given below. They are translated from those presented in Appendix C of [1]. Note that the consistent element mass matrix is symmetric. Therefore, only the non-zero elements of the upper-right triangular matrix are presented. These non-zero elements or entries of the mass matrix are:

$$m_{1,1} = \frac{\rho A h}{6}, \quad m_{1,6} = \frac{\rho A h}{30} a_{31}, \quad m_{1,7} = \frac{\rho A h}{12}, \quad m_{1,12} = -\frac{\rho A h}{60} a_{23},$$

$$m_{1,13} = \frac{\rho A h}{12}, \quad m_{1,18} = -\frac{\rho A h}{60}(2 a_{31} - a_{23}),$$

$$m_{2,2} = \frac{\rho A h}{6}, \quad m_{2,6} = \frac{\rho A h}{30}(-b_{12} + b_{31}), \quad m_{2,8} = \frac{\rho A h}{12},$$

$$m_{2,12} = \frac{\rho A h}{60}(2 b_{12} - b_{23}), \quad m_{2,14} = \frac{\rho A h}{12}, \quad m_{2,18} = -\frac{\rho A h}{60}(2 b_{31} - b_{23}),$$

$$m_{3,3} = \frac{\rho A h}{6}, \quad m_{3,4} = -\frac{\rho A h}{30} a_{31}, \quad m_{3,5} = -\frac{\rho A h}{30}(-b_{12} + b_{31}),$$

$$m_{3,9} = \frac{\rho A h}{12}, \quad m_{3,10} = \frac{\rho A h}{60} a_{23}, \quad m_{3,11} = -\frac{\rho A h}{60}(2 b_{12} - b_{23}),$$

$$m_{3,15} = \frac{\rho A h}{12}, \quad m_{3,16} = \frac{\rho A h}{60}(2 a_{31} - a_{23}), \quad m_{3,17} = \frac{\rho A h}{60}(2 b_{31} - b_{23}),$$

$$m_{4,4} = \frac{\rho A h}{90}(15 J_h + a_{31}^2), \quad m_{4,5} = \frac{\rho A h}{180} a_{31}(2 b_{31} - b_{12}), \quad m_{4,9} = -\frac{\rho A h}{60} a_{31},$$

$$m_{4,10} = -\frac{\rho A h}{180}(-15 J_h + a_{31} a_{23}), \quad m_{4,11} = \frac{\rho A h}{180} a_{31}(b_{12} - b_{23}),$$

$$m_{4,15} = -\frac{\rho A h}{30} a_{31}, \quad m_{4,16} = -\frac{\rho A h}{180}(-15 J_h + 2 a_{31}^2 - a_{31} a_{23}),$$

$$m_{4,17} = -\frac{\rho A h}{180} a_{31}(2b_{31} - b_{23}), \quad m_{5,5} = \frac{\rho A h}{90}\left(15J_h + b_{31}^2 + b_{12}^2 - b_{31}b_{12}\right),$$

$$m_{5,9} = -\frac{\rho A h}{60}\left(-2b_{12} + b_{31}\right), \quad m_{5,10} = -\frac{\rho A h}{180} a_{23}\left(-b_{12} + b_{31}\right),$$

$$m_{5,11} = \frac{\rho A h}{180}\left(15J_h - 2b_{12}^2 + b_{31}b_{12} + b_{12}b_{23} - b_{31}b_{23}\right),$$

$$m_{5,15} = \frac{\rho A h}{60}\left(b_{12} - 2b_{31}\right),$$

$$m_{5,16} = -\frac{\rho A h}{180}\left(2a_{31}b_{31} - a_{31}b_{12} + b_{12}a_{23} - b_{31}a_{23}\right),$$

$$m_{5,17} = -\frac{\rho A h}{180}\left(-15J_h + 2b_{31}^2 - b_{31}b_{12} + b_{12}b_{23} - b_{31}b_{23}\right),$$

$$m_{6,6} = \frac{\rho A h}{90}\left(15J_o + a_{31}^2 + b_{31}^2 + b_{12}^2 - b_{31}b_{12}\right),$$

$$m_{6,7} = \frac{\rho A h}{60} a_{31}, \quad m_{6,8} = \frac{\rho A h}{60}\left(-2b_{12} + b_{31}\right),$$

$$m_{6,12} = -\frac{\rho A h}{180}\left(-15J_o + 2b_{12}^2 - b_{31}b_{12} - b_{12}b_{23} + a_{31}a_{23} + b_{31}b_{23}\right),$$

$$m_{6,13} = \frac{\rho A h}{30} a_{31}, \quad m_{6,14} = \frac{\rho A h}{60}\left(2b_{31} - b_{12}\right),$$

$$m_{6,18} = -\frac{\rho A h}{180}\left(-15J_o + 2a_{31}^2 + 2b_{31}^2 - b_{31}b_{12} + b_{12}b_{23} - a_{31}a_{23} - b_{31}b_{23}\right),$$

$$m_{7,7} = \frac{\rho A h}{6}, \quad m_{7,12} = -\frac{\rho A h}{30} a_{23}, \quad m_{7,13} = \frac{\rho A h}{12},$$

$$m_{7,18} = -\frac{\rho A h}{60}\left(-2a_{23} + a_{31}\right),$$

$$m_{8,8} = \frac{\rho A h}{6}, \qquad m_{8,12} = \frac{\rho A h}{30}\left(b_{12} - b_{23}\right),$$

$$m_{8,14} = \frac{\rho A h}{12}, \quad m_{8,18} = -\frac{\rho A h}{60}\left(-2b_{23} + b_{31}\right), \quad m_{9,9} = \frac{\rho A h}{6},$$

$$m_{9,10} = \frac{\rho A h}{30}a_{23}, \quad m_{9,11} = -\frac{\rho A h}{30}\left(b_{12} - b_{23}\right), \quad m_{9,15} = \frac{\rho A h}{12},$$

$$m_{9,16} = \frac{\rho A h}{60}\left(-2a_{23} + a_{31}\right), \quad m_{9,17} = \frac{\rho A h}{60}\left(-2b_{23} + b_{31}\right),$$

$$m_{10,10} = \frac{\rho A h}{90}\left(15 J_h + a_{23}^2\right), \quad m_{10,11} = -\frac{\rho A h}{180}a_{23}\left(-2b_{23} + b_{12}\right),$$

$$m_{10,15} = \frac{\rho A h}{30}a_{23}, \qquad m_{10,16} = \frac{\rho A h}{180}\left(15 J_h - 2a_{23}^2 + a_{31}a_{23}\right),$$

$$m_{10,17} = \frac{\rho A h}{180}a_{23}\left(-2b_{23} + b_{31}\right),$$

$$m_{11,11} = \frac{\rho A h}{90}\left(15 J_h + b_{12}^2 + b_{23}^2 - b_{12}b_{23}\right), \quad m_{11,15} = -\frac{\rho A h}{60}\left(-2b_{23} + b_{12}\right),$$

$$m_{11,16} = -\frac{\rho A h}{180}\left(2a_{23}b_{23} + a_{31}b_{12} - b_{12}a_{23} - b_{23}a_{31}\right),$$

$$m_{11,17} = -\frac{\rho A h}{180}\left(-15 J_h + 2b_{23}^2 + b_{31}b_{12} - b_{12}b_{23} - b_{31}b_{23}\right),$$

$$m_{12,12} = \frac{\rho A h}{90}\left(15 J_o + b_{12}^2 + a_{23}^2 + b_{23}^2 - b_{12}b_{23}\right),$$

$$m_{12,13} = -\frac{\rho A h}{30}a_{23}, \quad m_{12,14} = \frac{\rho A h}{60}\left(-2b_{23} + b_{12}\right),$$

$$m_{12,18} = \frac{\rho A h}{180}\left(15 J_o - 2a_{23}^2 - 2b_{23}^2 - b_{31}b_{12} + b_{12}b_{23} + a_{31}a_{23} + b_{31}b_{23}\right),$$

$$m_{13,13} = \frac{\rho A h}{6}, \qquad m_{13,18} = -\frac{\rho A h}{30}\left(a_{31} - a_{23}\right),$$

$$m_{14,14} = \frac{\rho A h}{6}, \qquad m_{14,18} = -\frac{\rho A h}{30}\left(b_{31} - b_{23}\right),$$

$$m_{15,15} = \frac{\rho A h}{6}, \qquad m_{15,16} = \frac{\rho A h}{30}\left(a_{31} - a_{23}\right),$$

$$m_{15,17} = \frac{\rho A h}{30}\left(b_{31} - b_{23}\right), \quad m_{16,16} = \frac{\rho A h}{90}\left(15 J_h + a_{31}^2 + a_{23}^2 - a_{31}a_{23}\right),$$

$$m_{16,17} = \frac{\rho A h}{180}\left(2 a_{31}b_{31} + 2 a_{23}b_{23} - b_{23}a_{31} - b_{31}a_{23}\right),$$

$$m_{17,17} = \frac{\rho A h}{90}\left(15 J_h + b_{31}^2 + b_{23}^2 - b_{31}b_{23}\right),$$

$$m_{18,18} = \frac{\rho A h}{90}\left(15 J_o + a_{31}^2 + b_{31}^2 + a_{23}^2 + b_{23}^2 - a_{31}a_{23} - b_{31}b_{23}\right),$$

where ρ, A, and h are respectively the density of the material, area, and thickness of the triangular shell element at configuration C^t. In other words, ρ should be ρ^t, for example. But for conciseness, the superscript t for all the quantities has been disregarded in the foregoing explicit expressions. Furthermore, in the foregoing, $J_h = h^2/(12)$ is the moment of inertia per unit cross-sectional area, and $J_o = [r_2{}^2 + s_3{}^2 + r_3(r_3 - r_2)]/(18)$ is the polar moment of inertia per unit area of the triangular element and is about an axis through its centroid. The coefficients a_{ij} and b_{ij} are given below Eq. (1.84).

Reference

[1] Liu, M.L. (1993). Response statistics of shell structures with geometrical and material nonlinearities, Ph.D. Thesis, University of Western Ontario, Canada.

Appendix 2A   Eigenvalue Solution

The eigenvalue solution of a linear general mdof system consists of the evaluation of the eigenvalues and eigenvectors of the following matrix equation of motion

$$M\ddot{x} + C\dot{x} + Kx = 0 . \qquad (2A.1)$$

This is Eq. (2.14) with the applied force vector on the rhs being set to zero. The eigenvalue solution of this general linear mdof system has been presented by Ma and associates [1]. However, many eigenvalue solution algorithms in the FE packages available in the commercial and educational sectors are based on the following mdof system that disregards the damping forcing term. The latter approach is followed in this book for two main reasons. First, the amount of algebraic manipulation and analytical work is considerably smaller compared with that required in the solution of Ma and associates. Second, the computer programs written in Fortran employed in the computational work of the eigenvalue solution reported in this book are based on the undamped system. The following is an outline and more discussion on various approaches and theories can be found in books such as [2].

Consider the linear matrix equation of motion for the undamped mdof system

$$M\ddot{x} + Kx = 0 . \qquad (2A.2)$$

For harmonic motion, this equation can be written as

$$(K - \lambda M)x = 0 , \qquad \text{since } \ddot{x} = -\lambda x , \qquad \text{where } \lambda = \omega^2 . \qquad (2A.3)$$

Analytically, the *characteristic or frequency* equation of the system is defined by

$$\det[K - \lambda M] = |K - \lambda M| = 0 . \qquad (2A.4)$$

The roots λ_i of the characteristic equation are called *eigenvalues*. By substituting λ_i into Eq. (2A.3) one obtains the corresponding mode shape x_i which is called the *eigenvector*. The mode shape x_i is identified as $\Psi^{(i)}$ in Eq. (2.16).

In Eq. (2A.3) K is a real symmetric matrix, which is either positive definite, or positive semi-definite. The matrix M is also a real symmetric matrix and is positive definite. As a consequence of these matrix properties, all the eigenvalues are real and greater than, or equal to, zero.

Computationally, there are many ways of finding the eigenvalues and eigenvectors of Eq. (2A.3). However, in this book only two approaches have been employed through the applications of SPADAS [3] and [4] where the subspace iteration technique was applied. For brevity, only the steps of eigenvalue solution in SPADAS are outlined in what follows. For a detailed presentation and discussion the readers are referred to [3].

In SPADAS, Eq. (2A.3) is first cast into the standard form by employing the Cholesky's symmetric decomposition [5-7]. Subsequently, the resulting matrix is reduced, using Householder's method [8], to the tridiagonal form whose eigenvalues are obtained with the bisection method and Gershgorin theorem [5] to determine the upper and lower bounds on all the eigenvalues. The eigenvectors of the tridiagonal matrix are obtained by a method of inverse iteration [9]. Having found the eigenvectors of the tridiagonal matrix, the eigenvectors of Eq. (2A.3) are determined by backward substitution.

References

[1] Ma, F., Morzfeld, M., and Imam, A. (2010). The decoupling of damped linear systems in free and forced vibration, *J. of Sound and Vibration*, **329**, 3182-3202.

[2] Meirovitch, L. (1980). *Computational Methods in Structural Dynamics*, Sijthoff and Noordhoff, Alphen aan den Rijn, The Netherlands.

[3] Petyt, M. (1974). *SPADAS I: Theoretical Manual*, I.S.V.R, University of Southampton, United Kingdom.

[4] Bathe, K.J., Wilson, E.L., and Iding, R.H. (1974). NONSAP, A structural analysis program for static and dynamic response of nonlinear systems, Report No. SESM 74-3, Struct. Eng. Lab., U. of California, Berkeley.

[5] Bishop, R.E.D., Gladwell, G.M.L., and Michaelson, S. (1965). *The Matrix Analysis of Vibration*, Cambridge University Press, Cambridge.

[6] Fox, L. (1964). *An Introduction to Numerical Linear Algebra*, Clarendon Press, Oxford.

[7] Wilkinson, J.H. (1965). *The Algebraic Eigenvalue Problem*, Oxford University Press, Oxford.

[8] Householder, A.S., and Bauer, F.L. (1959). On certain methods for expanding the characteristic polynomial, *Numerische Mathematik*, **9**, 386-393.

[9] Wilkinson, J.H. (1962). Calculation of the eigenvectors of a symmetric tridiagonal matrix by inverse iteration, *Numerische Mathematik*, **4**, 368-376.

Appendix 2B   Derivation of Evolutionary Spectral Densities and Variances of Displacements

Before one can derive the evolutionary spectral densities and variances of displacements for a mdof system under uniformly modulated random excitations, it is relatively more direct to begin with the derivation of the evolutionary spectral densities of a mdof system under exponentially decaying random excitations. These results will be applied to the derivation for the uniformly modulated random excitation case. While this approach may seem to be rather tedious, it will enable one to relatively easily identify the individual nodal couplings and modal couplings contributing to the response being analyzed. This is the route followed in the present appendix.

2B.1   Evolutionary Spectral Densities Due to Exponentially Decaying Random Excitations

The derivation of evolutionary spectral densities of a mdof system under exponentially decaying random excitations consists of two parts. The first part is concerned with the derivation of the evolutionary auto-spectral densities or simply referred to as evolutionary spectral densities for conciseness. The second part is to do with the derivation of the evolutionary cross-spectral densities.

In the first part of the derivation, one begins with the application of Eq. (2.28) and remembering that instead of a uniformly modulated deterministic function it is the exponentially decaying function one is concerned here, such that the evolutionary auto-spectral densities become

$$S_{jj}(t,\omega) = S_{jj}^{u}(t,\omega) + S_{jj}^{c}(t,\omega) , \tag{2B.1}$$

in which the subscript j is the nodal number and

$$S_{jj}^{u}(t,\omega) = \sum_{r=1}^{L} \left(\mu_{jr} \Omega_r \right)^2 S_{rr}(\omega) \, e^{-2\zeta_r \omega_r t} \frac{p_{rr}^{u}(t,\omega)}{|H_r(\omega)|^2} , \tag{2B.2}$$

where the subscript r denotes the mode number and modal couplings are

$$S_{jj}^{c}(t,\omega) = 2 \sum_{r=2}^{L} \sum_{s=r}^{L} \left(\mu_{uj} \right)\left(\mu_{sj} \right) S_{us}(\omega) \, e^{-\left(\zeta_u \omega_u + \zeta_s \omega_s \right)t} \frac{p_{us}^{c}(t,\omega)}{D_{us}(\omega)} , \tag{2B.3}$$

$$u = r - 1 , \quad \Omega_r = \omega_r \sqrt{1 - \zeta_r^2} , \quad \mu_{jr} = \left(E_r / \Omega_r \right)\left(\psi_{jr} \right),$$

$$H_r(\omega) = \varepsilon_r^2 + \Omega_r^2 - \omega^2 + i2\varepsilon_r\omega \, , \qquad \varepsilon_r = \omega_r\left(\zeta_r - \alpha_r/\omega_r\right) ,$$

$$\left\langle \eta_r(t)\eta_s(t) \right\rangle = \int_{-\infty}^{\infty} S_{rs}(\omega)\, d\omega \, ,$$

with the assumption that $S_{rs}(\omega) = S_{sr}(\omega)$, and the uncoupled modal contributions to the response at node j are

$$p_{rr}^u(t,\omega) = A_{1r} + A_{2r}\omega^2 + A_{3r}\cos\omega t + A_{4r}\omega\sin\omega t \, , \qquad (2B.4)$$

where the coefficients are defined as

$$A_{1r} = 1 + e^{2\varepsilon_r t} + \left(\varepsilon_r^2 - \Omega_r^2\right)\left(\frac{s_r^2}{\Omega_r^2}\right) + \left(\frac{\varepsilon_r}{\Omega_r}\right)\sin 2\Omega_r t \, , \qquad A_{2r} = \left(\frac{s_r}{\Omega_r}\right)^2 ,$$

$$A_{3r} = -\left(\frac{2}{\Omega_r}\right)e^{\varepsilon_r t}\left(\Omega_r c_r + \varepsilon_r s_r\right) , \qquad A_{4r} = -2\,e^{\varepsilon_r t}\left(\frac{s_r}{\Omega_r}\right) ,$$

$$s_r = \sin\left(\Omega_r t\right) , \qquad c_r = \cos\left(\Omega_r t\right) , \qquad D_{us} = |H_u(\omega)|^2\,|H_s(\omega)|^2 \, ,$$

and part of the modal couplings in Eq. (3B.3) is through the following expressions

$$p_{us}^c(t,\omega) = \sigma_{1us} + \sigma_{2us}\,\omega^2 + \sigma_{3us}\,\omega^4 + \sigma_{4us}\,\omega^6$$

$$+\; \sigma_{5us}\cos\omega t + \sigma_{6us}\,\omega\sin\omega t + \sigma_{7us}\,\omega^2\cos\omega t \qquad (2B.5)$$

$$+\; \sigma_{8us}\,\omega^3\sin\omega t + \sigma_{9us}\,\omega^4\cos\omega t + \sigma_{10us}\,\omega^5\sin\omega t \, ,$$

where the coefficients are

$$\sigma_{1us} = B_{1us}K_{2us} \, , \qquad \sigma_{2us} = B_{4us}K_{2us} - B_{1us}K_{1us} - K_{3us}D_{3us} \, ,$$

$$\sigma_{3us} = B_{1us} - B_{4us}K_{1us} - K_{4us}D_{3us} \, , \qquad \sigma_{4us} = B_{4us} \, ,$$

$$\sigma_{5us} = B_{2us} K_{2us} , \qquad \sigma_{6us} = B_{3us} K_{2us} - K_{3us} D_{1us} ,$$

$$\sigma_{7us} = -\left(B_{2us} K_{1us} + K_{3us} D_{2us}\right) , \qquad \sigma_{8us} = -\left(B_{3us} K_{1us} + K_{4us} D_{1us}\right) ,$$

$$\sigma_{9us} = B_{2us} - K_{4us} D_{2us} , \qquad \sigma_{10us} = B_{3us} ,$$

$$B_{1us} = \Omega_u \Omega_s \left[e^{(\varepsilon_u + \varepsilon_s)t} + c_u c_s\right] + s_u s_s \varepsilon_u \varepsilon_s + \Omega_u c_u s_s \varepsilon_s + \Omega_s c_s s_u \varepsilon_u ,$$

$$B_{2us} = -\left[\Omega_u e^{\varepsilon_u t} s_s \varepsilon_s + \Omega_s e^{\varepsilon_s t} s_u \varepsilon_u + \Omega_u \Omega_s \left(c_s e^{\varepsilon_u t} + c_u e^{\varepsilon_s t}\right)\right] ,$$

$$B_{3us} = -\left(\Omega_u e^{\varepsilon_u t} s_s + \Omega_s e^{\varepsilon_s t} s_u\right) , \qquad B_{4us} = s_u s_s ,$$

$$D_{1us} = -\left[\Omega_u e^{\varepsilon_u t} s_s \varepsilon_s - \Omega_s e^{\varepsilon_s t} s_u \varepsilon_u + \Omega_u \Omega_s \left(c_s e^{\varepsilon_u t} - c_u e^{\varepsilon_s t}\right)\right] ,$$

$$D_{2us} = -\left[\Omega_s e^{\varepsilon_s t} s_u - \Omega_u e^{\varepsilon_u t} s_s\right] , \quad D_{3us} = \Omega_s c_s s_u - \Omega_u c_u s_s - s_u s_s \left(\varepsilon_u - \varepsilon_s\right) ,$$

$$K_{1us} = G_u^2 + G_s^2 - 4\varepsilon_u \varepsilon_s , \qquad K_{2us} = \left(G_u G_s\right)^2 ,$$

$$K_{3us} = 2\left(\varepsilon_s G_u^2 - \varepsilon_u G_s^2\right) , \qquad K_{4us} = 2\left(\varepsilon_u - \varepsilon_s\right) ,$$

$$G_u^2 = \varepsilon_u^2 + \Omega_u^2 , \qquad G_s^2 = \varepsilon_s^2 + \Omega_s^2 .$$

Substituting Eqs. (2B.4) and (2B.5) into Eqs. (2B.2) and (2B.3), respectively, the evolutionary spectral densities defined in Eq. (2B.1) can be obtained.

In the second part of the derivation the evolutionary cross-spectral densities of a mdof system under exponentially decaying random excitations are obtained. The evolutionary cross-spectral densities are defined as

$$S_{jk}(t,\omega) = S_{jk}^u(t,\omega) + S_{jk}^c(t,\omega) , \qquad (2B.6)$$

where the first term on the rhs denotes nodal couplings for all the individual modes while the second term on the rhs of Eq. (2B.6) represents nodal couplings and modal couplings in the response.

The first term on the rhs of Eq. (2B.6) is associated with the responses of the individual modes and is identical to Eq. (2B.2) except that the term ($\mu_{jr}\Omega_r)^2$ inside the summation sign is replaced by $\mu_{jr}\mu_{kr}\Omega_r^2$. The second term on the rhs of Eq. (2B.6) is due to the nodal couplings and modal couplings. It is given by

$$S_{jk}^c = \sum_{r=2}^{L}\sum_{s=r}^{L} e^{-(\zeta_u\omega_u + \zeta_s\omega_s)t} S_{us}(\omega)\left[\frac{\eta_{pk}P_{us}^c(t,\omega) + i\eta_{mk}q_{us}^c(t,\omega)}{|H_u(\omega)|^2 |H_s(\omega)|^2}\right], \quad (2B.7)$$

where all the terms in Eq. (2B.7) have been defined in the foregoing except

$$\eta_{pk} = \left(\mu_{ju}\right)\left(\mu_{ks}\right) + \left(\mu_{js}\right)\left(\mu_{ku}\right), \quad \eta_{mk} = \left(\mu_{ju}\right)\left(\mu_{ks}\right) - \left(\mu_{js}\right)\left(\mu_{ku}\right), \quad (2B.8)$$

$$q_{us}^c(t,\omega) = \beta_{1us}\,\omega + \beta_{2us}\,\omega^3 + \beta_{3us}\,\omega^5$$
$$+ \beta_{4us}\sin\omega t + \beta_{5us}\omega\cos\omega t + \beta_{6us}\,\omega^2\sin\omega t \quad (2B.9)$$
$$+ \beta_{7us}\,\omega^3\cos\omega t + \beta_{8us}\,\omega^4\sin\omega t + \beta_{9us}\,\omega^5\cos\omega t\,,$$

in which the coefficients are defined by

$$\beta_{1us} = B_{1us}K_{3us} + D_{3us}K_{2us}\,, \quad \beta_{2us} = B_{1us}K_{4us} + B_{4us}K_{3us} - D_{3us}K_{1us}\,,$$

$$\beta_{3us} = B_{4us}K_{4us} + D_{3us}\,, \quad \beta_{4us} = D_{1us}K_{2us}\,,$$

$$\beta_{5us} = B_{2us}K_{3us} + D_{2us}K_{2us}\,, \quad \beta_{6us} = B_{3us}K_{3us} - D_{1us}K_{1us}\,,$$

$$\beta_{7us} = B_{2us}K_{4us} - D_{2us}K_{1us}\,, \quad \beta_{8us} = B_{3us}K_{4us} + D_{1us}\,, \quad \beta_{9us} = D_{2us}\,,$$

and the remaining symbols have already been defined previously.

2B.2 Evolutionary Spectral Densities Due to Uniformly Modulated Random Excitations

The derivation of the evolutionary spectral densities of a mdof system under uniformly modulated random excitations is similarly to that presented in the last section except that considerably more algebraic manipulation is required. By applying Eq. (2.27) and after a lengthy algebraic manipulation one obtains Eq. (2.28) as

$$S_{jk}(t,\omega) = \sum_{i=1}^{5} S_{jk}^{(i)}(t,\omega) ,$$

where the terms on the rhs are defined as

$$S_{jk}^{(1)}(t,\omega) = \sum_{r=1}^{L} (\mu_{jr})(\mu_{kr})\Omega_r^2 e^{-2\zeta_r\omega_r t} S_{rr}(\omega) \frac{p_{rr}^{\alpha u}(t,\omega)}{|H_{r1}(\omega)|^2} ,$$

$$S_{jk}^{(2)}(t,\omega) = \sum_{r=2}^{L}\sum_{s=r}^{L} e^{-(\zeta_u\omega_u + \zeta_s\omega_s)t} S_{us}(\omega)\left[\frac{\eta_{pk}p_{us}^{\alpha c}(t,\omega) + i\eta_{mk}q_{us}^{\alpha c}(t,\omega)}{|H_{u1}(\omega)|^2 |H_{s1}(\omega)|^2}\right],$$

$$S_{jk}^{(3)}(t,\omega) = \sum_{r=1}^{L} (\mu_{jr})(\mu_{kr})\Omega_r^2 e^{-2\zeta_r\omega_r t} S_{rr}(\omega) \frac{p_{rr}^{\beta u}(t,\omega)}{|H_{r2}(\omega)|^2} ,$$

$$S_{jk}^{(4)}(t,\omega) = \sum_{r=2}^{L}\sum_{s=r}^{L} e^{-(\zeta_u\omega_u + \zeta_s\omega_s)t} S_{us}(\omega)\left[\frac{\eta_{pk}p_{us}^{\beta c}(t,\omega) + i\eta_{mk}q_{us}^{\beta c}(t,\omega)}{|H_{u2}(\omega)|^2 |H_{s2}(\omega)|^2}\right],$$

$$S_{jk}^{(5)}(t,\omega) = -\sum_{r=1}^{L}\sum_{s=1}^{L} e^{-(\zeta_r\omega_r + \zeta_s\omega_s)t} S_{rs}(\omega)\left[\frac{\eta_{pk}p_{rs}^{\alpha\beta}(t,\omega) + i\eta_{mk}q_{rs}^{\alpha\beta}(t,\omega)}{|H_{r1}(\omega)|^2 |H_{s2}(\omega)|^2}\right],$$

in which when $j = k$ the above expressions reduce to the evolutionary auto-spectral densities. When $j \neq k$, Eq. (2.28) gives the evolutionary cross-spectral densities.

Equations (2B.4), (2B.5), and (2B.9) are applied to generate all the terms in Eq. (2.28). In the latter, $p_{rr}^{\alpha u}(t,\omega)$ and $p_{us}^{\alpha c}(t,\omega)$ represent $p_{rr}^{u}(t,\omega)$ and $p_{us}^{c}(t,\omega)$ associated with α_{r1} in Eqs. (2B.4) and (2B.5), respectively. In other words, $p_{rr}^{\alpha u}(t,\omega)$ and $p_{us}^{\alpha c}(t,\omega)$ are obtained by substituting ε_r, ε_u and ε_s with ε_{r1}, ε_{u1} and ε_{s1}, respectively.

Likewise, in Eq. (2.28), $p_{rr}^{\beta u}(t,\omega)$ and $p_{us}^{\beta c}(t,\omega)$ represent $p_{rr}^{u}(t,\omega)$ and $p_{us}^{c}(t,\omega)$ associated with α_{r2} in Eqs. (2B.4) and (2B.5), respectively. Thus, $p_{rr}^{\beta u}(t,\omega)$ and $p_{us}^{\beta c}(t,\omega)$ are obtained by substituting ε_r, ε_u and ε_s with ε_{r2}, ε_{u2} and ε_{s2}, respectively.

Also, $p_{rs}^{\alpha\beta}(t,\omega)$ refers to $p_{rs}^{c}(t,\omega)$ associated with α_{r1} and α_{s2} in Eq. (2B.5). That is, $p_{rs}^{\alpha\beta}(t,\omega)$ is obtained by substituting ε_u and ε_s with ε_{r1} and ε_{s2}, respectively, while the subscript u is replaced by r.

Furthermore, $q_{us}^{\alpha c}(t,\omega)$ represents $q_{us}^{c}(t,\omega)$ associated with α_{u1} in Eq. (2B.9). That is, $q_{us}^{\alpha c}(t,\omega)$ is obtained by substituting ε_u and ε_s with ε_{u1} and ε_{s1}, respectively in Eq. (2B.9). On the other hand, the term $q_{us}^{\beta c}(t,\omega)$ denotes $q_{us}^{c}(t,\omega)$ associated with α_{u2} in Eq. (2B.9). Therefore, $q_{us}^{\beta c}(t,\omega)$ is obtained by substituting ε_u and ε_s with ε_{u2} and ε_{s2}, respectively in Eq. (2B.9).

The term $q_{rs}^{\alpha\beta}(t,\omega)$ represents $q_{us}^{c}(t,\omega)$ in Eq. (2B.9) associated with α_{r1} and α_{s2}. Thus, $q_{rs}^{\alpha\beta}(t,\omega)$ is obtained by replacing ε_u and ε_s with ε_{r1} and ε_{s2}, respectively.

Finally, H_{r1} in Eq. (2.28), for example, is H_r with ε_r being replaced by ε_{r1} in which α_r, is, in turn, replaced by α_{r1} such that $\varepsilon_{r1} = \omega_r (\zeta_r - \alpha_{r1}/\omega_r)$ and so on.

Before leaving this section it should be pointed out that when the rhs of Eq. (2.26) is set to unity, the excitations become Gaussian white noises.

2B.3 Variances of Displacements

A detailed closed form solution for Eq. (2.27) when the nodal numbers j and k are equal is considered in this section. When the nodal numbers j and k are not equal, similar statistics are included in Appendices 2C through 2E.

By making use of Eq. (2.27) and (2.28), the variances of displacement responses can be written as

$$\langle x_j^2(t) \rangle = \langle x_j^2(t) \rangle^u + \langle x_j^2(t) \rangle^c , \tag{2B.10}$$

where the terms on the rhs are defined by

$$\langle x_j^2(t) \rangle^u = \int_{-\infty}^{\infty} S_{jj}^u(t,\omega)\,d\omega , \quad \langle x_j^2(t) \rangle^c = \int_{-\infty}^{\infty} S_{jj}^c(t,\omega)\,d\omega . \tag{2B.11a, b}$$

Applying Eq. (2B.11a) for the discretized system under nonstationary random excitations, each of which is a product of a zero-mean Gaussian white noise and an exponentially decaying deterministic modulating function α_r, and Eq. (2.28) in which now the second subscript $k = j$ and $\alpha_{r1} = \alpha_r$ while the second exponential term associated with α_{r2} is set to zero, one can show that

$$\langle x_j^2(t) \rangle^u = \langle x_{jj}^2 \rangle^u = \sum_{r=1}^{L} \left(\mu_{jr}\Omega_r\right)^2 e^{-2\zeta_r\omega_r t} \int_{-\infty}^{\infty} \frac{S_{rr}(\omega)\,p_{rr}^u(t,\omega)}{|H_r(\omega)|^2}\,d\omega . \tag{2B.12}$$

By analogy with Laplace's method, if $S_{rr}(\omega)$ are smooth functions of ω with no sharp peaks, and the damping ratios ζ_r are small, with separate modes, then a good approximation may be obtained by setting the excitation spectral densities $S_{rr}(\omega)$ equal to their values at $\omega = \omega_r$, such that $S_{rr}(\omega)$ can be taken outside the integral. This technique has been employed by Caughey and Stumpf [1], and Corotis and Marshall [2] for sdof systems.

If $S_{rr}(\omega) = S_0$ is independent of ω, then Eq. (2B.11) becomes

$$\left\langle x_{jj}^2 \right\rangle^u = \sum_{r=1}^{L} \left(\mu_{jr} \Omega_r \right)^2 e^{-2\zeta_r \omega_r t} S_0 \int_{-\infty}^{\infty} \frac{p_{rr}^u(t,\omega)}{|H_r(\omega)|^2} \, d\omega \ . \tag{2B.13}$$

To perform the integration in this equation, complex contour integration is required. With the real variable ω replaced by the complex variable

$$\Omega = \omega + i\lambda \ , \tag{2B.14}$$

the integral on the rhs of Eq. (2B.13) then becomes

$$I^u(\Omega) = \oint \frac{p_{rr}^u(t,\omega)}{|H_r(\omega)|^2} \, d\omega$$

where the symbol $\oint$ indicates that the closed contour is in the upper half of the complex plane Ω. The contour of integration is shown in Figure 2B.1.

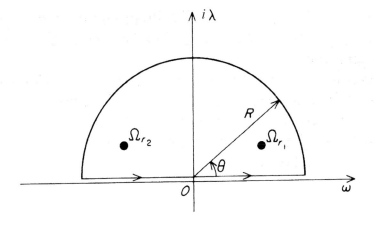

Figure 2B.1   Contour of integration.

The square of the modulus of receptance on the rhs of the last equation is

$$| H_r(\Omega)|^2 = \left(G_r^2 - \Omega^2\right)^2 + \left(2\xi_r G_r\Omega\right)^2 , \tag{2B.15}$$

in which the parameters

$$\xi_r = \varepsilon_r / G_r , \quad \varepsilon_r = \omega_r \zeta_r - \alpha_r , \quad G_r^2 = \varepsilon_r^2 + \Omega_r^2 , \quad \Omega_r = \omega_r \sqrt{1 - \zeta_r^2} .$$

The rhs of the contour integral has four singularities for every r. However, only two for every r inside the contour of integration with $\xi_r < 1$ are required. They are

$$\Omega_{r1} = G_r\left(\sqrt{1 - \xi_r^2} + i\xi_r \right) , \quad \Omega_{r2} = G_r\left(-\sqrt{1 - \xi_r^2} + i\xi_r \right) . \tag{2B.16a, b}$$

The operation of Eq. (2B.13) can be performed symbolically as follow

$$\int_{-\infty}^{\infty} = \lim_{R \to \infty} \left[\oint - \int \right] .$$

By Jordan's lemma, the integral over the semicircle vanishes as R approaches infinity. Consequently,

$$\int_{-\infty}^{\infty} = \lim_{R \to \infty} \oint = \lim_{R \to \infty} I^u(\Omega) = 2\pi i \sum_{\ell=1}^{m} \rho_{r\ell},$$

where $\rho_{r\ell}$ designates the residues of the function. The residues of the two poles defined in Eq. (2B.16) are

$$\sum_{\ell=1}^{2} \rho_{r\ell} = \sum_{\ell=1}^{2} \lim_{\Omega \to \Omega_{r\ell}} \left(\Omega - \Omega_{r\ell}\right) \frac{p_{rr}^u(t,\Omega)}{D_r(\Omega)} = \sum_{\ell=1}^{2} \lim_{\Omega \to \Omega_{r\ell}} \frac{p_{rr}^u(t,\Omega)}{dD_r(\Omega)/d\Omega} , \tag{2B.17}$$

in which the denominator term

$$D_r(\Omega) = | H_r(\Omega)|^2 = \left(G_r^2 - \Omega^2\right)^2 + \left(2\xi_r G_r\Omega\right)^2 .$$

Thus, one has its derivative with respect to Ω,

$$\frac{dD_r(\Omega)}{d\Omega} = 4\left[\left(2\xi_r^2 - 1\right)G_r^2\Omega + \Omega^3\right] . \tag{2B.18}$$

The residues are therefore given as

$$\rho_{r\ell} = \frac{p_{rr}^{u}(t,\Omega)\big|_{\Omega=\Omega_{r\ell}}}{i8G_r^3\xi_r\sqrt{1-\xi_r^2}\left(\sqrt{1-\xi_r^2}+i\xi_r\right)}, \qquad \ell = 1,2. \qquad (2B.19)$$

Substituting Eq. (2B.19) into the integral gives

$$\int_{-\infty}^{\infty} = \frac{\pi}{4G_r^3\xi_r}\left[\left(\sqrt{1-\xi_r^2}+i\xi_r\right)p_{rr}^{u}(t,\Omega)\big|_{\Omega=\Omega_{r2}}\right.$$

$$\left. -\left(-\sqrt{1-\xi_r^2}+i\xi_r\right)p_{rr}^{u}(t,\Omega)\big|_{\Omega=\Omega_{r1}}\right].$$

With the last equation and $p_{rr}^{u}(t,\omega)$ defined in Eq. (2B.4), after some lengthy algebraic manipulation and simplifying, and finally substituting the results into Eq. (2B.13), one can show that

$$\left\langle x_j^2\right\rangle^{u} = \sum_{r=1}^{L}\Phi_{or}^{u}\sum_{\delta=1}^{4}A_{\delta r}Q_{\delta r}, \qquad (2B.20)$$

where

$$\Phi_{or}^{u} = \frac{\pi\left(\mu_{jr}\Omega_r\right)^2 e^{-2\zeta_r\omega_r t}S_o}{G_r^3 y_r}, \qquad Q_{1r} = \sqrt{1-\xi_r^2}, \qquad Q_{2r} = G_r\gamma_r,$$

$$Q_{4r} = e^{-\xi_r G_r t}G_r\sin\gamma_r t, \qquad \gamma_r = G_r\sqrt{1-\xi_r^2}, \qquad y_r = 2\xi_r\sqrt{1-\xi_r^2},$$

$$Q_{3r} = e^{-\xi_r G_r t}\left(\xi_r\sin\gamma_r t + \sqrt{1-\xi_r^2}\cos\gamma_r t\right),$$

and the remaining symbols are already defined in Eqs. (2B.4), (2B.5), and (2B.9).
    Next, Eq. (2B.11b) is considered. The latter equation is the contributions of the modal couplings of the system to the variances of nodal displacements

$$\left\langle x_j^2(t)\right\rangle^{c} = \sum_{r=1}^{L}\sum_{s=1}^{L}\left(\mu_{jr}\right)\left(\mu_{js}\right)e^{-\left(\zeta_r\omega_r+\zeta_s\omega_s\right)t}I_{rs}, \qquad r\neq s, \qquad (2B.21)$$

in which the integrals

$$I_{rs} = \int_{-\infty}^{\infty} \left[\frac{p_{rs}^{c}(t,\omega) + i\, q_{rs}^{c}(t,\omega)}{D_{rs}(\omega)} \right] S_{rs}(\omega)\, d\omega \;, \tag{2B.22}$$

where $p_{rs}^{c}(t,\omega)$ can be obtained from Eq. (2B.5) in which the first subscript u is to be replaced with r, and similarly, $q_{rs}^{c}(t,\omega)$ from Eq. (2B.9) with u replaced by r.

Again, by analogy with Laplace's method and following the same reasoning used previously in this section, and assuming that the excitations associated with modes r and s are directly correlated, Eq. (2B.21) gives

$$\left\langle x_{j}^{2}(t) \right\rangle^{c} = 2 \sum_{r=2}^{L} \sum_{s=2}^{L} \left(\mu_{ju} \right)\left(\mu_{js} \right) S_{o}\, e^{-(\zeta_{u}\omega_{u} + \zeta_{s}\omega_{s})t}\, I_{us}, \quad u = r - 1, \tag{2B.23}$$

where the integrals on the rhs are defined by

$$I_{us} = \int_{-\infty}^{\infty} \frac{p_{us}^{c}(t,\omega)}{D_{us}(\omega)}\, d\omega \;. \tag{2B.24}$$

This integral is evaluated by contour integration, a procedure similar to that used above. The integrand has eight poles for every pair of integers u and s. However, only four of the eight poles are inside the contour of integration with $\xi_{r} < 1$, are required. The four poles are

$$\Omega_{u1} = G_{u}\left(\sqrt{1 - \xi_{u}^{2}} + i\xi_{u} \right), \quad \Omega_{u2} = G_{u}\left(-\sqrt{1 - \xi_{u}^{2}} + i\xi_{u} \right), \tag{2B.25a, b}$$

$$\Omega_{s1} = G_{s}\left(\sqrt{1 - \xi_{s}^{2}} + i\xi_{s} \right), \quad \Omega_{s2} = G_{s}\left(-\sqrt{1 - \xi_{s}^{2}} + i\xi_{s} \right). \tag{2B.25c, d}$$

With similar reasoning to the results in Eq. (2B.17), one has

$$\lim_{R\to\infty} I_{us}(\Omega) = 2\pi i \sum_{\ell=1}^{m} \left(\rho_{u\ell} + \rho_{s\ell} \right), \tag{2B.26}$$

where the residues of the four poles are

$$\sum_{\ell=1}^{2} \rho_{u\ell} = \sum_{\ell=1}^{2} \lim_{\Omega\to\Omega_{u\ell}} \left(\Omega - \Omega_{u\ell} \right) \frac{p_{us}^{c}(t,\Omega)}{D_{us}(\Omega)},$$

which gives

$$\sum_{\ell=1}^{2} \rho_{u\ell} = \sum_{\ell=1}^{2} \lim_{\Omega \to \Omega_{u\ell}} \frac{p_{us}^{c}(t,\Omega)}{dD_{us}(\Omega)/d\Omega}.$$ (2B.27a)

Similarly,

$$\sum_{\ell=1}^{2} \rho_{s\ell} = \sum_{\ell=1}^{2} \lim_{\Omega \to \Omega_{s\ell}} \frac{p_{us}^{c}(t,\Omega)}{dD_{us}(\Omega)/d\Omega}.$$ (2B.27b)

In the above equations,

$$D_{us}(\Omega) = \left[\left(G_{u}^{2} - \Omega^{2} \right)^{2} + \left(2\xi_{u} G_{u}\Omega \right)^{2} \right] \left[\left(G_{s}^{2} - \Omega^{2} \right)^{2} + \left(2\xi_{s} G_{s}\Omega \right)^{2} \right].$$

Thus, one has

$$\frac{dD_{us}(\Omega)}{d\Omega} = |H_{s}(\Omega)|^{2} \frac{d|H_{u}(\Omega)|^{2}}{d\Omega} + |H_{u}(\Omega)|^{2} \frac{d|H_{s}(\Omega)|^{2}}{d\Omega}.$$ (2B.28)

Applying Eq. (2B.18), for $\Omega = \Omega_{u\ell}$, $\ell = 1, 2$, it gives

$$\left. \frac{dD_{us}(\Omega)}{d\Omega} \right|_{\Omega=\Omega_{u\ell}} = 4|H_{s}(\Omega)| \left[\left(2\xi_{u}^{2} - 1 \right) G_{u}^{2}\Omega + \Omega^{3} \right] \Big|_{\Omega=\Omega_{u\ell}}.$$

Likewise, for $\Omega = \Omega_{s\ell}$, $\ell = 1, 2$,

$$\left. \frac{dD_{us}(\Omega)}{d\Omega} \right|_{\Omega=\Omega_{s\ell}} = 4|H_{u}(\Omega)| \left[\left(2\xi_{s}^{2} - 1 \right) G_{s}^{2}\Omega + \Omega^{3} \right] \Big|_{\Omega=\Omega_{s\ell}}.$$

Substituting for $\Omega_{u\ell}$, where $\ell = 1, 2$ into the first of the last two equations results

$$\left. \frac{dD_{us}(\Omega)}{d\Omega} \right|_{\Omega=\Omega_{u1}} = Z_{11} Z_{12}, \qquad \left. \frac{dD_{us}(\Omega)}{d\Omega} \right|_{\Omega=\Omega_{u2}} = Z_{21} Z_{22},$$ (2B.29a, b)

where the complex numbers on the rhs are defined as

$$Z_{11} = d_{1us} + i d_{2us}, \qquad Z_{12} = i 4 G_{u}^{3} y_{u} \left(\sqrt{1 - \xi_{u}^{2}} + i\xi_{u} \right),$$

$$Z_{21} = -\left(d_{1us} - i d_{2us} \right), \qquad Z_{22} = i 4 G_{u}^{3} y_{u} \left(-\sqrt{1 - \xi_{u}^{2}} + i\xi_{u} \right),$$

$$d_{1us} = G_s^4 + 2G_u^2 G_s^2 \left(2\xi_s^2 - 1\right) w_u + G_u^4 u_u \, ,$$

$$d_{2us} = 2G_u^2 y_u \left[G_s^2 \left(2\xi_s^2 - 1\right) + G_u^2 w_u \right] ,$$

$$u_u = 1 - 8\xi_u^2 + 8\xi_u^4 \, , \qquad w_u = 1 - 2\xi_u^2 \, ,$$

and the remaining symbols have already been defined above except that the subscript r is replaced by u.

Substituting Eqs. (2B.29a) and (2B.29b) into Eq. (2B.27a), and applying Eqs. (2B.26) and (2B.23) in turn, one has

$$J_1(t,us) = 2 \sum_{r=2}^{L} \sum_{s=2}^{L} \left(\mu_{ju}\right)\left(\mu_{js}\right) S_o \, e^{-\left(\zeta_u \omega_u + \zeta_s \omega_s\right)t} \, J(us) \, ,$$

where the term $J(us)$ on the rhs is defined by

$$J(us) = 2\pi i \left[\frac{p_{us}^c(t,\Omega)\big|_{\Omega = \Omega_{u1}}}{Z_{11} Z_{12}} + \frac{p_{us}^c(t,\Omega)\big|_{\Omega = \Omega_{u2}}}{Z_{21} Z_{22}} \right] .$$

Upon application of Eq. (2B.23) and the last equation, and after some lengthy algebraic manipulation one can show that

$$J_1(t,us) = \sum_{r=2}^{L} \sum_{s=r}^{L} \Phi_{1us}^c \sum_{\zeta=1}^{10} \sigma_{\zeta us} p_{\zeta us}^c \, , \qquad (2B.30)$$

in which the terms on the rhs are given as

$$\Phi_{1us}^c = \frac{-\pi\left(\mu_{ju} \mu_{js}\right) e^{-\left(\zeta_u \omega_u + \zeta_s \omega_s\right)t} S_o}{G_u^3 y_u \left[\left(d_{1us}\right)^2 + \left(d_{2us}\right)^2\right]} \, ,$$

$$p_{1us}^c = z_{1us} \, , \qquad p_{2us}^c = G_u^2 \left(z_{1us} w_u - z_{2us} y_u\right) , \qquad (2B.31a,\, b)$$

$$p_{3us}^c = G_u^4 \left(z_{1us} u_u - z_{2us} v_u\right) , \qquad (2B.31c)$$

$$p_{4us}^c = G_u^6 \left[z_{1us}\left(u_u w_u - v_u y_u\right) - z_{2us}\left(u_u y_u + v_u w_u\right)\right] , \qquad (2B.31d)$$

$$p_{5us}^{c} = e^{-\xi_u G_u t}\left(z_{1us}\cos\gamma_u t - z_{2us}\sin\gamma_u t\right) , \qquad (2B.31e)$$

$$p_{6us}^{c} = G_u e^{-\xi_u G_u t}\left[\left(\xi_u z_{1us} + z_{2us}\sqrt{1-\xi_u^2}\right)\cos\gamma_u t\right.$$
$$\left. + \left(z_{1us}\sqrt{1-\xi_u^2} - \xi_u z_{2us}\right)\sin\gamma_u t\right] , \qquad (2B.31f)$$

$$p_{7us}^{c} = G_u^2 e^{-\xi_u G_u t}\left[\left(w_u z_{1us} - y_u z_{2us}\right)\cos\gamma_u t\right.$$
$$\left. - \left(w_u z_{2us} + y_u z_{1us}\right)\sin\gamma_u t\right] , \qquad (2B.31g)$$

$$p_{8us}^{c} = G_u^3 e^{-\xi_u G_u t}\left(f_{1us}\sin\gamma_u t + f_{2us}\cos\gamma_u t\right) , \qquad (2B.31h)$$

$$p_{9us}^{c} = G_u^4 e^{-\xi_u G_u t}\left(f_{3us}\cos\gamma_u t - f_{4us}\sin\gamma_u t\right) , \qquad (2B.31i)$$

$$p_{10us}^{c} = G_u^5 e^{-\xi_u G_u t}\left(f_{5us}\sin\gamma_u t + f_{6us}\cos\gamma_u t\right) . \qquad (2B.31j)$$

The parameters inside the above relations are

$$v_u = 2 w_u y_u , \qquad z_{1us} = \xi_u d_{2us} - d_{1us}\sqrt{1-\xi_u^2} , \qquad (2B.32a, b)$$

$$z_{2us} = \xi_u d_{1us} + d_{2us}\sqrt{1-\xi_u^2} , \qquad (2B.32c)$$

$$f_{1us} = z_{1us}\left(w_u\sqrt{1-\xi_u^2} - \xi_u y_u\right)$$
$$- z_{2us}\left(\xi_u w_u + y_u\sqrt{1-\xi_u^2}\right) , \qquad (2B.32d)$$

$$f_{2us} = z_{1us}\left(y_u\sqrt{1-\xi_u^2} + \xi_u w_u\right)$$
$$+ z_{2us}\left(w_u\sqrt{1-\xi_u^2} - \xi_u y_u\right) , \qquad (2B.32e)$$

$$f_{3us} = z_{1us}\left(w_u^2 - y_u^2\right) - 2z_{2us}\,w_u\,y_u \ , \tag{2B.32f}$$

$$f_{4us} = z_{2us}\left(w_u^2 - y_u^2\right) + 2z_{1us}\,w_u\,y_u \ , \tag{2B.32g}$$

$$f_{5us} = \sqrt{1 - \xi_u^2}\left[z_{1us}\left(w_u^2 - y_u^2\right) - 2z_{2us}\,w_u\,y_u\right]$$
$$- \xi_u\left[\left(w_u^2 - y_u^2\right)z_{2us} + 2z_{1us}\,w_u\,y_u\right] , \tag{2B.32h}$$

$$f_{6us} = \sqrt{1 - \xi_u^2}\left[z_{2us}\left(w_u^2 - y_u^2\right) + 2z_{1us}\,w_u\,y_u\right]$$
$$+ \xi_u\left[\left(w_u^2 - y_u^2\right)z_{1us} - 2z_{2us}\,w_u\,y_u\right] . \tag{2B.32i}$$

Similarly, by applying Eqs. (2B.27b), (2B.26) and (2B.23), the part of the variance of the displacement response associated with the poles Ω_{s1} and Ω_{s2} becomes

$$J_2(t,us) = J_1(t,su) \ , \tag{2B.33}$$

where $J_1(t,su)$ is $J_1(t,us)$ with u and s replaced by s and u, respectively.

Substituting the above equations and Eq. (2B.20) into (2B.10), the variance of displacement x_j becomes

$$\left\langle x_j^2 \right\rangle = \sum_{r=1}^{L} \Phi_{or}^u \sum_{\delta=1}^{4} A_{\delta r}\,Q_{\delta r} + J_1(t,us) + J_2(t,us) \ . \tag{2B.34}$$

Recall that Eq. (2B.34) is for nonstationary random excitations defined as products of zero-mean Gaussian white noises and exponentially decaying deterministic modulating functions with modulating parameters $\alpha_{r1} = \alpha_r$.

For envelope modulating functions as defined by Eq. (2.26), Eq. (2B.34) is applied similarly to the case with modulating functions parameters, $\alpha_{r2} = \alpha_r$ with due consideration of the coupling terms. The outline and results are given in Eq. (2.28) and Appendix 2C.

Finally, it should be noted that when $L = 1$, that is for sdof systems, the result given by Eq. (2B.34) tallies with that obtained by Shinozuka [3]. Setting $L = 1$ and the exponentially decaying deterministic modulating function to unity, Eq. (2B.34) reduces to the corresponding result derived by Caughey and Stumpf [1].

References

[1] Caughey, T.K., and Stumpf, H.J. (1961). Transient response of a dynamic system under random excitation, *Trans. A.S.M.E. Journal of Applied Mechanics*, **28**, 563-566.

[2] Corotis, R.B., and Marshall, T.A. (1977). Oscillator response to modulated random excitation, *Proc. A.S.C.E. Journal of Engineering Mechanics*, **103**, 501-513.

[3] Shinozuka, M. (1964). Probability of structural failure under random loading, *Proc. A.S.C.E. Journal of Engineering Mechanics*, **90(EM5)**, 147-170.

Appendix 2C   Time-dependent Covariances of Displacements

With the results in Appendix 2B the covariances of displacements can be obtained as expressed in Eq. (2.39) in which the terms on the rhs are given by

$$N_1^{(1)}(x_j x_k) = \sum_{r=1}^{L} \Phi_{or}^{\alpha u} \sum_{\delta=1}^{4} A_{\delta r}^{\alpha} Q_{\delta r}^{\alpha},$$

$$N_2^{(1)}(x_j x_k) = \sum_{r=2}^{L} \sum_{s=r}^{L} \Phi_{1us}^{\alpha c} \sum_{\zeta=1}^{10} \sigma_{\zeta us}^{\alpha} P_{\zeta us}^{\alpha c},$$

$$N_3^{(1)}(x_j x_k) = \sum_{r=2}^{L} \sum_{s=r}^{L} \Phi_{1su}^{\alpha c} \sum_{\zeta=1}^{10} \sigma_{\zeta su}^{\alpha} P_{\zeta su}^{\alpha c},$$

$$N_4^{(1)}(x_j x_k) = \sum_{r=1}^{L} \Phi_{or}^{\beta u} \sum_{\delta=1}^{4} A_{\delta r}^{\beta} Q_{\delta r}^{\beta},$$

$$N_5^{(1)}(x_j x_k) = \sum_{r=2}^{L} \sum_{s=r}^{L} \Phi_{1us}^{\beta c} \sum_{\zeta=1}^{10} \sigma_{\zeta us}^{\beta} P_{\zeta us}^{\beta c},$$

$$N_6^{(1)}(x_j x_k) = \sum_{r=2}^{L} \sum_{s=r}^{L} \Phi_{1su}^{\beta c} \sum_{\zeta=1}^{10} \sigma_{\zeta su}^{\beta} P_{\zeta su}^{\beta c},$$

$$N_7^{(1)}(x_j x_k) = -\sum_{r=1}^{L} \sum_{s=1}^{L} \Phi_{1rs}^{\alpha \beta} \sum_{\zeta=1}^{10} \sigma_{\zeta rs}^{\alpha \beta} P_{\zeta rs}^{\alpha \beta},$$

$$N_8^{(1)}(x_j x_k) = -\sum_{r=1}^{L} \sum_{s=1}^{L} \Phi_{1sr}^{\alpha \beta} \sum_{\zeta=1}^{10} \sigma_{\zeta sr}^{\alpha \beta} P_{\zeta sr}^{\alpha \beta},$$

$$N_9^{(1)}(x_j x_k) = \sum_{r=2}^{L} \sum_{s=r}^{L} \Phi_{2us}^{\alpha c} \sum_{\eta=1}^{3} \beta_{\eta us}^{\alpha} q_{\eta us}^{\alpha c},$$

$$N_{10}^{(1)}(x_j x_k) = \sum_{r=2}^{L} \sum_{s=r}^{L} \Phi_{2su}^{\alpha c} \sum_{\eta=1}^{3} \beta_{\eta su}^{\alpha} q_{\eta su}^{\alpha c},$$

$$N_{11}^{(1)}(x_j x_k) = \sum_{r=2}^{L} \sum_{s=r}^{L} \Phi_{2us}^{\beta c} \sum_{\eta=1}^{3} \beta_{\eta us}^{\beta} q_{\eta us}^{\beta c},$$

$$N_{12}^{(1)}(x_j x_k) = \sum_{r=2}^{L} \sum_{s=r}^{L} \Phi_{2su}^{\beta c} \sum_{\eta=1}^{3} \beta_{\eta su}^{\beta} q_{\eta su}^{\beta c},$$

$$N_{13}^{(1)}(x_j x_k) = -\sum_{r=1}^{L} \sum_{s=1}^{L} \Phi_{2rs}^{\alpha\beta} \sum_{\eta=1}^{3} \beta_{\eta rs}^{\alpha\beta} q_{\eta rs}^{\alpha\beta},$$

$$N_{14}^{(1)}(x_j x_k) = -\sum_{r=1}^{L} \sum_{s=1}^{L} \Phi_{2sr}^{\alpha\beta} \sum_{\eta=1}^{3} \beta_{\eta sr}^{\alpha\beta} q_{\eta sr}^{\alpha\beta},$$

in which the factors and terms are included in Appendix 2B and the following.

In the foregoing the terms $N_1^{(1)}(x_j\,x_k)$ through $N_8^{(1)}(x_j\,x_k)$ are due to the evolutionary coincident spectral density function and $N_9^{(1)}(x_jx_k)$ through $N_{14}^{(1)}(x_jx_k)$ constitute the effect of the evolutionary quadrature spectral density function. When the subcripts j and k are equal, the evolutionary quadrature spectral density function becomes zero, Eq. (2.39) reduces to the variance of displacement at node j.

The factors or coefficients of the covariances of displacements defined by Eq. (2.39) are presented now. Operating on the integral and going through some lengthy algebraic manipulation as outlined in Appendix 2B and simplifying, one can show that the factors are

$$\Phi_{or}^{\alpha u} = \frac{\pi \mu_{jr} \mu_{kr} \Omega_r^2 e^{-2\zeta_r \omega_r t} S_o}{G_{r1}^3 y_{r1}}, \tag{2C.1a}$$

$$\Phi_{1us}^{\alpha c} = \frac{-\pi(\eta_{pk})e^{-(\zeta_u \omega_u + \zeta_s \omega_s)t} S_o}{G_{u1}^3 y_{u1}\left[(d_{1us}^{\alpha})^2 + (d_{2us}^{\alpha})^2\right]}, \tag{2C.1b}$$

$$\Phi_{1su}^{\alpha c} = \frac{-\pi(\eta_{pk})e^{-(\zeta_s \omega_s + \zeta_u \omega_u)t} S_o}{G_{s1}^3 y_{s1}\left[(d_{1su}^{\alpha})^2 + (d_{2su}^{\alpha})^2\right]}, \tag{2C.1c}$$

$$\Phi_{or}^{\beta u} = \frac{\pi \mu_{jr} \mu_{kr} \Omega_r^2 e^{-2\zeta_r \omega_r t} S_o}{G_{r2}^3 y_{r2}}, \tag{2C.1d}$$

$$\Phi_{1us}^{\beta c} = \frac{-\pi\left(\eta_{pk}\right)e^{-\left(\zeta_u\omega_u + \zeta_s\omega_s\right)t}S_o}{G_{u2}^3 y_{u2}\left[\left(d_{1us}^\beta\right)^2 + \left(d_{2us}^\beta\right)^2\right]} \quad , \qquad (2C.1e)$$

$$\Phi_{1su}^{\beta c} = \frac{-\pi\left(\eta_{pk}\right)e^{-\left(\zeta_u\omega_u + \zeta_s\omega_s\right)t}S_o}{G_{s2}^3 y_{s2}\left[\left(d_{1su}^\beta\right)^2 + \left(d_{2su}^\beta\right)^2\right]} \quad , \qquad (2C.1f)$$

$$\Phi_{1rs}^{\alpha\beta} = \frac{-\pi\left(\eta_{pk}\right)e^{-\left(\zeta_r\omega_r + \zeta_s\omega_s\right)t}S_o}{G_{r1}^3 y_{r1}\left[\left(d_{1rs}^{\alpha\beta}\right)^2 + \left(d_{2rs}^{\alpha\beta}\right)^2\right]} \quad , \qquad (2C.1g)$$

$$\Phi_{1sr}^{\alpha\beta} = \frac{-\pi\left(\eta_{pk}\right)e^{-\left(\zeta_r\omega_r + \zeta_s\omega_s\right)t}S_o}{G_{s2}^3 y_{s2}\left[\left(d_{1sr}^{\alpha\beta}\right)^2 + \left(d_{2sr}^{\alpha\beta}\right)^2\right]} \quad , \qquad (2C.1h)$$

while the factors containing the subscript 2 are similar to those containing the subscript 1 except that in the latter η_{pk} is being replaced with η_{mk}. Thus,

$$\Phi_{2us}^{\alpha c} = \frac{-\pi\left(\eta_{mk}\right)e^{-\left(\zeta_u\omega_u + \zeta_s\omega_s\right)t}S_o}{G_{u1}^3 y_{u1}\left[\left(d_{1us}^\alpha\right)^2 + \left(d_{2us}^\alpha\right)^2\right]} \quad , \qquad (2C.2a)$$

$$\Phi_{2su}^{\alpha c} = \frac{-\pi\left(\eta_{mk}\right)e^{-\left(\zeta_s\omega_s + \zeta_u\omega_u\right)t}S_o}{G_{s1}^3 y_{s1}\left[\left(d_{1su}^\alpha\right)^2 + \left(d_{2su}^\alpha\right)^2\right]} \quad , \qquad (2C.2b)$$

$$\Phi_{2us}^{\beta c} = \frac{-\pi\left(\eta_{mk}\right)e^{-\left(\zeta_u\omega_u + \zeta_s\omega_s\right)t}S_o}{G_{u2}^3 y_{u2}\left[\left(d_{1us}^\beta\right)^2 + \left(d_{2us}^\beta\right)^2\right]} \quad , \qquad (2C.2c)$$

$$\Phi_{2su}^{\beta c} = \frac{-\pi\left(\eta_{mk}\right)e^{-\left(\zeta_u\omega_u + \zeta_s\omega_s\right)t}S_o}{G_{s2}^3 y_{s2}\left[\left(d_{1su}^\beta\right)^2 + \left(d_{2su}^\beta\right)^2\right]} \quad , \qquad (2C.2d)$$

$$\Phi_{2rs}^{\alpha\beta} = \frac{-\pi\left(\eta_{mk}\right)e^{-\left(\zeta_r\omega_r + \zeta_s\omega_s\right)t}S_o}{G_{r1}^3 y_{r1}\left[\left(d_{1rs}^{\alpha\beta}\right)^2 + \left(d_{2rs}^{\alpha\beta}\right)^2\right]},$$

(2C.2e)

$$\Phi_{2sr}^{\alpha\beta} = \frac{-\pi\left(\eta_{mk}\right)e^{-\left(\zeta_r\omega_r + \zeta_s\omega_s\right)t}S_o}{G_{s2}^3 y_{s2}\left[\left(d_{1sr}^{\alpha\beta}\right)^2 + \left(d_{2sr}^{\alpha\beta}\right)^2\right]}.$$

(2C.2f)

The terms inside the square brackets in the denominator of Eq. (2C.2a) are

$$d_{1us}^{\alpha} = G_{s1}^4 + 2G_{u1}^2 G_{s1}^2 \left(2\xi_{s1}^2 - 1\right)w_{u1} + G_{u1}^4 u_{u1},$$

(2C.3a)

$$d_{2us}^{\alpha} = 2G_{u1}^2 y_{u1}\left[G_{s1}^2\left(2\xi_{s1}^2 - 1\right) + G_{u1}^2 w_{u1}\right].$$

(2C.3b)

Similarly, the terms with the superscript β can be obtained by applying the last two relations with appropriate change of superscript and subscripts to give

$$d_{1us}^{\beta} = G_{s2}^4 + 2G_{u2}^2 G_{s2}^2 \left(2\xi_{s2}^2 - 1\right)w_{u2} + G_{u2}^4 u_{u2},$$

(2C.3c)

$$d_{2us}^{\beta} = 2G_{u2}^2 y_{u2}\left[G_{s2}^2\left(2\xi_{s2}^2 - 1\right) + G_{u2}^2 w_{u2}\right].$$

(2C.3d)

The terms with the superscript $\alpha\beta$ can similarly be found to be

$$d_{1rs}^{\alpha\beta} = G_{s2}^4 + 2G_{r1}^2 G_{s2}^2 \left(2\xi_{s2}^2 - 1\right)w_{r1} + G_{r1}^4 u_{r1},$$

(2C.3e)

$$d_{2rs}^{\alpha\beta} = 2G_{r1}^2 y_{r1}\left[G_{s2}^2\left(2\xi_{s2}^2 - 1\right) + G_{r1}^2 w_{r1}\right],$$

(2C.3f)

in addition to those defined in Appendix 2B. One also has

$$\xi_{si} = \frac{\varepsilon_{si}}{G_{si}}, \quad u_{ui} = 1 - 8\xi_{ui}^2 + 8\xi_{ui}^4,$$

(2C.4a, b)

$$w_{ui} = 1 - 2\xi_{ui}^2, \qquad y_{si} = 2\xi_{si}\sqrt{1 - \xi_{si}^2}, \qquad i = 1, 2, \qquad \text{(2C.4c, d)}$$

$$Q_{1r}^{\alpha} = \sqrt{1 - \xi_{r1}^2}, \qquad Q_{2r}^{\alpha} = G_{r1}\gamma_{r1}, \qquad \text{(2C.4e, f)}$$

$$Q_{3r}^{\alpha} = e^{-\xi_{r1} G_{r1} t}\left(\xi_{r1}\sin\gamma_{r1}t + \sqrt{1 - \xi_{r1}^2}\cos\gamma_{r1}t\right), \qquad \text{(2C.4g)}$$

$$Q_{4r}^{\alpha} = e^{-\xi_{r1} G_{r1} t} G_{r1}\sin\gamma_{r1}t . \qquad \text{(2C.4h)}$$

The four corresponding expressions containing the superscript β in Eq. (2.39) can be obtained by replacing the superscript α, on the lhs and the subscript $r1$ on the rhs of the last four equations, with the superscript β and subscript $r2$, respectively. Thus,

$$Q_{1r}^{\beta} = \sqrt{1 - \xi_{r2}^2}, \qquad Q_{2r}^{\beta} = G_{r2}\gamma_{r2}, \qquad \text{(2C.5a, b)}$$

$$Q_{3r}^{\beta} = e^{-\xi_{r2} G_{r2} t}\left(\xi_{r2}\sin\gamma_{r2}t + \sqrt{1 - \xi_{r2}^2}\cos\gamma_{r2}t\right), \qquad \text{(2C.5c)}$$

$$Q_{4r}^{\beta} = e^{-\xi_{r2} G_{r2} t} G_{r2}\sin\gamma_{r2}t . \qquad \text{(2C.5d)}$$

Now, returning to Eq. (2.39), the terms $A_{\delta r}^{\alpha}$ are $A_{\delta r}$ in Eq. (2B.4) with the subscript r being replaced by $r1$. Similarly, the terms $A_{\delta r}^{\beta}$ in Eq. (2.39) are $A_{\delta r}$ in Eq. (2B.4) with the subscript r being replaced by $r2$. The $p_{\zeta us}^{\alpha c}$ terms in Eq. (2.39) are now defined in the following,

$$p_{1us}^{\alpha c} = z_{1us}^{\alpha}, \qquad p_{2us}^{\alpha c} = G_{u1}^2\left(z_{1us}^{\alpha}w_{u1} - z_{2us}^{\alpha}y_{u1}\right), \qquad \text{(2C.6a, b)}$$

$$p_{3us}^{\alpha c} = G_{u1}^4\left(z_{1us}^{\alpha}u_{u1} - z_{2us}^{\alpha}v_{u1}\right), \qquad \text{(2C.6c)}$$

$$p_{4us}^{\alpha c} = G_{u1}^6\left[z_{1us}^{\alpha}\left(u_{u1}w_{u1} - v_{u1}y_{u1}\right) - z_{2us}^{\alpha}\left(u_{u1}y_{u1} + v_{u1}w_{u1}\right)\right], \qquad \text{(2C.6d)}$$

$$p_{5us}^{\alpha c} = e^{-\xi_{u1}G_{u1}t}\left(z_{1us}^{\alpha}\cos\gamma_{u1}t - z_{2us}^{\alpha}\sin\gamma_{u1}t\right), \qquad (2C.6e)$$

$$p_{6us}^{\alpha c} = G_{u1}e^{-\xi_{u1}G_{u1}t}\left[\left(\xi_{u1}z_{1us}^{\alpha} + z_{2us}^{\alpha}\sqrt{1-\xi_{u1}^{2}}\right)\cos\gamma_{u1}t\right.$$
$$\left. + \left(z_{1us}^{\alpha}\sqrt{1-\xi_{u1}^{2}} - \xi_{u1}z_{2us}^{\alpha}\right)\sin\gamma_{u1}t\right], \qquad (2C.6f)$$

$$p_{7us}^{\alpha c} = G_{u1}^{2}e^{-\xi_{u1}G_{u1}t}\left[\left(w_{u1}z_{1us}^{\alpha} - y_{u1}z_{2us}^{\alpha}\right)\cos\gamma_{u1}t\right.$$
$$\left. - \left(w_{u1}z_{2us}^{\alpha} + y_{u1}z_{1us}^{\alpha}\right)\sin\gamma_{u1}t\right], \qquad (2C.6g)$$

$$p_{8us}^{\alpha c} = G_{u1}^{3}e^{-\xi_{u1}G_{u1}t}\left(f_{1us}^{\alpha}\sin\gamma_{u1}t + f_{2us}^{\alpha}\cos\gamma_{u1}t\right), \qquad (2C.6h)$$

$$p_{9us}^{\alpha c} = G_{u1}^{4}e^{-\xi_{u1}G_{u1}t}\left(f_{3us}^{\alpha}\cos\gamma_{u1}t - f_{4us}^{\alpha}\sin\gamma_{u1}t\right), \qquad (2C.6i)$$

$$p_{10us}^{\alpha c} = G_{u1}^{5}e^{-\xi_{u1}G_{u1}t}\left(f_{5us}^{\alpha}\sin\gamma_{u1}t + f_{6us}^{\alpha}\cos\gamma_{u1}t\right), \qquad (2C.6j)$$

$$v_{u1} = 2w_{u1}y_{u1}, \qquad z_{1us}^{\alpha} = \xi_{u1}d_{2us}^{\alpha} - d_{1us}^{\alpha}\sqrt{1-\xi_{u1}^{2}}, \qquad (2C.7a, b)$$

$$z_{2us}^{\alpha} = \xi_{u1}d_{1us}^{\alpha} + d_{2us}^{\alpha}\sqrt{1-\xi_{u1}^{2}}, \qquad (2C.7c)$$

$$f_{1us}^{\alpha} = z_{1us}^{\alpha}\left(w_{u1}\sqrt{1-\xi_{u1}^{2}} - \xi_{u1}y_{u1}\right)$$
$$- z_{2us}^{\alpha}\left(\xi_{u1}w_{u1} + y_{u1}\sqrt{1-\xi_{u1}^{2}}\right), \qquad (2C.7d)$$

$$f_{2us}^{\alpha} = z_{1us}^{\alpha}\left(y_{u1}\sqrt{1-\xi_{u1}^{2}} + \xi_{u1}w_{u1}\right)$$
$$+ z_{2us}^{\alpha}\left(w_{u1}\sqrt{1-\xi_{u1}^{2}} - \xi_{u1}y_{u1}\right), \qquad (2C.7e)$$

$$f_{3us}^{\alpha} = z_{1us}^{\alpha}\left(w_{u1}^2 - y_{u1}^2\right) - 2z_{2us}^{\alpha} w_{u1} y_{u1} , \tag{2C.7f}$$

$$f_{4us}^{\alpha} = z_{2us}^{\alpha}\left(w_{u1}^2 - y_{u1}^2\right) + 2z_{1us}^{\alpha} w_{u1} y_{u1} , \tag{2C.7g}$$

$$f_{5us}^{\alpha} = \sqrt{1 - \xi_{u1}^2}\left[z_{1us}^{\alpha}\left(w_{u1}^2 - y_{u1}^2\right) - 2z_{2us}^{\alpha} w_{u1} y_{u1}\right]$$
$$- \xi_{u1}\left[\left(w_{u1}^2 - y_{u1}^2\right)z_{2us}^{\alpha} + 2z_{1us}^{\alpha} w_{u1} y_{u1}\right] , \tag{2C.7h}$$

$$f_{6us}^{\alpha} = \sqrt{1 - \xi_{u1}^2}\left[z_{2us}^{\alpha}\left(w_{u1}^2 - y_{u1}^2\right) + 2z_{1us}^{\alpha} w_{u1} y_{u1}\right]$$
$$+ \xi_{u1}\left[\left(w_{u1}^2 - y_{u1}^2\right)z_{1us}^{\alpha} - 2z_{2us}^{\alpha} w_{u1} y_{u1}\right] . \tag{2C.7i}$$

In Eqs. (2C.6) and (2C.7) all the expressions with the superscript α indicate that they are related to the modulating function parameter α_{u1} of the random excitation. Therefore, one can obtain $p_{\zeta us}^{\beta c}$ from $p_{\zeta us}^{\alpha c}$ in Eqs. (2C.6a) through (2C.6j) by replacing α_{u1} with α_{u2}. On the other hand, one can obtain $p_{\zeta rs}^{\alpha \beta}$ in Eq. (2.39) from $p_{\zeta us}^{\alpha c}$ in Eqs. (2C.6a) through (2C.6j) by replacing the appropriate modulating function parameter α_{u1} with α_{r1}. Thus, these expressions are

$$p_{1rs}^{\alpha\beta} = z_{1rs}^{\alpha\beta} , \qquad p_{2rs}^{\alpha\beta} = G_{r1}^2\left(z_{1rs}^{\alpha\beta} w_{r1} - z_{2rs}^{\alpha\beta} y_{r1}\right) , \tag{2C.8a, b}$$

$$p_{3rs}^{\alpha\beta} = G_{r1}^4\left(z_{1rs}^{\alpha\beta} u_{r1} - z_{2rs}^{\alpha\beta} v_{r1}\right) ,$$
$$p_{4rs}^{\alpha\beta} = G_{r1}^6\left[z_{1rs}^{\alpha\beta}\left(u_{r1} w_{r1} - v_{r1} y_{r1}\right) - z_{2rs}^{\alpha\beta}\left(u_{r1} y_{r1} + v_{r1} w_{r1}\right)\right] , \tag{2C.8c, d}$$

$$p_{5rs}^{\alpha\beta} = e^{-\xi_{r1} G_{r1} t}\left(z_{1rs}^{\alpha\beta} \cos\gamma_{r1} t - z_{2rs}^{\alpha\beta} \sin\gamma_{r1} t\right) , \tag{2C.8e}$$

$$p_{6rs}^{\alpha\beta} = G_{r1} e^{-\xi_{r1} G_{r1} t}\left[\left(\xi_{r1} z_{1rs}^{\alpha\beta} + z_{2rs}^{\alpha\beta}\sqrt{1 - \xi_{r1}^2}\right)\cos\gamma_{r1} t\right.$$
$$\left. + \left(z_{1rs}^{\alpha\beta}\sqrt{1 - \xi_{r1}^2} - \xi_{r1} z_{2rs}^{\alpha}\right)\sin\gamma_{r1} t\right] , \tag{2C.8f}$$

$$p_{7rs}^{\alpha\beta} = G_{r1}^2 e^{-\xi_{r1}G_{r1}t} \left[\left(w_{r1} z_{1rs}^{\alpha\beta} - y_{r1} z_{2rs}^{\alpha\beta} \right) \cos\gamma_{r1}t \right.$$
$$\left. - \left(w_{r1} z_{2rs}^{\alpha\beta} + y_{r1} z_{1rs}^{\alpha\beta} \right) \sin\gamma_{r1}t \right], \qquad (2C.8g)$$

$$p_{8rs}^{\alpha\beta} = G_{r1}^3 e^{-\xi_{r1}G_{r1}t} \left(f_{1rs}^{\alpha\beta} \sin\gamma_{r1}t + f_{2rs}^{\alpha\beta} \cos\gamma_{r1}t \right), \qquad (2C.8h)$$

$$p_{9rs}^{\alpha\beta} = G_{r1}^4 e^{-\xi_{r1}G_{r1}t} \left(f_{3rs}^{\alpha\beta} \cos\gamma_{r1}t - f_{4rs}^{\alpha\beta} \sin\gamma_{r1}t \right), \qquad (2C.8i)$$

$$p_{10rs}^{\alpha\beta} = G_{r1}^5 e^{-\xi_{r1}G_{r1}t} \left(f_{5rs}^{\alpha\beta} \sin\gamma_{r1}t + f_{6rs}^{\alpha\beta} \cos\gamma_{r1}t \right), \qquad (2C.8j)$$

in which the coefficients are defined as

$$v_{r1} = 2 w_{r1} y_{r1}, \qquad z_{1rs}^{\alpha\beta} = \xi_{r1} d_{2rs}^{\alpha\beta} - d_{1rs}^{\alpha\beta} \sqrt{1 - \xi_{r1}^2}, \qquad (2C.9a, b)$$

$$z_{2rs}^{\alpha\beta} = \xi_{r1} d_{1rs}^{\alpha\beta} + d_{2rs}^{\alpha\beta} \sqrt{1 - \xi_{r1}^2}, \qquad (2C.9c)$$

$$f_{1rs}^{\alpha\beta} = z_{1rs}^{\alpha\beta} \left(w_{r1} \sqrt{1 - \xi_{r1}^2} - \xi_{r1} y_{r1} \right) - z_{2rs}^{\alpha\beta} \left(\xi_{r1} w_{r1} + y_{r1} \sqrt{1 - \xi_{r1}^2} \right), \qquad (2C.9d)$$

$$f_{2rs}^{\alpha\beta} = z_{1rs}^{\alpha\beta} \left(y_{r1} \sqrt{1 - \xi_{r1}^2} + \xi_{r1} w_{r1} \right) + z_{2rs}^{\alpha\beta} \left(w_{r1} \sqrt{1 - \xi_{r1}^2} - \xi_{r1} y_{r1} \right), \qquad (2C.9e)$$

$$f_{3rs}^{\alpha\beta} = z_{1rs}^{\alpha\beta} \left(w_{r1}^2 - y_{r1}^2 \right) - 2 z_{2rs}^{\alpha\beta} w_{r1} y_{r1}, \qquad (2C.9f)$$

$$f_{4rs}^{\alpha\beta} = z_{2rs}^{\alpha\beta} \left(w_{r1}^2 - y_{r1}^2 \right) + 2 z_{1rs}^{\alpha\beta} w_{r1} y_{r1}, \qquad (2C.9g)$$

$$f_{5rs}^{\alpha\beta} = \sqrt{1 - \xi_{r1}^2} \left[z_{1rs}^{\alpha\beta} \left(w_{r1}^2 - y_{r1}^2 \right) - 2 z_{2rs}^{\alpha\beta} w_{r1} y_{r1} \right]$$
$$- \xi_{r1} \left[\left(w_{r1}^2 - y_{r1}^2 \right) z_{2rs}^{\alpha\beta} + 2 z_{1rs}^{\alpha\beta} w_{r1} y_{r1} \right], \qquad (2C.9h)$$

$$f_{6rs}^{\alpha\beta} = \sqrt{1 - \xi_{r1}^2} \left[z_{2rs}^{\alpha\beta} \left(w_{r1}^2 - y_{r1}^2 \right) + 2z_{1rs}^{\alpha\beta} w_{r1} y_{r1} \right]$$
$$+ \xi_{r1} \left[\left(w_{r1}^2 - y_{r1}^2 \right) z_{1rs}^{\alpha\beta} - 2z_{2rs}^{\alpha\beta} w_{r1} y_{r1} \right]. \tag{2C.9i}$$

Also, $p_{\zeta sr}^{\alpha\beta}$ can be obtained from $p_{\zeta us}^{\alpha c}$ in Eqs. (2C.6a) through (2C.6j) by replacing α_{u1} with α_{s2}. For instance,

$$p_{4sr}^{\alpha\beta} = G_{s2}^6 \left[z_{1sr}^{\alpha\beta} \left(u_{s2} w_{s2} - v_{s2} y_{s2} \right) - z_{2sr}^{\alpha\beta} \left(u_{s2} y_{s2} + v_{s2} w_{s2} \right) \right], \tag{2C.10}$$

and so on. The terms associated with $p_{\zeta us}^{\alpha c}$, $p_{\zeta us}^{\beta c}$, $p_{\zeta rs}^{\alpha\beta}$, and $p_{\zeta sr}^{\alpha\beta}$ can be similarly obtained. For brevity, they are not presented here. However, it should be mentioned that the generating functions for these terms are defined in Eq. (2B.5). That is, for example, the terms associated with $p_{\zeta us}^{\alpha c}$ are

$$\sigma_{1us}^{\alpha} = B_{1us}^{\alpha} K_{2us}^{\alpha}, \quad \sigma_{2us}^{\alpha} = B_{4us}^{\alpha} K_{2us}^{\alpha} - B_{1us}^{\alpha} K_{1us}^{\alpha} - K_{3us}^{\alpha} D_{3us}^{\alpha}, \tag{2C.11a-d}$$

$$\sigma_{3us}^{\alpha} = B_{1us}^{\alpha} - B_{4us}^{\alpha} K_{1us}^{\alpha} - K_{4us}^{\alpha} D_{3us}^{\alpha}, \qquad \sigma_{4us}^{\alpha} = B_{4us}^{\alpha},$$

$$\sigma_{5us}^{\alpha} = B_{2us}^{\alpha} K_{2us}^{\alpha}, \quad \sigma_{6us}^{\alpha} = B_{3us}^{\alpha} K_{2us}^{\alpha} - K_{3us}^{\alpha} D_{1us}^{\alpha}, \tag{2C.11e-g}$$

$$\sigma_{7us}^{\alpha} = - \left(B_{2us}^{\alpha} K_{1us}^{\alpha} + K_{3us}^{\alpha} D_{2us}^{\alpha} \right),$$

$$\sigma_{8us}^{\alpha} = - \left(B_{3us}^{\alpha} K_{1us}^{\alpha} + K_{4us}^{\alpha} D_{1us}^{\alpha} \right), \tag{2C.11h-j}$$

$$\sigma_{9us}^{\alpha} = B_{2us}^{\alpha} - K_{4us}^{\alpha} D_{2us}^{\alpha}, \qquad \sigma_{10us}^{\alpha} = B_{3us}^{\alpha},$$

and the other terms in the above equation can be generated by applying Eq. (2B.5) with appropriate addition to the superscript and change of subscripts. For example,

$$K_{1us}^{\alpha} = G_{u1}^2 + G_{s1}^2 - 4\varepsilon_{u1} \varepsilon_{s1}, \quad K_{2us}^{\alpha} = \left(G_{u1} G_{s1} \right)^2,$$

$$K_{3us}^{\alpha} = 2 \left(\varepsilon_{s1} G_{u1}^2 - \varepsilon_{u1} G_{s1}^2 \right), \quad K_{4us}^{\alpha} = 2 \left(\varepsilon_{u1} - \varepsilon_{s1} \right), \tag{2C.12a-d}$$

$$G_{u1}^2 = \varepsilon_{u1}^2 + \Omega_u^2, \quad G_{s1}^2 = \varepsilon_{s1}^2 + \Omega_s^2, \qquad \text{(2C.13a, b)}$$

and so on. For another illustration of the use of generating functions in Eq. (2B.5), consider the first two terms associated with $p_{\zeta rs}^{\alpha\beta}$,

$$\sigma_{1rs}^{\alpha\beta} = B_{1rs}^{\alpha\beta} K_{2rs}^{\alpha\beta},$$

$$\text{(2C.14a, b)}$$

$$\sigma_{2rs}^{\alpha\beta} = B_{4rs}^{\alpha\beta} K_{2rs}^{\alpha\beta} - B_{1rs}^{\alpha\beta} K_{1rs}^{\alpha\beta} - K_{3rs}^{\alpha\beta} D_{3rs}^{\alpha\beta},$$

in which

$$K_{1rs}^{\alpha\beta} = G_{r1}^2 + G_{s2}^2 - 4\varepsilon_{r1}\varepsilon_{s2}, \qquad K_{2rs}^{\alpha\beta} = \left(G_{r1} G_{s2}\right)^2,$$

$$\text{(2C.15a-d)}$$

$$K_{3rs}^{\alpha\beta} = 2\left(\varepsilon_{s2} G_{r1}^2 - \varepsilon_{r1} G_{s2}^2\right), \qquad K_{4rs}^{\alpha\beta} = 2\left(\varepsilon_{r1} - \varepsilon_{s2}\right),$$

$$G_{r1}^2 = \varepsilon_{r1}^2 + \Omega_r^2, \quad G_{s2}^2 = \varepsilon_{s2}^2 + \Omega_s^2, \qquad \text{(2C.16a, b)}$$

The foregoing terms are the contribution of the evolutionary coincident spectral density to the response. The terms involved with the contribution of the evolutionary quadrature spectral density function, that is, the last six terms on the rhs of Eq. (2.39), to the response is now considered in the following. First, the terms $q_{\eta us}^{\alpha c}$ are derived as

$$q_{1us}^{\alpha c} = -G_{u1}\left(\xi_{u1} z_{1us}^\alpha + z_{2us}^\alpha \sqrt{1 - \xi_{u1}^2}\right), \qquad \text{(2C.17a)}$$

$$q_{2us}^{\alpha c} = -G_{u1}^3\left[\xi_{u1} z_{1us}^\alpha\left(3 - 4\xi_{u1}^2\right) + z_{2us}^\alpha \sqrt{1 - \xi_{u1}^2}\left(1 - 4\xi_{u1}^2\right)\right], \qquad \text{(2C.17b)}$$

$$q_{3us}^{\alpha c} = -G_{u1}^5 f_{6us}^\alpha. \qquad \text{(2C.17c)}$$

The terms associated with Eq. (2C.17) are generated from those in Eq. (2B.9) with appropriate addition to the superscript and change of subscripts as

$$\beta_{1us}^\alpha = B_{1us}^\alpha K_{3us}^\alpha + D_{3us}^\alpha K_{2us}^\alpha,$$

$$\text{(2C.18a, b)}$$

$$\beta_{2us}^\alpha = B_{1us}^\alpha K_{4us}^\alpha + B_{4us}^\alpha K_{3us}^\alpha - D_{3us}^\alpha K_{1us}^\alpha,$$

$$\beta_{3us}^{\alpha} = B_{4us}^{\alpha} K_{4us}^{\alpha} + D_{3us}^{\alpha} . \qquad (2C.18c)$$

The other expressions in the evolutioary quadrature spectral density contribution to the response can be similarly generated by making use of Eqs. (2C.17) and (2C.18). For instance,

$$q_{1rs}^{\alpha\beta} = - G_{r1}\left(\xi_{r1} z_{1rs}^{\alpha\beta} + z_{2rs}^{\alpha\beta} \sqrt{1 - \xi_{r1}^2} \right) , \qquad (2C.19a)$$

$$q_{2rs}^{\alpha\beta} = - G_{r1}^3 \left[\xi_{r1} z_{1rs}^{\alpha\beta}\left(3 - 4\xi_{r1}^2\right) + z_{2rs}^{\alpha\beta} \sqrt{1 - \xi_{r1}^2}\left(1 - 4\xi_{r1}^2\right) \right], \qquad (2C.19b)$$

$$q_{3rs}^{\alpha\beta} = - G_{r1}^5 f_{6rs}^{\alpha\beta} . \qquad (2C.19c)$$

The coefficients associated with the above expressions are

$$\beta_{1rs}^{\alpha\beta} = B_{1rs}^{\alpha\beta} K_{3rs}^{\alpha\beta} + D_{3rs}^{\alpha\beta} K_{2rs}^{\alpha\beta} ,$$

$$\beta_{2rs}^{\alpha\beta} = B_{1rs}^{\alpha\beta} K_{4rs}^{\alpha\beta} + B_{4rs}^{\alpha\beta} K_{3rs}^{\alpha\beta} - D_{3rs}^{\alpha\beta} K_{1rs}^{\alpha\beta} , \qquad (2C.20)$$

$$\beta_{3rs}^{\alpha\beta} = B_{4rs}^{\alpha\beta} K_{4rs}^{\alpha\beta} + D_{3rs}^{\alpha\beta} .$$

Appendix 2D   Covariances of Displacements and Velocities

From Eq. (2.41) the covariances of displacements at node j and velocities at node k are given by

$$\left\langle x_j(t)\dot{x}_k(t)\right\rangle = \sum_{i=1}^{14} N_i^{(12)}\left(x_j\dot{x}_k\right),$$

where the terms on the rhs are defined as

$$N_1^{(12)}\left(x_j\dot{x}_k\right) = \sum_{r=1}^{L} \Phi_{1r}^{\alpha u} A_{2r}^{\alpha}, \qquad N_2^{(12)}\left(x_j\dot{x}_k\right) = \sum_{r=2}^{L}\sum_{s=r}^{L} \Phi_{pus}^{\alpha c} \sum_{\delta=1}^{4} \sigma_{\delta us}^{\alpha} P_{p\delta us}^{\alpha c},$$

$$N_3^{(12)}\left(x_j\dot{x}_k\right) = \sum_{r=2}^{L}\sum_{s=r}^{L} \Phi_{psu}^{\alpha c} \sum_{\delta=1}^{4} \sigma_{\delta su}^{\alpha} P_{p\delta su}^{\alpha c}, \qquad N_4^{(12)}\left(x_j\dot{x}_k\right) = \sum_{r=1}^{L} \Phi_{1r}^{\beta u} A_{2r}^{\beta},$$

$$N_5^{(12)}\left(x_j\dot{x}_k\right) = \sum_{r=2}^{L}\sum_{s=r}^{L} \Phi_{pus}^{\beta c} \sum_{\delta=1}^{4} \sigma_{\delta us}^{\beta} P_{p\delta us}^{\beta c},$$

$$N_6^{(12)}\left(x_j\dot{x}_k\right) = \sum_{r=2}^{L}\sum_{s=r}^{L} \Phi_{psu}^{\beta c} \sum_{\delta=1}^{4} \sigma_{\delta su}^{\beta} P_{p\delta su}^{\beta c},$$

$$N_7^{(12)}\left(x_j\dot{x}_k\right) = -\sum_{r=1}^{L}\sum_{s=1}^{L} \Phi_{prs}^{\alpha\beta} \sum_{\delta=1}^{4} \sigma_{\delta rs}^{\alpha\beta} P_{p\delta rs}^{\alpha\beta},$$

$$N_8^{(12)}\left(x_j\dot{x}_k\right) = -\sum_{r=1}^{L}\sum_{s=1}^{L} \Phi_{psr}^{\alpha\beta} \sum_{\delta=1}^{4} \sigma_{\delta sr}^{\alpha\beta} P_{p\delta sr}^{\alpha\beta},$$

$$N_9^{(12)}\left(x_j\dot{x}_k\right) = \sum_{r=2}^{L}\sum_{s=r}^{L} \Phi_{qus}^{\alpha c} \sum_{\lambda=1}^{9} \beta_{\lambda us}^{\alpha} q_{q\lambda us}^{\alpha c},$$

$$N_{10}^{(12)}\left(x_j\dot{x}_k\right) = \sum_{r=2}^{L}\sum_{s=r}^{L} \Phi_{qsu}^{\alpha c} \sum_{\lambda=1}^{9} \beta_{\lambda su}^{\alpha} q_{q\lambda su}^{\alpha c},$$

$$N_{11}^{(12)}\left(x_j\dot{x}_k\right) = \sum_{r=2}^{L}\sum_{s=r}^{L} \Phi_{qus}^{\beta c} \sum_{\lambda=1}^{9} \beta_{\lambda us}^{\beta} q_{q\lambda us}^{\beta c},$$

$$N_{12}^{(12)}(x_j \dot{x}_k) = \sum_{r=2}^{L} \sum_{s=r}^{L} \Phi_{qsu}^{\beta c} \sum_{\lambda=1}^{9} \beta_{\lambda su}^{\beta} q_{q\lambda su}^{\beta c},$$

$$N_{13}^{(12)}(x_j \dot{x}_k) = -\sum_{r=1}^{L} \sum_{s=1}^{L} \Phi_{qrs}^{\alpha\beta} \sum_{\lambda=1}^{9} \beta_{\lambda rs}^{\alpha\beta} q_{q\lambda rs}^{\alpha\beta},$$

$$N_{14}^{(12)}(x_j \dot{x}_k) = -\sum_{r=1}^{L} \sum_{s=1}^{L} \Phi_{qsr}^{\alpha\beta} \sum_{\lambda=1}^{9} \beta_{\lambda sr}^{\alpha\beta} q_{q\lambda sr}^{\alpha\beta},$$

in which the first eight terms are due to the evolutionary coincident spectral density function, while the last six constitute the effect of the evolutionary quadrature spectral density function.

In the above derivation the assumption is that the modulating functions of the nonstationary random excitations are smooth functions varying slowly in time t.

Following similar steps for the derivation of covariances of displacements in Appendix 2C, one can obtain factors and terms in Eq. (2.41) as in the following

$$\Phi_{1r}^{\alpha u} = \pi \mu_{jr} \mu_{kr} \Omega_r^2 e^{-2\zeta_r \omega_r t} S_o, \tag{2D.1a}$$

$$\Phi_{pus}^{\alpha c} = \frac{-\pi(\eta_{pk}) e^{-(\zeta_u \omega_u + \zeta_s \omega_s)t} S_o}{G_{u1}^2 y_{u1} \left[\left(d_{1us}^{\alpha} \right)^2 + \left(d_{2us}^{\alpha} \right)^2 \right]}, \tag{2D.1b}$$

$$\Phi_{psu}^{\alpha c} = \frac{-\pi(\eta_{pk}) e^{-(\zeta_s \omega_s + \zeta_u \omega_u)t} S_o}{G_{s1}^2 y_{s1} \left[\left(d_{1su}^{\alpha} \right)^2 + \left(d_{2su}^{\alpha} \right)^2 \right]}, \tag{2D.1c}$$

$$\Phi_{1r}^{\beta u} = \pi \mu_{jr} \mu_{kr} \Omega_r^2 e^{-2\zeta_r \omega_r t} S_o, \tag{2D.1d}$$

$$\Phi_{pus}^{\beta c} = \frac{-\pi(\eta_{pk}) e^{-(\zeta_u \omega_u + \zeta_s \omega_s)t} S_o}{G_{u2}^2 y_{u2} \left[\left(d_{1us}^{\beta} \right)^2 + \left(d_{2us}^{\beta} \right)^2 \right]}, \tag{2D.1e}$$

$$\Phi_{psu}^{\beta c} = \frac{-\pi(\eta_{pk}) e^{-(\zeta_u \omega_u + \zeta_s \omega_s)t} S_o}{G_{s2}^2 y_{s2} \left[\left(d_{1su}^{\beta} \right)^2 + \left(d_{2su}^{\beta} \right)^2 \right]}, \tag{2D.1f}$$

$$\Phi_{prs}^{\alpha\beta} = \frac{-\pi\left(\eta_{pk}\right)e^{-\left(\zeta_r\omega_r + \zeta_s\omega_s\right)t}S_o}{G_{r1}^2 y_{r1}\left[\left(d_{1rs}^{\alpha\beta}\right)^2 + \left(d_{2rs}^{\alpha\beta}\right)^2\right]}, \tag{2D.1g}$$

$$\Phi_{psr}^{\alpha\beta} = \frac{-\pi\left(\eta_{pk}\right)e^{-\left(\zeta_r\omega_r + \zeta_s\omega_s\right)t}S_o}{G_{s2}^2 y_{s2}\left[\left(d_{1sr}^{\alpha\beta}\right)^2 + \left(d_{2sr}^{\alpha\beta}\right)^2\right]}, \tag{2D.1h}$$

$$\Phi_{qus}^{\alpha c} = \frac{-\pi\left(\eta_{mk}\right)e^{-\left(\zeta_u\omega_u + \zeta_s\omega_s\right)t}S_o}{G_{u1}^3 y_{u1}\left[\left(d_{1us}^{\alpha}\right)^2 + \left(d_{2us}^{\alpha}\right)^2\right]}, \tag{2D.1i}$$

where the factors are related to Eqs. (2C.1a) through (2C.1h) by simple factors such as G_{u1} and G_{r1}. The factors containing the first subscript q are similar to those containing the first subscript p except that in the latter, η_{pk} is replaced by η_{mk}.

The terms constituting the contribution of the evolutionary coincident spectral density to the response such as $p_{p\delta us}^{\alpha c}$ in Eq. (2.41) are obtained as

$$p_{p1us}^{\alpha c} = -\left(\xi_{u1} z_{1us}^{\alpha} + z_{2us}^{\alpha}\sqrt{1 - \xi_{u1}^2}\right), \tag{2D.2a}$$

$$p_{p2us}^{\alpha c} = -G_{u1}^2\left[\xi_{u1} z_{1us}^{\alpha}\left(3 - 4\xi_{u1}^2\right) + z_{2us}^{\alpha}\sqrt{1 - \xi_{u1}^2}\left(1 - 4\xi_{u1}^2\right)\right], \tag{2D.2b}$$

$$p_{p3us}^{\alpha c} = -G_{u1}^4 f_{6us}^{\alpha}, \quad p_{p4us}^{\alpha c} = -G_{u1}^6\left(y_{u1} f_{5us}^{\alpha} + f_{6us}^{\alpha} w_{u1}\right). \tag{2D.2c, d}$$

The parameters on the rhs of Eq. (2D.2) are defined in Eqs. (2C.3a), (2C.3c, d), (2C.4b,c), (2C.5h,i), and so on.

Similar to those in Appendix 2C, the remaining terms in the contribution of the evolutionary coincident spectral density in Eq. (2.41) can be derived. For instance, the terms $p_{p\delta us}^{\beta c}$ can be obtained as

$$p_{p1us}^{\beta c} = -\left(\xi_{u2} z_{1us}^{\beta} + z_{2us}^{\beta}\sqrt{1 - \xi_{u2}^2}\right), \tag{2D.3a}$$

$$p_{p2us}^{\beta c} = -G_{u2}^2\left[\xi_{u2} z_{1us}^{\beta}\left(3 - 4\xi_{u2}^2\right) + z_{2us}^{\beta}\sqrt{1 - \xi_{u2}^2}\left(1 - 4\xi_{u2}^2\right)\right], \tag{2D.3b}$$

$$p^{\beta c}_{p3us} = - G^4_{u2} f^\beta_{6us} \, , \quad p^{\beta c}_{p4us} = - G^6_{u2} \left(y_{u2} f^\beta_{5us} + f^\beta_{6us} w_{u2} \right) , \quad \text{(2D.3c, d)}$$

whereas $p^{\alpha\beta}_{p\delta rs}$ can be shown to be

$$p^{\alpha\beta}_{p1rs} = - \left(\xi_{r1} z^{\alpha\beta}_{1rs} + z^{\alpha\beta}_{2rs} \sqrt{1 - \xi^2_{r1}} \right) , \quad \text{(2D.4a)}$$

$$p^{\alpha\beta}_{p2rs} = - G^2_{r1} \left[\xi_{r1} z^{\alpha\beta}_{1rs} \left(3 - 4\xi^2_{r1} \right) + z^{\alpha\beta}_{2rs} \sqrt{1 - \xi^2_{r1}} \left(1 - 4\xi^2_{r1} \right) \right], \quad \text{(2D.4b)}$$

$$p^{\alpha\beta}_{p3rs} = - G^4_{r1} f^{\alpha\beta}_{6rs} \, , \quad p^{\alpha\beta}_{p4rs} = - G^6_{r1} \left(y_{r1} f^{\alpha\beta}_{5rs} + f^{\alpha\beta}_{6rs} w_{r1} \right) . \quad \text{(2D.4c, d)}$$

Now, the terms constituting the contribution of the evolutionary quadrature spectral density function to the response are considered. Firstly, the terms $q^{\alpha c}_{q\lambda us}$ are obtained as

$$q^{\alpha c}_{q1us} = - G_{u1} \left(z^\alpha_{1us} w_{u1} - z^\alpha_{2us} y_{u1} \right) , \quad \text{(2D.5a)}$$

$$q^{\alpha c}_{q2us} = - G^3_{u1} \left(z^\alpha_{1us} u_{u1} - z^\alpha_{2us} v_{u1} \right) , \quad \text{(2D.5b)}$$

$$q^{\alpha c}_{q3us} = - G^5_{u1} \left[z^\alpha_{1us} \left(u_{u1} w_{u1} - v_{u1} y_{u1} \right) - z^\alpha_{2us} \left(u_{u1} y_{u1} + v_{u1} w_{u1} \right) \right], \quad \text{(2D.5c)}$$

$$q^{\alpha c}_{q4us} = - e^{-\xi_{u1} G_{u1} t} \left[\left(\xi_{u1} z^\alpha_{1us} + z^\alpha_{2us} \sqrt{1 - \xi^2_{u1}} \right) \cos\gamma_{u1} t \right.$$
$$\left. + \left(z^\alpha_{1us} \sqrt{1 - \xi^2_{u1}} - \xi_{u1} z^\alpha_{2us} \right) \sin\gamma_{u1} t \right] , \quad \text{(2D.5d)}$$

$$q^{\alpha c}_{q5us} = - G_{u1} e^{-\xi_{u1} G_{u1} t} \left[\left(w_{u1} z^\alpha_{1us} - y_{u1} z^\alpha_{2us} \right) \cos\gamma_{u1} t \right.$$
$$\left. - \left(w_{u1} z^\alpha_{2us} + y_{u1} z^\alpha_{1us} \right) \sin\gamma_{u1} t \right] , \quad \text{(2D.5e)}$$

$$q^{\alpha c}_{q6us} = - G^2_{u1} e^{-\xi_{u1} G_{u1} t} \left(f^\alpha_{1us} \sin\gamma_{u1} t + f^\alpha_{2us} \cos\gamma_{u1} t \right) , \quad \text{(2D.5f)}$$

$$q_{q7us}^{\alpha c} = -G_{u1}^3 e^{-\xi_{u1} G_{u1} t} \left(f_{3us}^{\alpha} \cos\gamma_{u1} t - f_{4us}^{\alpha} \sin\gamma_{u1} t \right),$$ (2D.5g)

$$q_{q8us}^{\alpha c} = -G_{u1}^4 e^{-\xi_{u1} G_{u1} t} \left(f_{5us}^{\alpha} \sin\gamma_{u1} t + f_{6us}^{\alpha} \cos\gamma_{u1} t \right),$$ (2D.5h)

$$q_{q9us}^{\alpha c} = -G_{u1}^5 e^{-\xi_{u1} G_{u1} t} \left[\left(f_{5us}^{\alpha} \sqrt{1 - \xi_{u1}^2} - \xi_{u1} f_{6us}^{\alpha} \right) \cos\gamma_{u1} t \right.$$
$$\left. - \left(f_{6us}^{\alpha} \sqrt{1 - \xi_{u1}^2} + \xi_{u1} f_{5us}^{\alpha} \right) \sin\gamma_{u1} t \right].$$ (2D.5i)

Similarly, the terms $q_{q\lambda rs}^{\alpha\beta}$ can be found as

$$q_{q1rs}^{\alpha\beta} = -G_{r1} \left(z_{1rs}^{\alpha\beta} w_{r1} - z_{2rs}^{\alpha\beta} y_{r1} \right),$$ (2D.6a)

$$q_{q2rs}^{\alpha\beta} = -G_{r1}^3 \left(z_{1rs}^{\alpha\beta} u_{r1} - z_{2rs}^{\alpha\beta} v_{r1} \right),$$ (2D.6b)

$$q_{q3rs}^{\alpha\beta} = -G_{r1}^5 \left[z_{1rs}^{\alpha\beta} \left(u_{r1} w_{r1} - v_{r1} y_{r1} \right) \right.$$
$$\left. - z_{2rs}^{\alpha\beta} \left(u_{r1} y_{r1} + v_{r1} w_{r1} \right) \right],$$ (2D.6c)

$$q_{q4rs}^{\alpha\beta} = -e^{-\xi_{r1} G_{r1} t} \left[\left(\xi_{r1} z_{1rs}^{\alpha\beta} + z_{2rs}^{\alpha\beta} \sqrt{1 - \xi_{r1}^2} \right) \cos\gamma_{r1} t \right.$$
$$\left. + \left(z_{1rs}^{\alpha\beta} \sqrt{1 - \xi_{r1}^2} - \xi_{r1} z_{2rs}^{\alpha\beta} \right) \sin\gamma_{r1} t \right],$$ (2D.6d)

$$q_{q5rs}^{\alpha\beta} = -G_{r1} e^{-\xi_{r1} G_{r1} t} \left[\left(w_{r1} z_{1rs}^{\alpha\beta} - y_{r1} z_{2rs}^{\alpha\beta} \right) \cos\gamma_{r1} t \right.$$
$$\left. - \left(w_{r1} z_{2rs}^{\alpha\beta} + y_{r1} z_{1rs}^{\alpha\beta} \right) \sin\gamma_{r1} t \right],$$ (2D.6e)

$$q_{q6rs}^{\alpha\beta} = -G_{r1}^2 e^{-\xi_{r1} G_{r1} t} \left(f_{1rs}^{\alpha\beta} \sin\gamma_{r1} t + f_{2rs}^{\alpha\beta} \cos\gamma_{r1} t \right),$$ (2D.6f)

$$q_{q7rs}^{\alpha\beta} = -G_{r1}^3 e^{-\xi_{r1} G_{r1} t} \left(f_{3rs}^{\alpha\beta} \cos\gamma_{r1} t - f_{4rs}^{\alpha\beta} \sin\gamma_{r1} t \right),$$ (2D.6g)

$$q_{q8rs}^{\alpha\beta} = -G_{r1}^4 e^{-\xi_{r1}G_{r1}t}\left(f_{5rs}^{\alpha\beta}\sin\gamma_{r1}t + f_{6rs}^{\alpha\beta}\cos\gamma_{r1}t\right), \qquad (2D.6h)$$

$$q_{q9rs}^{\alpha\beta} = -G_{r1}^5 e^{-\xi_{r1}G_{r1}t}\left[\left(f_{5rs}^{\alpha\beta}\sqrt{1-\xi_{r1}^2} - \xi_{r1}f_{6rs}^{\alpha\beta}\right)\cos\gamma_{r1}t \right.$$
$$\left. - \left(f_{6rs}^{\alpha\beta}\sqrt{1-\xi_{r1}^2} + \xi_{r1}f_{5rs}^{\alpha\beta}\right)\sin\gamma_{r1}t\right]. \qquad (2D.6i)$$

Also, for example, the terms associated with $q_{q\lambda us}^{\alpha c}$ are generated from Eq. (2B.9) in a similar manner to those presented in Eq. (2C.15). Of course, in addition to those included in the latter equation the other six terms are

$$\beta_{4us}^{\alpha} = D_{1us}^{\alpha}K_{2us}^{\alpha}, \qquad \beta_{5us}^{\alpha} = B_{2us}^{\alpha}K_{3us}^{\alpha} + D_{2us}^{\alpha}K_{2us}^{\alpha},$$

$$\beta_{6us}^{\alpha} = B_{3us}^{\alpha}K_{3us}^{\alpha} - D_{1us}^{\alpha}K_{1us}^{\alpha}, \qquad (2D.7a\text{-}f)$$

$$\beta_{7us}^{\alpha} = B_{2us}^{\alpha}K_{4us}^{\alpha} - D_{2us}^{\alpha}K_{1us}^{\alpha},$$

$$\beta_{8us}^{\alpha} = B_{3us}^{\alpha}K_{4us}^{\alpha} + D_{1us}^{\alpha}, \qquad \beta_{9us}^{\alpha} = D_{2us}^{\alpha}.$$

Finally, it should be mentioned that the remaining terms in Eq. (2.41) can be similarly obtained as those presented above.

Appendix 2E   Time-dependent Covariances of Velocities

From Eq. (2.43) the covariances of velocities at node j and k are given by

$$\left\langle \dot{x}_j(t) \dot{x}_k(t) \right\rangle = \sum_{i=1}^{14} N_i^{(2)}\left(\dot{x}_j \dot{x}_k\right), \qquad (2.43)$$

where the terms on the rhs are defined as

$$N_1^{(2)}\left(\dot{x}_j \dot{x}_k\right) = \sum_{r=1}^{L} \Phi_{yr}^{\alpha u} \sum_{\delta=1}^{4} A_{\delta r}^{\alpha} Q_{y\delta r}^{\alpha},$$

$$N_2^{(2)}\left(\dot{x}_j \dot{x}_k\right) = \sum_{r=2}^{L} \sum_{s=r}^{L} \Phi_{yus}^{\alpha c} \sum_{\zeta=1}^{10} \sigma_{\zeta us}^{\alpha} P_{y\zeta us}^{\alpha c},$$

$$N_3^{(2)}\left(\dot{x}_j \dot{x}_k\right) = \sum_{r=2}^{L} \sum_{s=r}^{L} \Phi_{ysu}^{\alpha c} \sum_{\zeta=1}^{10} \sigma_{\zeta su}^{\alpha} P_{y\zeta su}^{\alpha c},$$

$$N_4^{(2)}\left(\dot{x}_j \dot{x}_k\right) = \sum_{r=1}^{L} \Phi_{yr}^{\beta u} \sum_{\delta=1}^{4} A_{\delta r}^{\beta} Q_{y\delta r}^{\beta},$$

$$N_5^{(2)}\left(\dot{x}_j \dot{x}_k\right) = \sum_{r=2}^{L} \sum_{s=r}^{L} \Phi_{yus}^{\beta c} \sum_{\zeta=1}^{10} \sigma_{\zeta us}^{\beta} P_{y\zeta us}^{\beta c},$$

$$N_6^{(2)}\left(\dot{x}_j \dot{x}_k\right) = \sum_{r=2}^{L} \sum_{s=r}^{L} \Phi_{ysu}^{\beta c} \sum_{\zeta=1}^{10} \sigma_{\zeta su}^{\beta} P_{y\zeta su}^{\beta c},$$

$$N_7^{(2)}\left(\dot{x}_j \dot{x}_k\right) = -\sum_{r=1}^{L} \sum_{s=1}^{L} \Phi_{yrs}^{\alpha\beta} \sum_{\zeta=1}^{10} \sigma_{\zeta rs}^{\alpha\beta} P_{y\zeta rs}^{\alpha\beta},$$

$$N_8^{(2)}\left(\dot{x}_j \dot{x}_k\right) = -\sum_{r=1}^{L} \sum_{s=1}^{L} \Phi_{ysr}^{\alpha\beta} \sum_{\zeta=1}^{10} \sigma_{\zeta sr}^{\alpha\beta} P_{y\zeta sr}^{\alpha\beta},$$

$$N_9^{(2)}\left(\dot{x}_j \dot{x}_k\right) = \sum_{r=2}^{L} \sum_{s=r}^{L} \Phi_{zus}^{\alpha c} \sum_{\eta=1}^{3} \beta_{\eta us}^{\alpha} q_{z\eta us}^{\alpha c},$$

$$N_{10}^{(2)}\left(\dot{x}_j\,\dot{x}_k\right) = \sum_{r=2}^{L}\sum_{s=r}^{L} \Phi_{zsu}^{\alpha c} \sum_{\eta=1}^{3} \beta_{\eta su}^{\alpha} q_{z\eta su}^{\alpha c},$$

$$N_{11}^{(2)}\left(\dot{x}_j\,\dot{x}_k\right) = \sum_{r=2}^{L}\sum_{s=r}^{L} \Phi_{zus}^{\beta c} \sum_{\eta=1}^{3} \beta_{\eta us}^{\beta} q_{z\eta us}^{\beta c},$$

$$N_{12}^{(2)}\left(\dot{x}_j\,\dot{x}_k\right) = \sum_{r=2}^{L}\sum_{s=r}^{L} \Phi_{zsu}^{\beta c} \sum_{\eta=1}^{3} \beta_{\eta su}^{\beta} q_{z\eta su}^{\beta c},$$

$$N_{13}^{(2)}\left(\dot{x}_j\,\dot{x}_k\right) = -\sum_{r=1}^{L}\sum_{s=1}^{L} \Phi_{zrs}^{\alpha\beta} \sum_{\eta=1}^{3} \beta_{\eta rs}^{\alpha\beta} q_{z\eta rs}^{\alpha\beta},$$

$$N_{14}^{(2)}\left(\dot{x}_j\,\dot{x}_k\right) = -\sum_{r=1}^{L}\sum_{s=1}^{L} \Phi_{zsr}^{\alpha\beta} \sum_{\eta=1}^{3} \beta_{\eta sr}^{\alpha\beta} q_{z\eta sr}^{\alpha\beta},$$

in which the first eight terms on the lhs are the contribution of the evolutionary coincident spectral density functions whereas the remaining six terms on the lhs are due to the evolutionary quadrature spectral density functions. When the subscripts j and k are equal, the evolutionary quadrature spectral density functions become zero, and Eq. (2.43) reduces to the variances of velocity responses.

The steps for the derivation of covariances of velocities are similar to those for covariances of displacements in Appendix 2C, and as such one can show that the factors in $N_1^{(2)}(\bullet)$ through $N_8^{(2)}(\bullet)$, with the dot inside the brackets denoting the argument in Eq. (2.43) above, are

$$\Phi_{yr}^{\alpha u} = \frac{\pi\,\mu_{jr}\,\mu_{kr}\,\Omega_r^2\,e^{-2\zeta_r\omega_r t}\,S_o}{G_{r1}\,y_{r1}}, \tag{2E.1a}$$

$$\Phi_{yus}^{\alpha c} = \frac{-\pi\left(\eta_{pk}\right)e^{-\left(\zeta_u\omega_u + \zeta_s\omega_s\right)t}\,S_o}{G_{u1}\,y_{u1}\left[\left(d_{1us}^{\alpha}\right)^2 + \left(d_{2us}^{\alpha}\right)^2\right]}, \tag{2E.1b}$$

$$\Phi_{ysu}^{\alpha c} = \frac{-\pi\left(\eta_{pk}\right)e^{-\left(\zeta_s\omega_s + \zeta_u\omega_u\right)t}\,S_o}{G_{s1}\,y_{s1}\left[\left(d_{1su}^{\alpha}\right)^2 + \left(d_{2su}^{\alpha}\right)^2\right]}, \tag{2E.1c}$$

$$\Phi_{yr}^{\beta u} = \frac{\pi \mu_{jr} \mu_{kr} \Omega_r^2 e^{-2\zeta_r \omega_r t} S_o}{G_{r2} y_{r2}} , \qquad (2E.1d)$$

$$\Phi_{yus}^{\beta c} = \frac{-\pi (\eta_{pk}) e^{-(\zeta_u \omega_u + \zeta_s \omega_s)t} S_o}{G_{u2} y_{u2} \left[(d_{1us}^{\beta})^2 + (d_{2us}^{\beta})^2 \right]} , \qquad (2E.1e)$$

$$\Phi_{ysu}^{\beta c} = \frac{-\pi (\eta_{pk}) e^{-(\zeta_u \omega_u + \zeta_s \omega_s)t} S_o}{G_{s2} y_{s2} \left[(d_{1su}^{\beta})^2 + (d_{2su}^{\beta})^2 \right]} , \qquad (2E.1f)$$

$$\Phi_{yrs}^{\alpha\beta} = \frac{-\pi (\eta_{pk}) e^{-(\zeta_r \omega_r + \zeta_s \omega_s)t} S_o}{G_{r1} y_{r1} \left[(d_{1rs}^{\alpha\beta})^2 + (d_{2rs}^{\alpha\beta})^2 \right]} , \qquad (2E.1g)$$

$$\Phi_{ysr}^{\alpha\beta} = \frac{-\pi (\eta_{pk}) e^{-(\zeta_r \omega_r + \zeta_s \omega_s)t} S_o}{G_{s2} y_{s2} \left[(d_{1sr}^{\alpha\beta})^2 + (d_{2sr}^{\alpha\beta})^2 \right]} , \qquad (2E.1h)$$

where the factors are respectively related to Eqs. (2C.1a) through (2C.1h) by some simple factors such as G_{u1}^2 and G_{r1}^2. Specifically, for example, Eq. (2C.1e) is related to Eq. (2E.1e) through the equation $G_{u1}^2 \times (2C.1e) = (2E.1e)$ and Eq. (2C.1g) is related to Eq. (2E.1g) by $G_{r1}^2 \times (2C.1g) = (2E.1g)$, and so on.

The factors containing the first subscript z in $N_9^{(2)}(\bullet)$ through $N_{14}^{(2)}(\bullet)$ in Eq. (2.43) above are similar to those containing the first subscript y in Eqs. (2E.1a) through (2E.1h) except that in the latter equations η_{pk} is replaced with η_{mk}. Thus, for example, the factor in $N_9^{(2)}(\bullet)$ is obtained by applying the factor in $N_2^{(2)}(\bullet)$ as defined in Eq. (2E.1b) in which η_{pk} is replaced by η_{mk} to give

$$\Phi_{zus}^{\alpha c} = \frac{-\pi (\eta_{mk}) e^{-(\zeta_u \omega_u + \zeta_s \omega_s)t} S_o}{G_{u1} y_{u1} \left[(d_{1us}^{\alpha})^2 + (d_{2us}^{\alpha})^2 \right]} . \qquad (2E.1i)$$

Similarly, the factor in $N_{12}^{(2)}(\bullet)$ is obtained by applying the factor in $N_5^{(2)}(\bullet)$ as defined in Eq. (2E.1e) in which η_{pk} is replaced by η_{mk} to give

$$\Phi_{zus}^{\beta c} = \frac{-\pi\left(\eta_{mk}\right)e^{-\left(\zeta_u \omega_u + \zeta_s \omega_s\right)t} S_o}{G_{u2} y_{u2}\left[\left(d_{1us}^{\beta}\right)^2 + \left(d_{2us}^{\beta}\right)^2\right]} . \qquad (2E.1j)$$

As pointed out in Appendix 2C, the terms $A_{\delta r}^{\alpha}$ are $A_{\delta r}$, where the first subscript $\delta = 1, 2, 3, 4$, in Eq. (2B.4) with the subscript r in the latter equation being replaced with $r1$. Similarly, the terms $A_{\delta r}^{\beta}$ are $A_{\delta r}$ in Eq. (2B.4) with the subscript r in the latter equation being replaced with $r2$, respectively.

The terms associated with $A_{\delta r}^{\alpha}$ are defined as follows

$$Q_{y1r}^{\alpha} = \sqrt{1 - \xi_{r1}^2}, \qquad Q_{y2r}^{\alpha} = G_{r1}^2 Q_{y1r}^{\alpha}\left(1 - 4\xi_{r1}^2\right), \qquad (2E.2a, b)$$

$$Q_{y3r}^{\alpha} = e^{-\xi_{r1} G_{r1} t}\left(\sqrt{1 - \xi_{r1}^2} \cos\gamma_{r1}t - \xi_{r1} \sin\gamma_{r1}t\right), \qquad (2E.2c)$$

$$Q_{y4r}^{\alpha} = G_{r1} e^{-\xi_{r1} G_{r1} t}\left(y_{r1} \cos\gamma_{r1}t + w_{r1} \sin\gamma_{r1}t\right). \qquad (2E.2d)$$

Now, consider the remaining terms that constitute the contribution of the evolutionary coincident spectral density to the response. For instance, the terms $p_{y\zeta us}^{\alpha c}$ can be obtained as

$$p_{y1us}^{\alpha c} = z_{1us}^{\alpha} w_{u1} - z_{2us}^{\alpha} y_{u1}, \qquad (2E.3a)$$

$$p_{y2us}^{\alpha c} = G_{u1}^2\left(z_{1us}^{\alpha} u_{u1} - z_{2us}^{\alpha} v_{u1}\right), \qquad (2E.3b)$$

$$p_{y3us}^{\alpha c} = G_{u1}^4\left[z_{1us}^{\alpha}\left(u_{u1} w_{u1} - v_{u1} y_{u1}\right) - z_{2us}^{\alpha}\left(u_{u1} y_{u1} + v_{u1} w_{u1}\right)\right], \qquad (2E.3c)$$

$$p_{y4us}^{\alpha c} = G_{u1}^6\left[\sqrt{1 - \xi_{u1}^2}\left(w_{u1} f_{5us}^{\alpha} - y_{u1} f_{6us}^{\alpha}\right) - \xi_{u1}\left(w_{u1} f_{6us}^{\alpha} + y_{u1} f_{5us}^{\alpha}\right)\right], \qquad (2E.3d)$$

$$p_{y5us}^{\alpha c} = e^{-\xi_{u1} G_{u1} t}\left[\left(w_{u1} z_{1us}^{\alpha} - y_{u1} z_{2us}^{\alpha}\right)\cos\gamma_{u1}t \right.$$
$$\left. - \left(w_{u1} z_{2us}^{\alpha} + y_{u1} z_{1us}^{\alpha}\right)\sin\gamma_{u1}t\right], \qquad (2E.3e)$$

$$p_{y6us}^{\alpha c} = G_{u1} e^{-\xi_{u1} G_{u1} t} \left(f_{1us}^{\alpha} \sin\gamma_{u1} t + f_{2us}^{\alpha} \cos\gamma_{u1} t \right), \tag{2E.3f}$$

$$p_{y7us}^{\alpha c} = G_{u1}^2 e^{-\xi_{u1} G_{u1} t} \left(f_{3us}^{\alpha} \cos\gamma_{u1} t - f_{4us}^{\alpha} \sin\gamma_{u1} t \right), \tag{2E.3g}$$

$$p_{y8us}^{\alpha c} = G_{u1}^3 e^{-\xi_{u1} G_{u1} t} \left(f_{5us}^{\alpha} \sin\gamma_{u1} t + f_{6us}^{\alpha} \cos\gamma_{u1} t \right), \tag{2E.3h}$$

$$p_{y9us}^{\alpha c} = G_{u1}^4 e^{-\xi_{u1} G_{u1} t} \left[\left(f_{5us}^{\alpha} \sqrt{1 - \xi_{u1}^2} - \xi_{u1} f_{6us}^{\alpha} \right) \cos\gamma_{u1} t \right.$$
$$\left. - \left(f_{6us}^{\alpha} \sqrt{1 - \xi_{u1}^2} + \xi_{u1} f_{5us}^{\alpha} \right) \sin\gamma_{u1} t \right], \tag{2E.3i}$$

$$p_{y10us}^{\alpha c} = G_{u1}^5 e^{-\xi_{u1} G_{u1} t} \left[\left(f_{5us}^{\alpha} w_{u1} - y_{u1} f_{6us}^{\alpha} \right) \sin\gamma_{u1} t \right.$$
$$\left. + \left(w_{u1} f_{6us}^{\alpha} + y_{u1} f_{5us}^{\alpha} \right) \cos\gamma_{u1} t \right]. \tag{2E.3j}$$

The terms $p_{y\zeta su}^{\alpha c}$, $p_{y\zeta us}^{\beta c}$, $p_{y\zeta su}^{\beta c}$, $p_{y\zeta rs}^{\alpha\beta}$, and $p_{y\zeta sr}^{\alpha\beta}$ can be similarly derived.

The contribution of the evolutionary quadrature spectral density to the response is included in the following. First, the terms $q_{z\zeta us}^{\alpha c}$ are obtained as

$$q_{z1us}^{\alpha c} = -G_{u1} \left[\xi_{u1} z_{1us}^{\alpha} \left(3 - 4\xi_{u1}^2 \right) + z_{2us}^{\alpha} \sqrt{1 - \xi_{u1}^2} \left(1 - 4\xi_{u1}^2 \right) \right], \tag{2E.4a}$$

$$q_{z2us}^{\alpha c} = -G_{u1}^3 f_{6us}^{\alpha}, \qquad q_{z3us}^{\alpha c} = -G_{u1}^5 \left(y_{u1} f_{5us}^{\alpha} + f_{6us}^{\alpha} w_{u1} \right). \tag{2E.4b, c}$$

The remaining terms of the contribution of the evolutionary quadrature spectral density to the response can similarly be obtained.

Before leaving this appendix, it suffices to note that the terms associated with $p_{y\zeta us}^{\alpha c}$ and those associated with $q_{z\zeta us}^{\alpha c}$, for example, have already been considered and presented in Appendix 2C.

Appendix 2F   Cylindrical Shell Element Matrices

With reference to the truncated conical shell in Figure 2.14, the cylindrical shell element is a special case in which the angle $\varphi = 0$, the element length ℓ, radius of the cross-section of the element R, thickness h, and circumferential wave number n which is n_c in Sub-section 2.5.2. The material properties of the shell element are: density ρ, Young's modulus E, and Poisson's ratio ν. The element nodal displacement vector is

$$q = \begin{bmatrix} u_1 & v_1 & w_1 & \phi_1 & u_2 & v_2 & w_2 & \phi_2 \end{bmatrix}^T, \qquad \text{where } \phi_i = \frac{\partial w_i}{\partial x}, \qquad i = 1,2.$$

The consistent element mass matrix of the cylindrical shell is given by [1]

$$m = \pi \ell h \rho \left[m_{ij} \right]_{8 \times 8},$$

where $[m_{ij}]$ is symmetric and the non-zero entries of the upper triangular matrix are:

$$m_{11} = \frac{R}{3} = m_{22} = m_{55} = m_{66}, \qquad m_{15} = (\tfrac{1}{2})\, m_{11}, \qquad m_{26} = m_{15},$$

$$m_{33} = \frac{13}{15}R, \qquad m_{34} = \frac{11}{210}R\ell, \qquad m_{37} = \frac{9}{70}R, \qquad m_{38} = -\frac{13}{420}R\ell,$$

$$m_{44} = \frac{R\ell^2}{105}, \qquad m_{47} = -m_{38}, \qquad m_{48} = -\frac{R\ell^2}{140},$$

$$m_{77} = \frac{13R}{35}, \qquad m_{78} = -m_{34}, \qquad m_{88} = m_{44}.$$

The consistent element stiffness matrix of the cylindrical shell is defined by [1]

$$k = \pi \left[k_{ij} \right]_{8 \times 8},$$

where $[k_{ij}]$ is symmetric and the entries of the upper triangular matrix are:

$$k_{11} = \frac{RD_{11}}{\ell} + \frac{n^2 \ell D_{33}}{3R}, \qquad k_{12} = \frac{n}{2}\left(D_{12} - D_{33} \right),$$

$$k_{13} = -\frac{D_{12}}{2}, \qquad k_{14} = -\frac{D_{12}\ell}{12}, \qquad k_{15} = -\frac{D_{11}R}{\ell} + \frac{n^2 \ell D_{33}}{6R},$$

$$k_{16} = \frac{n}{2}\left(D_{12} + D_{33}\right), \quad k_{17} = k_{13}, \quad k_{18} = -k_{14},$$

$$k_{22} = \frac{n^2 \ell}{3R}\left(D_{22} + \frac{D_{55}}{R^2}\right) + \frac{1}{\ell}\left(RD_{33} + \frac{4}{R}D_{66}\right),$$

$$k_{23} = -\frac{7n\ell}{20R}\left(D_{22} + \frac{n^2 D_{55}}{R^2}\right) - \frac{n}{R\ell}\left(D_{45} + 4D_{66}\right),$$

$$k_{24} = -\frac{n\ell^2}{20R}\left(D_{22} + \frac{n^2 D_{55}}{R^2}\right) - \frac{nD_{45}}{R}, \qquad k_{25} = -k_{16},$$

$$k_{26} = \frac{n^2 \ell}{6R}\left(D_{22} + \frac{D_{55}}{R^2}\right) - \frac{1}{\ell}\left(RD_{33} + \frac{4D_{66}}{R}\right),$$

$$k_{27} = -\frac{3n\ell}{20R}\left(D_{22} + \frac{n^2 D_{55}}{R^2}\right) + \frac{n}{R\ell}\left(D_{45} + 4D_{66}\right),$$

$$k_{28} = \frac{n\ell^2}{30R}\left(D_{22} + \frac{n^2 D_{55}}{R^2}\right),$$

$$k_{33} = \frac{13\ell}{35R}\left(D_{22} + \frac{n^4 D_{55}}{R^2}\right) + \frac{12n^2}{5R\ell}\left(D_{45} + 2D_{66}\right) + \frac{12RD_{44}}{\ell^3},$$

$$k_{34} = \frac{11\ell^2}{210R}\left(D_{22} + \frac{n^4 D_{55}}{R^2}\right) + \frac{2n^2}{5R}\left(3D_{45} + D_{66}\right) + \frac{6RD_{44}}{\ell^2},$$

$$k_{35} = -k_{13}, \quad k_{36} = k_{27},$$

$$k_{37} = \frac{9\ell}{70R}\left(D_{22} + \frac{n^4 D_{55}}{R^2}\right) - \frac{12n^2}{5R\ell}\left(D_{45} + 2D_{66}\right) - \frac{12RD_{44}}{\ell^3},$$

$$k_{38} = -\frac{13\ell^2}{420R}\left(D_{22} + \frac{n^4 D_{55}}{R^2}\right) + \frac{n^2}{5R}\left(D_{45} + 2D_{66}\right) + \frac{6RD_{44}}{\ell^2} ,$$

$$k_{44} = \frac{\ell^3}{105R}\left(D_{22} + \frac{n^4 D_{55}}{R^2}\right) + \frac{4n^2\ell}{15R}\left(D_{45} + 2D_{66}\right) + \frac{4RD_{44}}{\ell} ,$$

$$k_{45} = -k_{14} , \quad k_{46} = -k_{28} , \quad k_{47} = -k_{38} ,$$

$$k_{48} = -\frac{\ell^3}{140R}\left(D_{22} + \frac{n^4 D_{55}}{R^2}\right) - \frac{n^2\ell}{15R}\left(D_{45} + 2D_{66}\right) + \frac{2RD_{44}}{\ell} ,$$

$$k_{55} = k_{11} , \quad k_{56} = -k_{12} , \quad k_{57} = -k_{13} , \quad k_{58} = k_{14} , \quad k_{66} = k_{22} ,$$

$$k_{67} = k_{23} , \quad k_{68} = -k_{24} , \quad k_{77} = k_{33} , \quad k_{78} = -k_{34} , \quad k_{88} = k_{44} ,$$

in which D_{ij} are the elements of the elastic matrix and they are defined as

$$D_{11} = D_{22} = \frac{Eh}{1 - v^2} , \quad D_{12} = D_{21} = vD_{11} , \quad D_{33} = \frac{Eh}{2(1 + v)} ,$$

$$D_{44} = D_{55} = \frac{Eh^3}{12(1 - v^2)} , \quad D_{45} = D_{54} = vD_{44} , \quad D_{66} = \frac{Eh^3}{24(1 + v)} .$$

The explicit element matrices of a truncated conical shell element can be found in Ref. [1]. They are not included here for brevity.

Reference

[1] To, C.W.S. and Wang, B. (1991). An axisymmetric thin shell finite element for vibration analysis, *Computers and Structures*, **40(3)**, 555-568.

Appendix 3A Deterministic Newmark Family of Algorithms

In this appendix the main objective is to present the steps of derivation of Eq. (3.37). As pointed out at the beginning of Sub-section 3.5.1, by making use of Eqs. (3.34) through (3.36) and a combination of these latter equations one is able to achieve the objective. The combination of Eqs. (3.34) through (3.36) applied, symbolically, is

$$(\Delta t)^2 \beta \left[(3.34) + (3.36) \right] + (\Delta t)^2 (1 - 2\beta) \left[(3.35) \right]$$

$$+ (\Delta t)^2 \left(\gamma - \frac{1}{2} \right) \left[(3.35) - (3.34) \right] .$$

Consider the terms associated with M, C, K, and the sum of forcing functions, respectively for the above combination as in the followings.

Firstly, for the terms associated with the mass matrix M, one has

$$(\Delta t)^2 \left[\beta a_{s+1} + \beta a_{s-1} + (1 - 2\beta) a_s + \left(\gamma - \frac{1}{2} \right) (a_s - a_{s-1}) \right]$$

$$= (\Delta t)^2 \left\{ \beta a_{s+1} + \left(\frac{1}{2} - \beta \right) a_s - \left[\beta a_s + \left(\frac{1}{2} - \beta \right) a_{s-1} \right] \right. \quad (3A.1)$$

$$\left. + \frac{1}{2} (a_s + a_{s-1}) + \left(\gamma - \frac{1}{2} \right) (a_s - a_{s-1}) \right\} .$$

By making use of Eq. (3.24), the rhs of the last equation becomes

$$(\Delta t)^2 \left\{ \beta a_{s+1} + \left(\frac{1}{2} - \beta \right) a_s - \left[\beta a_s + \left(\frac{1}{2} - \beta \right) a_{s-1} \right] \right.$$

$$\left. + \frac{1}{2} (a_s + a_{s-1}) + \left(\gamma - \frac{1}{2} \right) (a_s - a_{s-1}) \right\} \quad (3A.2)$$

$$= d_{s+1} - 2 d_s - (\Delta t)(v_s - v_{s-1}) + d_a ,$$

where the last term on the rhs is defined as

$$d_a = d_{s-1} + \frac{1}{2} (\Delta t)^2 (a_s + a_{s-1}) + (\Delta t)^2 \left(\gamma - \frac{1}{2} \right) (a_s - a_{s-1}) .$$

By Eq. (3.25), the bracketed terms on the rhs of Eq. (3A.2) become zero and therefore the terms associated with M reduces to

$$(\Delta t)^2 \left[\beta a_{s+1} + \beta a_{s-1} + (1 - 2\beta) a_s + \left(\gamma - \frac{1}{2} \right) (a_s - a_{s-1}) \right]$$

$$= (d_{s+1} - 2d_s + d_{s-1}) . \tag{3A.3}$$

Secondly, for the terms associated with the damping matrix C, one obtains

$$(\Delta t)^2 \beta (v_{s+1} + v_{s-1}) + (\Delta t)^2 (1 - 2\beta) v_s + (\Delta t)^2 \left(\gamma - \frac{1}{2} \right) (v_s - v_{s-1})$$

$$= (\Delta t)^2 \left[v_{s-1} + \beta (v_{s+1} - v_s) + (v_s - v_{s-1}) \left(\gamma - \beta + \frac{1}{2} \right) \right] . \tag{3A.4}$$

From Eq. (3.24),

$$(\Delta t) v_s = d_{s+1} - d_s - (\Delta t)^2 \left(\frac{1}{2} - \beta \right) a_s - \beta (\Delta t)^2 a_{s+1} . \tag{3A.5}$$

Therefore, the rhs of Eq. (3A.4) can be shown to be

$$\text{rhs of (3A.4)} = (\Delta t) \left(\sum_{i=1}^{5} d^{(i)} \right) , \tag{3A.6}$$

where the terms on the rhs of Eq. (3A.4) are defined by

$$d^{(1)} = \frac{1}{2} (d_{s+1} - d_s) - \frac{1}{2} (\Delta t)^2 \left(\frac{1}{2} - \beta \right) a_s ,$$

$$d^{(2)} = - \frac{1}{2} \left[(\Delta t)^2 \beta a_{s+1} - d_s + d_{s-1} \right],$$

$$d^{(3)} = - \frac{1}{2} (\Delta t)^2 \left[\left(\frac{1}{2} - \beta \right) a_{s-1} + \beta a_s \right],$$

$$d^{(4)} = (\Delta t) \left[\beta \left(v_{s+1} - v_s \right) + \left(v_s - v_{s-1} \right) \left(\frac{1}{2} - \beta \right) \right],$$

$$d^{(5)} = (\Delta t) \left(v_s - v_{s-1} \right) \left(\gamma - \frac{1}{2} \right).$$

From Eq. (3.25),

$$v_{s+1} - v_s = (\Delta t) \left[(1 - \gamma) a_s + \gamma a_{s+1} \right]. \tag{3A.7}$$

Similarly,

$$v_s - v_{s-1} = (\Delta t) \left[(1 - \gamma) a_{s-1} + \gamma a_s \right]. \tag{3A.8}$$

Substituting Eqs. (3A.7) and (3A.8) into the rhs of Eq. (3A.4), and after rearranging, one can show that

$$\text{rhs of (3A.4)} = (\Delta t) \left(d^{(5)} + \sum_{i=6}^{8} d^{(i)} \right), \tag{3A.9}$$

in which the terms

$$d^{(6)} = \frac{1}{2} \left(d_{s+1} - d_s \right) - \frac{1}{2} (\Delta t)^2 \left(\frac{1}{2} - \beta \right) \left(a_s + a_{s-1} \right),$$

$$d^{(7)} = (\Delta t)^2 \beta \left\{ -\frac{1}{2} \left(a_{s+1} + a_s \right) + \left[(1 - \gamma) a_s + \gamma a_{s+1} \right] \right\},$$

$$d^{(8)} = (\Delta t)^2 \left(\frac{1}{2} - \beta \right) \left[(1 - \gamma) a_{s-1} + \gamma a_s \right].$$

Rearranging and simplifying the rhs of Eq. (3A.9), it reduces to

$$\text{rhs of (3A.9)} = (\Delta t) \left[d^{(5)} + \frac{1}{2} \left(d_{s+1} - d_s \right) + \left(\gamma - \frac{1}{2} \right) \beta_1 \right], \tag{3A.10}$$

where

$$\beta_1 = (\Delta t)^2 \left[\left(\frac{1}{2} - \beta \right) a_s + \beta a_{s+1} - \left(\frac{1}{2} - \beta \right) a_{s-1} - \beta a_s \right].$$

From Eq. (3.24), one has

$$(\Delta t)^2 \left[\left(\frac{1}{2} - \beta \right) a_s + \beta a_{s+1} \right] = d_{s+1} - d_s - (\Delta t) v_s . \tag{3A.11}$$

Applying this equation to the rhs of Eq. (3A.10), and simplifying Eq. (3A.4) finally reduces to

$$(\Delta t)^2 \beta \left(v_{s+1} + v_{s-1} \right) + (\Delta t)^2 (1 - 2\beta) v_s$$

$$+ (\Delta t)^2 \left(\gamma - \frac{1}{2} \right) \left(v_s - v_{s-1} \right) \tag{3A.12}$$

$$= (\Delta t) \left[\gamma d_{s+1} + (1 - 2\gamma) d_s + (\gamma - 1) d_{s-1} \right].$$

Thirdly, for the terms associated with the stiffness matrix K one has

$$(\Delta t)^2 \beta \left(d_{s+1} + d_{s-1} \right) + (\Delta t)^2 (1 - 2\beta) d_s$$

$$+ (\Delta t)^2 \left(\gamma - \frac{1}{2} \right) \left(d_s - d_{s-1} \right) \tag{3A.13}$$

$$= (\Delta t)^2 \left[\beta d_{s+1} + \left(\frac{1}{2} + \gamma - 2\beta \right) d_s + \left(\frac{1}{2} + \beta - \gamma \right) d_{s-1} \right].$$

Finally, for the terms associated with the force vectors, one obtains

$$(\Delta t)^2 \beta \left(P_{s+1} + P_{s-1} \right) + (\Delta t)^2 (1 - 2\beta) P_s$$

$$+ \left(P_s - P_{s-1} \right) (\Delta t)^2 (\gamma - \frac{1}{2}) \tag{3A.14}$$

$$= (\Delta t)^2 \left[\beta P_{s+1} + (\frac{1}{2} + \gamma - 2\beta) P_s + (\frac{1}{2} + \beta - \gamma) P_{s-1} \right].$$

Applying Eqs. (3A.3) through (3A.14) and rearranging terms, it leads to

$$M_1 d_{s+1} = M_2 d_s - M_3 d_{s-1} + (\Delta t)^2 F_{es} , \qquad (3A.15)$$

where the coefficient matrices M_i with $i = 1, 2, 3$, and the force vector F_{es} are given below Eq. (3.37). Equation (3A.15) is Eq. (3.37) and it gives the displacement vector in the next time step d_{s+1} in terms of the displacement vector of the current time step d_s, the displacement vector of the past time step d_{s-1}, and the force vector F_{es}.

For a sdof system with small damping ratio ζ the algorithms with the stability condition of $2\beta \geq \gamma \geq 1/2$ are unconditional while that of $\gamma \geq 1/2$ and $\beta < \gamma/2$ are conditional. When $\beta = 0$ and $\gamma = 1/2$, it is the central difference method whose order of accuracy is 2 and critical time step size is

$$(\Delta t)_c = \frac{2}{\omega_1} ,$$

where ω_1 is the undamped natural frequency. More detailed discussion on the critical time step size for other members of the Newmark family of algorithms can be found in Chapter 9 of [1].

Reference

[1] Hughes, T.J.R. (1987). *The Finite Element Method: Linear Static and Dynamic Finite Element Analysis*, Prentice-Hall, Englewood Cliffs, New Jersey.

Index